METHODS IN MOLECULAR BIOLOGY

Series Editor
John M. Walker
School of Life and Medical Sciences
University of Hertfordshire
Hatfield, Hertfordshire, AL10 9AB, UK

For further volumes:
http://www.springer.com/series/7651

Angiogenesis Protocols

Third Edition

Edited by

Stewart G. Martin

*University of Nottingham, Academic Unit of Clinical Oncology, School of Medicine,
Nottingham University Hospitals NHS Trust, Nottingham, UK*

Peter W. Hewett

*Institute of Cardiovascular Science, College of Medical and Dental Sciences,
University of Birmingham, Birmingham, UK*

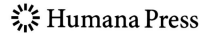 Humana Press

Editors
Stewart G. Martin
University of Nottingham
Academic Unit of Clinical Oncology
School of Medicine, Nottingham University
   Hospitals NHS Trust
Nottingham, UK

Peter W. Hewett
Institute of Cardiovascular Science
College of Medical and Dental Sciences
University of Birmingham
Birmingham, UK

ISSN 1064-3745   ISSN 1940-6029   (electronic)
Methods in Molecular Biology
ISBN 978-1-4939-3626-7   ISBN 978-1-4939-3628-1   (eBook)
DOI 10.1007/978-1-4939-3628-1

Library of Congress Control Number: 2016936584

Cover illustration: Elena I. Deryugina, Department of Cell and Molecular Biology, The Scripps Research Institute, La Jolla, CA, USA

Printed on acid-free paper

This Humana Press imprint is published by Springer Nature
The registered company is Springer Science+Business Media LLC New York

Preface

It is 15 years since the first edition of this book was published and 7 years since the second. The interest in angiogenesis, during the intervening time, continues to increase. It is a testament to the quality of the chapters included in the first two editions, and continued expansion of interest in angiogenesis research, that we now have a third edition of the book.

The original concept in writing this was, and still is, to provide angiogenesis researchers with a single source of relevant methodologies for cell isolation and assessing angiogenesis in vitro and in vivo. As always, inclusivity was key to this endeavor; techniques are described in detail and range in difficulty and resource requirements—this ensures that most, if not all, interested laboratories can participate in this exciting research area, irrespective of levels of resource and expertise.

As with the first two editions, the foundations remain firmly in place in the form of chapters on cell isolation, assessing angiogenesis in patient samples, and in vitro and in vivo assays, techniques that now form part of the canon of angiogenesis literature. A number of such chapters continue to be included in the current edition and have been updated to reflect changes that may have occurred since the previous edition. Our understanding of angiogenesis, and lymphangiogenesis, has moved on and our understanding of the complexities of the various processes involved is much more profound. As a result of this the current edition includes an expansion in the number of techniques, with many new chapters being included. There are now 27 chapters reflecting the diverse range of methodological approaches available to researchers.

There are new chapters on assessing leukocyte involvement in angiogenesis, lymphatic cell and pericyte isolation techniques, spheroid and ring based in vitro assays, and on pericyte involvement in angiogenesis. New in vivo based chapters include the chorioallantoic membrane models, corneal pocket assays to assess angiogenesis and lymphangiogenesis, models of muscle angiogenesis, and use of zebrafish embryos to study vascular angiogenesis and senescence.

By expanding the number of chapters we have, inevitably, had to sacrifice certain chapters from the second edition—very difficult decisions. We like to think, however, that this volume will continue to provide not only a practical handbook for key techniques but also an informative and enjoyable read for all those interested, no matter how directly, in angiogenesis.

Nottingham, UK   *Stewart G. Martin*
Birmingham, UK   *Peter W. Hewett*

Contents

Contributors

NARMEEN S. AHMAD • *University of Nottingham, Academic Unit of Clinical Oncology, School of Medicine, Nottingham University Hospitals NHS Trust, Nottingham, UK*

SILVIA P. ANDRADE • *Department of Physiology and Biophysics, Institute of Biological Sciences, Federal University of Minas Gerais, Belo Horizonte, MG, Brazil*

JOSEPH F. ARBOLEDA-VELASQUEZ • *Department of Ophthalmology, Schepens Eye Research Institute, Mass Eye and Ear Infirmary, Harvard Medical School, Boston, MA, USA*

DAVID O. BATES • *Cancer Biology, Division of Cancer and Stem Cells, School of Medicine, Queen's Medical Centre, University of Nottingham, Nottingham, UK*

ANDREW V. BENEST • *Cancer Biology, Division of Cancer and Stem Cells, School of Medicine, Queen's Medical Centre, University of Nottingham, Nottingham, UK*

ROY BICKNELL • *School of Immunity and Infection and Cancer Studies, Institute of Biomedical Research, College of Medical and Dental Sciences, University of Birmingham, Birmingham, UK*

RYAN M. BROWN • *Centre for Doctoral Training in Physical Sciences of Imaging in the Biomedical Sciences and School of Chemistry, University of Birmingham, Birmingham, UK*

LYNN M. BUTLER • *School of Clinical and Medical Sciences, College of Medical and Dental Sciences, The University of Birmingham, Birmingham, UK*

STEVEN CLASPER • *MRC Human Immunology Unit, Weatherall Institute of Molecular Medicine, John Radcliffe Hospital, Oxford, UK*

PATRICIA A. COOPER • *Institute of Cancer Therapeutics, University of Bradford, Bradford, UK*

WILLIAM COURT • *Cancer Research UK Cancer Therapeutics Unit, McElwain Laboratories, The Institute of Cancer Research, Surrey, UK*

VINCENT J. CUNNINGHAM • *Department of Oncology, The Medical School, University of Sheffield, Sheffield, UK*

ELENA I. DERYUGINA • *Department of Cell and Molecular Biology, The Scripps Research Institute, La Jolla, CA, USA*

YINDI DING • *Institute for Vascular Signalling, Centre for Molecular Medicine, Goethe University, Frankfurt, Germany*

SANDRA DONNINI • *Department of Life Sciences, University of Siena, Siena, Italy; Istituto Toscano Tumori (ITT), Florence, Italy*

JENNIFER T. DURHAM • *Department of Developmental, Molecular, and Chemical Biology, Graduate Program in Cell and Molecular Physiology, The Sackler School of Graduate Biomedical Sciences, Tufts University School of Medicine, Boston, MA, USA*

SUZANNE A. ECCLES • *Cancer Research UK Cancer Therapeutics Unit, McElwain Laboratories, The Institute of Cancer Research, Surrey, UK*

STUART EGGINTON • *School of Biomedical Science, University of Leeds, Leeds, UK*

MÔNICA A.N.D. FERREIRA • *Department of General Pathology, Institute of Biological Sciences, Federal University of Minas Gerais, Belo Horizonte, MG, Brazil*

INGRID FLEMING • *Institute for Vascular Signalling, Centre for Molecular Medicine, Goethe University, Frankfurt, Germany*

STEPHEN B. FOX • *Department of Pathology, Peter MacCallum Cancer Centre, Melbourne, VIC, Australia*

MELISSA V. GAMMONS • *MRC Laboratory of Molecular Biology, University of Cambridge, Cambridge, UK*

ALI HAFEZI-MOGHADAM • *Center for Excellence in Functional and Molecular Imaging, Brigham and Women's Hospital, Boston, MA, USA; Department of Radiology, Harvard Medical School, Boston, MA, USA*

VICTORIA L. HEATH • *School of Immunity and Infection, Institute of Biomedical Research, College of Medical and Dental Sciences, University of Birmingham, Birmingham, UK*

IRA M. HERMAN • *Department of Developmental, Molecular, and Chemical Biology, Graduate Program in Cell and Molecular Physiology, The Sackler School of Graduate Biomedical Sciences, Tufts University School of Medicine, Boston, MA, USA*

PETER W. HEWETT • *Institute of Cardiovascular Science, College of Medical and Dental Sciences, University of Birmingham, Birmingham, UK*

DAVID G. JACKSON • *MRC Human Immunology Unit, Weather all Institute of Molecular Medicine, John Radcliffe Hospital, Oxford, UK*

NICHOLAS JENE • *Department of Pathology, Peter MacCallum Cancer Centre, Melbourne, VIC, Australia*

THOMAS KORFF • *Department of Cardiovascular Research, Institute of Physiology and Pathophysiology, Heidelberg University, Heidelberg, Germany*

MICHAEL LEUNIG • *Department of Orthopaedic Surgery, Schulthess Klinik, Zurich, Switzerland*

ZERINA LOKMIC • *Murdoch Childrens Research Institute, Melbourne's Royal Children Hospital, The University of Melbourne, Parkville, VIC, Australia; Faculty of Medicine, Dentistry and Health Sciences, The University of Melbourne, Parkville, VIC, Australia*

STEWART G. MARTIN • *University of Nottingham, Academic Unit of Clinical Oncology, School of Medicine, Nottingham University Hospitals NHS Trust, Nottingham, UK*

HELEN M. MCGETTRICK • *School of Clinical and Medical Sciences, College of Medical and Dental Sciences, The University of Birmingham, Birmingham, UK*

CHRISTOPHER J. MEAH • *School of Immunity and Infection, Institute of Biomedical Research, College of Medical and Dental Sciences, University of Birmingham, Birmingham, UK*

HARRY MELLOR • *School of Biochemistry, Medical Sciences Building, University of Bristol, Bristol, UK*

LUCIA MORBIDELLI • *Department of Life Sciences, University of Siena, Siena, Italy*

SHINTARO NAKAO • *Center for Excellence in Functional and Molecular Imaging, Brigham and Women's Hospital, Boston, MA, USA; Department of Radiology, Harvard Medical School, Boston, MA, USA*

GERARD B. NASH • *School of Clinical and Medical Sciences, College of Medical and Dental Sciences, The University of Birmingham, Birmingham, UK*

JIA-MIN PANG • *Department of Pathology, Peter MacCallum Cancer Centre, Melbourne, VIC, Australia*

JESSAL J. PATEL • *Cardiovascular Division, School of Medicine, King's College London, London, UK*

LISA PATTERSON • *Cancer Research UK Cancer Therapeutics Unit, McElwain Laboratories, The Institute of Cancer Research, Surrey, UK*

LARISSA PFISTERER • *Department of Cardiovascular Research, Institute of Physiology and Pathophysiology, Heidelberg University, Heidelberg, Germany*

LUCA PRIMO • *Cancer Institute of Candiolo-FPO IRCCS and Department of Oncology, University of Torino, Turin, Italy*

VINCENT A. PRIMO • *Department of Ophthalmology, Schepens Eye Research Institute, Mass Eye and Ear Infirmary, Harvard Medical School, Boston, MA, USA*

VIVIEN E. PRISE • *Department of Oncology, The Medical School, University of Sheffield, Sheffield, UK*

MARK RICHARDS • *School of Biochemistry, Medical Sciences Building, University of Bristol, Bristol, UK*

EMMA RISTORI • *Department of Life Sciences, University of Siena, Siena, Italy*

DANIEL J. ROYSTON • *MRC Human Immunology Unit, Weatherall Institute of Molecular Medicine, John Radcliffe Hospital, Oxford, UK*

HEINRICH SAUER • *Department of Physiology, Faculty of Medicine, Justus-Liebig-University, Giessen, Germany*

AXEL SCKELL • *Department of Trauma and Reconstructive Surgery, University Medicine Greifswald, Greifswald, Germany*

GIORGIO SEANO • *Cancer Institute of Candiolo-FPO IRCCS and Department of Oncology, University of Torino, Turin, Italy*

FATEMEH SHARIFPANAH • *Department of Physiology, Faculty of Medicine, Justus-Liebig-University, Giessen, Germany*

ANTHONY R. SHEETS • *Department of Developmental, Molecular, and Chemical Biology, Graduate Program in Cell and Molecular Physiology, The Sackler School of Graduate Biomedical Sciences, Tufts University School of Medicine, Boston, MA, USA*

STEVEN D. SHNYDER • *Institute of Cancer Therapeutics, University of Bradford, Bradford, UK*

RICHARD C.M. SIOW • *Cardiovascular Division, School of Medicine, King's College London, London, UK*

SARAH J. STORR • *University of Nottingham, Academic Unit of Clinical Oncology, School of Medicine, Nottingham University Hospitals NHS Trust, Nottingham, UK*

IAIN B. STYLES • *School of Computer Science, University of Birmingham, Birmingham, UK*

GILLIAN M. TOZER • *Department of Oncology, The Medical School, University of Sheffield, Sheffield, UK*

MARINA ZICHE • *Department of Life Sciences, University of Siena, Siena, Italy; Istituto Toscano Tumori (ITT), Florence, Italy*

NINA ZIPPEL • *Institute for Vascular Signalling, Centre for Molecular Medicine, Goethe University, Frankfurt, Germany*

Part I

Microscopic Assessment

Chapter 1

Assessing Tumor Angiogenesis in Histological Samples

Jia-Min Pang, Nicholas Jene, and Stephen B. Fox

Abstract

Tumor neovascularization acquires their vessels through a number of processes including angiogenesis, vasculogenesis, vascular remodeling, intussusception, and possibly vascular mimicry in certain tumors. The end result of the tumor vasculature has been quantified by counting the number of immunohistochemically identified microvessels in areas of maximal vascularity, so-called hot spot. Other techniques have been developed such as Chalkley counting and the use of image analysis systems that are robust and reproducible as well as being more objective. Many of the molecular pathways that govern tumor neovascularization have been identified and many reagents are now available to study these tissue sections. These include angiogenic growth factors and their receptors and cell adhesion molecules, proteases, and markers of activated, proliferating, cytokine-stimulated, or angiogenic vessels, such as CD105. It is also possible to differentiate quiescent from active vessels. Other reagents that can identify proteins involved in microenvironmental influences such as hypoxia have also been generated. Although the histological assessment of tumor vascularity is used mostly in the research context, it may also have clinical applications if appropriate methodology and trained observers perform the studies.

Key words Tumor angiogenesis, Microvessel density, Chalkley counts, Angiogenic factors, Hypoxia, Vascular grading

1 Introduction

Although it has been recognized for many centuries that tumors are more vascular than their normal counterpart, it is only since Folkmans' hypothesis on anti-angiogenesis [1] that a more quantitative method for measuring the blood vasculature in tissue sections has been pursued. Folkman and colleagues recognized that quantitation of the tumor vasculature might play an important role in predicting tumor behavior and patient management and therefore developed a *m*icroscopic *a*ngiogenesis *g*rading *s*ystem, designated as the MAGS score. The score was calculated by measuring vessel number, endothelial cell hyperplasia, and cytology in tinctorially stained tissue sections [2]. It was designed to be an objective method for quantifying tumor angiogenesis that would yield important information on the relationship to other clinicopathological

Stewart G. Martin and Peter W. Hewett (eds.), *Angiogenesis Protocols*, Methods in Molecular Biology, vol. 1430,
DOI 10.1007/978-1-4939-3628-1_1, © Springer Science+Business Media New York 2016

tumor characteristics and aid the testing of anti-angiogenic therapies. However, although it was possible to classify tumors into "endothelial poor" or "rich," the technical limitations of sample selection, inter- and intra-observer variation, and conceptual biological problems were such that the technique could not be easily applied. Interest in grading tumor angiogenesis was rekindled in the 1980s and 1990s with the advent of nonspecific endothelial markers [3–5], but it has been only more recently with the advent of more specific endothelial markers that quantitation studies have been performed. Most investigators have highlighted endothelium using immunohistochemistry and based quantification on a method developed by Weidner et al. [6]. This method highlights the tumor vasculature with immunohistochemistry and counts individual vessels in the most vascular areas (so-called hot spots) of the tumor. These studies have shown that an increased microvessel density is a powerful prognostic tool in many human tumor types (reviewed in [7]). Nevertheless, due to limitations in capillary identification and quantitation, not all investigators have been able to confirm a relationship between tumor vascularity and prognosis (reviewed in [8]). This chapter will briefly discuss the considerations in quantifying tumor angiogenesis by microvessel quantitation in tissue sections and give the current optimal protocol for assessment. Since an increased understanding of how tumors acquire a neovasculature has emerged over the last few years along with the identification of the molecules involved in these processes, the pathways involved in tumor neovascularization and the microenvironmental influence such as hypoxia that profoundly affect the vascular program will also be discussed.

1.1 Mechanisms of Tumor Neovascularization

The hypothesis presented by Judah Folkman that tumors are angiogenesis dependent [9] is likely only to be partly the case, with the tumors using a variety of mechanisms to establish a blood supply. A non-angiogenic mechanism of tumor growth that occurs in lung carcinomas [10] and in secondary breast cancer metastasis to lung [11] is where tumor cells fill the existing structure within the lung, i.e., the alveolar spaces, whether primary or metastatic without destroying the underlying architecture with the tumor using the established blood vessels rather than generating new vessels as occurs during angiogenesis. Thus, there is no associated stromal desmoplasia, and unlike many solid tumors that are more vascular than their normal tissue counterpart, these tumors have a similar vascularity as normal lung. A similar co-option model has also been described in a brain tumor model whereby early in development, the tumor uses the existing blood vessels without eliciting an angiogenic response [12, 13]. The co-option model is also likely to occur in other tumor types such as colorectal hepatic metastases [14]. Thus in some circumstances, tumors are able to "parasitize" the normal stroma and sinusoidal vasculature for their metabolic needs.

Vasculogenesis, the de novo generation of blood vessels from endothelial progenitors, is another method exploited by tumors to generate neovessels. This has been reported in animal models and in some human tumors such as inflammatory subtype breast tumors [15–19]. The importance of this method of tumor vascularization may be more relevant in early tumor development since inhibition of stem cells or endothelial cell precursor mobilization prevents xenografts from inducing the initial angiogenic response [20].

Although currently there is a paucity of evidence, it has also been suggested that intussusception contributes to the establishment of a tumor blood supply. This is the process by which larger vessels within a tumor are divided by columns of tumor cells splitting an individual large vessel into two or more channels. Unlike conventional sprouting angiogenesis, endothelial cell proliferation is not a feature [21, 22]. This may be part of vascular remodeling which may be the dominant mechanism in the establishment of the tumor vascular bed [23–25].

Another somewhat contentious mechanism of neovascularization is vascular mimicry [26]. Vascular mimicry has been reported in aggressive ocular melanomas and ovarian tumors [26] and to be associated with poor survival in several cancers, including colorectal carcinoma, hepatocellular carcinoma, melanoma, non-small cell lung carcinomas, and sarcoma [27]. This is where the neoplastic cells rather than endothelial cells line the blood vessels and conduct the blood [28, 29]. The tumor cells acquire both the morphology and phenotype of endothelium and co-express some vascular markers. Partial lining of the blood vessels by tumor cells has been known for many years [30] and more recently reported in animal models using advanced techniques [31], but in our experience using morphology and double immunohistochemistry, this is not frequently observed in the common solid tumors (unpublished data).

2 Materials

1. Silane-coated or charged microscope slides (e.g., Superfrost Plus®).

2. Oven at 37 °C/hot plate at 70 °C.

3. Xylene or xylene substitute (e.g., Citroclear, Histolene).

4. Graded alcohols (100, 90, and 70 % ethanol).

5. 5 % H_2O_2 in methanol (methanol and 30 % hydrogen peroxide).

6. IHC reaction buffer: phosphate-buffered saline (PBS), Tris-buffered saline (TBS), or commercial preformulated buffer (e.g., Ventana reaction buffer).

7. 10 mM citrate pH 6.0 or commercial products: high-pH TRS pH 9.0 (Dako), Cell Conditioning Solution CC1 pH 8.0 (Ventana).

8. Pressure cooker or automated platform, e.g., Ventana Benchmark Ultra, Leica Bond.

9. Antibodies to CD31 (JC70; Dako M0823), CD34 (QBEND10; Dako M7165) for vascular endothelium, and D2-40 (D2-40; Dako M3619) and LYVE-1 (Abcam ab14917) for lymphatic endothelium. Double immunohistochemistry using markers of proliferation such as Ki-67 [32] and BrdU [23] can be considered if proliferating endothelium needs to be assessed.

10. Detection system: streptavidin-biotin kit or polymer detection systems, e.g., EnVision FLEX (Dako), Ultraview (Ventana).

11. Chromogens: diaminobenzidine (DAB), Fast Red.

12. Counterstain: Scott's Tap Water, Mayer's hematoxylin.

13. Mountant, e.g., Pertex, Ultramount.

14. Microscope cover slips.

15. Chalkley graticule (25 dot).

3 Methods

3.1 Highlighting the Vasculature in Tissue Sections

It should be emphasized that time should be devoted to optimization and validation of the immunohistochemical staining procedure since quality staining with little background greatly facilitates assessment. Consideration should be given to antibody clone choice, antigen retrieval methods, primary antibody titration, and the choice of detection system. Some chromogens form alcohol-soluble end products (e.g., AEC and BCIP/NBT) and care must be taken to use an aqueous mountant with these products. Many histopathology laboratories are well versed in immunohistochemistry necessitating only minor adjustments to the in-house protocol.

Most current diagnostic histopathology laboratories will perform all immunohistochemical staining using fully automated instrumentation. Several companies offer instruments capable of fully automated slide preparation including Ventana Benchmark, Leica Bond, and Dako Omnis. The main advantages of running immunohistochemistry with an automated platform include increased workflow, productivity, and greater consistency in slide staining. In busy laboratories, turnaround time and the ability to run slides overnight unattended are of high importance. Optimization of each individual antibody is initially performed in the laboratory using recommended detection reagents specific to the instrument. Most automated platforms provide reasonable flexibility in protocol optimization; however, most will only operate using detection reagents provided by the instrument's manufacturer. Once a protocol is validated, slides are sectioned and dried, and all other steps are performed onboard the instrument excluding clearing and cover slipping.

All steps performed on an automated system can be performed manually in the laboratory; some manual procedures are outlined below.

3.1.1 Manual Staining Procedure

1. Cut 3–4 μm formalin-fixed paraffin-embedded sections of the representative tumor block onto silane-coated or charged slides.

2. If using a fully automated platform, e.g., Ventana Benchmark Ultra, all steps following cutting of the tissue section may be performed onboard the instrument (excluding cover slipping).

3. Dry at 37 °C overnight in an incubator or melt for 15 min on 70 °C hot plate.

4. Dewax using xylene or substitute for 15 min before passing through graded alcohols (100 % ethanol, 100 % ethanol, and 70 % ethanol) into water.

5. Place in IHC reaction buffer for 5 min.

6. Perform antigen retrieval as required, either in a pressure cooker or water bath.

7. Apply primary antibody at room temperature as outlined in Table 1.

8. Block endogenous peroxidase if using a horseradish peroxidase (HRP)-detection system. Incubate the slides in 3–5 % hydrogen peroxide in distilled water or methanol for 10 min (methanol is gentler on sections that show poor adhesion to the slides).

9. Rinse three times in IHC running buffer for 5 min.

10. Apply the appropriate detection system as outlined below. Ensure that detection system takes into account the type of primary antibody, i.e., whether mouse monoclonal or rabbit polyclonal or rabbit monoclonal.

3.1.2 Detection Systems

The choice of detection system will depend upon the type of primary antibody and the sensitivity required for the assay.

Table 1
Primary antibodies and immunohistochemistry conditions for detection of vascular and lymphatic channels in tissues

Antibody	Antigen retrieval	Clone/source	Dilution	Incubation (min)
CD31 (monoclonal)	Ultra Cell Conditioning 1 (Ventana), 100 °C for 32 min	JC70a; Dako (M0823)	1/50	30
CD34 (monoclonal)	None	QBEND10: Immunotech (0786)	1/50	30
D2-40 (monoclonal)	Pressure cook @ 125 °C for 2 min in TRIS/EDTA pH 9	D2-40: Dako (M3619)	1/50	30
LYVE-1 (polyclonal)	Pressure cook @ 125 °C for 2 min in 10 mM citrate pH 6.0	Poly: Abcam (ab14917)	1/500	30

3.1.3 Chain Polymer-Conjugated Technology

This type of detection system uses an inert dextran backbone that has been labeled with multiple copies of HRP to enhance the detection system. Most major suppliers offer these detection systems (e.g., Dako EnVision FLEX, Ventana Ultraview). The main advantage of these systems over older biotin-based detection systems is that there is no possibility of binding with endogenous tissue biotin. These detection kits give good sensitivity and are more rapid than the alkaline phosphatase anti-alkaline phosphatase (APAAP) technique as they only require a single step for detection. Although some antibodies come already attached to the polymer allowing a single-step staining method, many antibodies are not available and thus utilize a two-step method where a primary antibody is followed by a mouse- or rabbit-labeled dextran-HRP complex. The sensitivity results from the conjugate having 100 s of enzyme molecules and multiple secondary anti-mouse (or rabbit) antibody molecules per backbone. The chromogen 3,3-diaminobenzidine (DAB) forms a brown insoluble end product when it is oxidized in the presence of hydrogen peroxide. Other chromogens may also be used with these kits, such as Naphthol Fast Red which forms a red alcohol-soluble precipitate. Detection kits contain all the necessary reagents for each step including blocking if required. If double immunohistochemical staining is being considered, the technology allows for discrimination using different secondary reagents and substrates.

1. Wash the slides in IHC running buffer after the primary antibody incubation.
2. Incubate the sections with the polymer solution for 30 min.
3. Rinse the sections three times with IHC running buffer.
4. Apply the chromogen substrate (e.g., DAB) for 10 min.
5. Rinse in tap water for 2–3 min.
6. Counterstain and then wash off excess in running tap water.
7. Dehydrate, clear, and mount.

3.1.4 StreptABC (Avidin-Biotin Complex)

This method uses the high affinity of streptavidin for biotin. It requires sequential application of a biotinylated secondary antibody followed by a tertiary antibody complex of streptavidin-biotin HRP. The open sites on the streptavidin complex to the biotin on the secondary antibody. StreptABC kits may come as predilute or ready-to-use reagents or may require titration of the secondary antibody. Ensure that the secondary antibody is appropriate for the primary antibody. A blocking step for endogenous biotin may be required if background binding is observed. Commercial kits are available for this purpose.

1. Apply the secondary antibody for 15–30 min.
2. Wash the slides in IHC running buffer.

3. Apply the streptavidin complex for 15–30 min.

4. Apply the chromogen substrate for 5–10 min.

5. Rinse the slides in tap water and counterstain if desired.

6. Mount in aqueous mountant if using an alcohol-soluble chromogen.
   Otherwise, dehydrate in graded alcohols, then clear in xylene, and mount with permanent mountant.

3.1.5 APAAP

This method uses a soluble enzyme anti-enzyme antibody complex (calf intestinal APAAP) to act on new fuchsin substrate. The primary and final antibody complex are bridged by excess rabbit anti-mouse antibody that binds to the primary mouse antibody with one Fab leaving a Fab site free to bind the tertiary complex. Repeated rounds of application with secondary and tertiary antibodies amplify the staining intensity. The enzyme hydrolyzes the naphthol esters in the substrate to phenols which couple to colorless diazonium salts in the chromogen to produce a red color. Endogenous alkaline phosphatase is inhibited by the addition of 5 mM levamisole that does not inhibit calf intestinal alkaline phosphatase.

1. Apply the secondary antibody for 30 min.

2. Wash the slides in IHC running buffer.

3. Apply the tertiary complex for 30 min.

4. Wash the slides in running tap water.

5. Counterstain if required.

6. Mount in aqueous mountant if using an alcohol-soluble chromogen. Otherwise, dehydrate in graded alcohols, then clear in xylene, and mount with permanent mountant.

3.2 Assessment of Tumor Blood Vascularity

Confirm satisfactory staining using normal entrapped vasculature as internal positive control. An optional parallel negative control section using an IgG_1 isotype antibody may be used. The three "hot spot" areas containing the maximum number of *discrete* microvessels should be identified by scanning the entire tumor at low power ($\times 40$ and $\times 100$) (Fig. 1). This is the most subjective step of the procedure. It has been demonstrated that the experience of the observer determines the success of identifying the relevant hot spots [33]. Poor selection will in turn lead to an inability to classify patients into different prognostic groups. Therefore it is recommended that inexperienced observers spend time in a laboratory where a period of training can be undertaken. Ideally comparisons between hot spots chosen by an experienced investigator and trainee should be performed and continued on different series until there is >90 % agreement. Training can be completed by assessing sections from a series which already contains prognostic information [32].

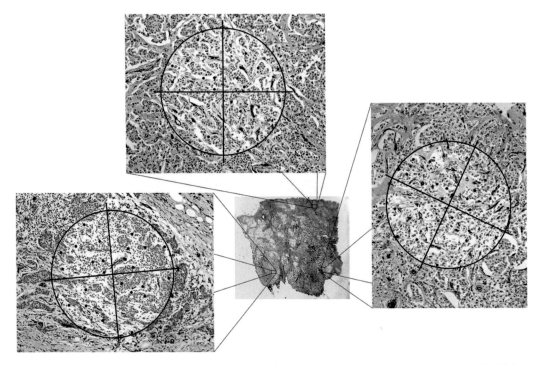

Fig. 1 The tumor is scanned at low power (×40–100) (*center*), and the three areas that contain the highest number of discrete microvessels are selected

Inexperienced observers tend to be drawn to areas with dilated vascular channels, often within the sclerotic body of the tumor. These central areas together with necrotic tumor should be ignored. Vascular lumina or the presence of erythrocytes is not a requirement to be considered a countable vessel, and indeed, many of the microvessels have a collapsed configuration. Although the hot spot areas can occur anywhere within the tumor, they are generally at the tumor periphery making it important to include the normal-tumor interface in the representative area to be assessed. Vessels outside the tumor margin by one ×200–250 field diameter and immediately adjacent benign tissues should not be counted. The procedure takes 2–5 min.

3.3 Chalkley Counting

Once selected, a 25-point Chalkley point eyepiece graticule [34] at ×200–250 should then be oriented over each hot spot region so that the maximum number of graticule points is on or within areas of highlighted vessels (Fig. 2). Particular care should be taken in the occasional case (<1 % breast cancers) where an intense plasma cell infiltrate can mimic a hot spot and also obscure the underlying tumor vasculature. Plasma cells can otherwise be disregarded on morphological grounds. The mean of the three Chalkley counts is then generated for each tumor and used for statistical analysis. The procedure takes 2–3 min.

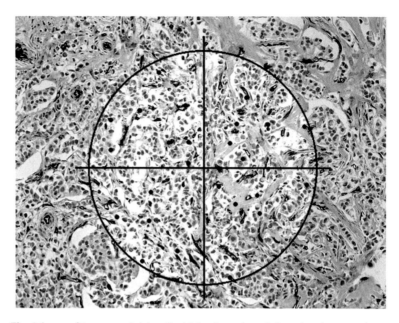

Fig. 2 Areas of tumor containing the highest number of discrete microvessels are examined at high power (×200–250), and the Chalkley point graticule is then rotated in the eyepiece so the maximum number of graticule dots coincides with the vessels or their lumens, which are then recorded

3.4 Intratumoral Microvessel Density

For this index any endothelial cell or endothelial cell cluster separate from adjacent microvessel, tumor cells, or matrix elements is considered a countable vessel. Those that appear to be derived from the same vessel if distinct should also be counted. Again vessel lumens and erythrocytes are not included in the criteria defining a microvessel. There is no cutoff for vessel caliber. The procedure takes 3–6 min.

3.5 Vascular Grading

To facilitate assessing angiogenesis in tissue sections, akin to semi-quantitative tumor grading, a vascular grading based on the subjective appraisal by trained observers over a conference microscope has been assessed [6, 35]. Significant correlations were demonstrated between vascular grade and both microvessel density ($p = 0.002$) and Chalkley count ($p = 0.0001$). Although the method is reproducible [36], delineating criteria is difficult due to the subjective nature of the system, and a considerable investment in time would be required to align the cutoffs required for multicenter studies. However, although there is some loss of power associated with translation of numerical to categorical data, the overall time savings engendered by this make it an attractive proposition. Nevertheless, further validation in a large series of randomized patients is warranted to determine its prognostic utility before its application can be considered in such studies.

3.6 Novel Angiogenic Antigens

Instead of highlighting all the tumor-associated endothelium, an alternative approach would be to selectively identify only the vasculature that is undergoing active neovascularization. This might be valuable not only in more accurately quantifying tumor angiogenesis but might also have important implications for anti-vascular targeting [37]. A number of antibodies have been identified which recognize antigens that have been reported to be upregulated in tumor-associated endothelium compared with normal tissues and include EN7/44, endosialin, endoglin (CD105), and nestin [32, 38].

3.6.1 CD105 (Endoglin)

CD105 is an accessory receptor for transforming growth factor beta, upregulated in proliferating endothelial cells and believed to be more specific to newly formed vessels compared with CD31 and CD34. CD105-derived microvascular density has been found to be more closely associated with prognosis in several forms of cancer, including astrocytoma, glioblastoma, upper urinary tract urothelial carcinoma, non-small cell lung cancer, hepatocellular carcinoma, and breast carcinoma, compared with CD31 and CD34 [39–42], and is more closely correlated with VEGF expression compared with CD31 [40].

CD105 can be expressed in tumor cells in addition to endothelial cells [43–45], complicating its use with automated image analysis.

Phase I and II clinical trials are in progress examining the use of TR105, an anti-CD105 antibody, in combination with bevacizumab (vascular endothelial growth factor (VEGF) monoclonal antibody) or VEGF receptor (VEGFR) tyrosine kinase inhibitors in several cancer types, including the relationship between CD105 expression and clinical response [46].

3.6.2 Nestin

Nestin is an intermediate filament protein, present in proliferating endothelial progenitor cells but not mature endothelial cells [47]. Nestin expression in endothelial cells has been reported to be correlated with expression of proliferation markers Ki-67 and PNCA [48–50] and to be more closely associated with prognosis than CD34 in colorectal carcinoma [48]. In breast cancer, co-expression of nestin and Ki-67, as a marker of microvessel proliferation, was associated with aggressive phenotypic features and poor survival [51]. As with CD105, nestin can also be expressed in tumor cells themselves [49] and expressed in association with inflammation.

3.7 Tumor Vascular Architecture

The vascular morphology of tumors is different within tumors of similar and different histological types [30]. It has been suggested that particular vascular patterns might both help distinguish benign from malignant lesions [52, 53] and be a prognostic marker; in ocular melanomas, a closed back-to-back loop vascular pattern was associated with death from metastasis [54], and in lung carcinoma distinct patterns of neovascularization might potentially respond differently to anticancer treatments [55]. In addition, it has been recently reported that the distribution of lymphatic vessels near the tumor edge in carcinoma of the cervix was associated with the presence of lymphovascular invasion and lymph node metastases [56].

3.8 Computer-Aided Image Analysis

We and others have attempted to automate the counting procedure by using computer image analysis systems [4, 33, 57–67]. These systems have several drawbacks not including the capital and running costs notwithstanding over those shared with manual methods. An endothelial marker that gives sensitive and specific capillary staining is essential to reduce background signal.

Computer simulation of Chalkley counting using ImageJ, a public-domain image processing program, has been demonstrated to show high correlation with traditional Chalkley counts, but this still requires operator input to select hot spots for analysis [68]. Currently, there is no software available that can identify hot spots but, when developed, the requirement for motorized stages will add further expense. Although partially automated systems with area and shape filters using defined color tolerances are available, most systems are not fully automated, require a high degree of operator interaction, and like manual counting suffer from observer bias. Currently computer image analysis is more costly, time consuming, and no more accurate than a trained observer, and these factors make it unsuitable to routine diagnostic practice. In addition, correlation between image analysis-determined microvascular density and manual microvascular density assessment and association of image analysis-determined microvascular density with clinical outcome are yet to be established. Hansen et al. reported poor correlation between image analysis and manual determination of microvascular density using miRNA-126 and CD34 [69], while Mohammed et al. reported acceptable correlation of microvascular density between the two methods, but the image analysis-derived microvascular densities were not associated with clinical outcome in patients with invasive breast cancer [70].

Computer image analysis systems do however have the advantage of being able to quantify the whole section rather than hot spots, but the prognostic implications of whole tumor vascularity will need to be established. Computer-aided image analysis also allows 3D-reconstructions of vessel systems, such as that performed by Asioli et al. who demonstrated the spatial organization of pre-lymphatics in breast tissue [71]. In addition, more complex calculations of vessel parameters are able to be determined than by the human eye alone. This is useful as vasculature, like most natural objects, have non-Euclidean geometry [72]. Computer-aided image analysis can be used to determine fractal parameters [72]; however, the utility of these as predictive and prognostic markers is yet to be fully determined.

Nevertheless, data from these studies have demonstrated that most vascular indices including microvessel density, vessel perimeter, and vascular area are significantly correlated suggesting that they are equivalent indices of angiogenesis [59]. However, previously it had been hypothesized that microvessel density might not be the most important vascular parameter, since a large vascular perimeter or area might be a better measure of angiogenesis since these may reflect the functional aspects of endothelial surface and volume of blood available for interaction with the tumor [59]. Open-source software, such as AngioTool [73], have been developed to determine morphological and spatial parameters of vascular networks and validated in preclinical models. In human tumors, high vessel complexity and large vessel size have been reported to be associated with poor survival in breast cancer patients, which the study authors hypothesize is related to the functional implications of these vessels [74].

4 Notes

1. Since endothelium is highly heterogeneous [75], the choice of antibody profoundly influences the number of microvessels available for assessment. Many such as those directed to vimentin [57], lectin [4, 76], alkaline phosphatase [3], and type IV collagen [58, 77] suffer from low specificity and are present on many non-endothelial elements. Others including antibodies to Factor VIII-related antigen and the marker used in most studies [6, 35, 78–83] identify only a proportion of capillaries and also detect lymphatic endothelium. The most specific and sensitive endothelial marker currently available is CD31 which is present in most capillaries and is a reliable epitope for immunostaining in routinely handled formalin-fixed paraffin-embedded tissues [84]. CD34 has been recommended by the Second International Consensus on the Methodology and Criteria of Evaluation of Angiogenesis Quantification in Solid Human Tumors. However, care is required when interpreting staining since some stromal cells also express this antigen, particularly in breast [32].

   Once the tumor vasculature has been immunohistochemically highlighted, the tumor is scanned at low magnification (×40–100) to identify angiogenic hot spots [6, 35]. The number of vessels is then quantified at high magnification (×200–400; field area 0.15–0.74 mm²) in these regions. These areas of high vascularity are chosen on the basis that these are the tumor regions which are likely to be biologically important. A high magnification (which will identify more microvessels by virtue of increased resolution [85]) used over a too small an area will always give a high vessel index, whereas a low magnification over too large an area will dilute out the "hot spot."

Further, tumors naturally have a limited number hot spots which would be diluted if too many were counted. Thus, although the number of hot spots assessed varies from 1 to 5 [6, 35, 78, 79, 86–88], most studies have examined three from a single representative tissue block. Nevertheless, both the magnification used and its corresponding tumor field area will determine the vessel number derived from each hot spot. It is thus recommended that three regions are examined using a microscope magnification of field area of between ×200 and 400 (corresponds to areas of approximately 0.15–0.74 mm^2 depending on the microscope type) [32, 59].

Although less subjective than identifying angiogenic hot spots [32], the process of counting vessels has also resulted in significant variation in published series. This has been emphasized in the study of Axelsson et al. [89] where the authors, after an initial training period with Weidner, who defined the criteria as to what constituted individual microvessels (*see* Subheading 3), did not observe a correlation between microvessel density and patient survival. Even experienced observers occasionally disagree as to what constitutes a microvessel. To overcome these problems, after selection of each hot spot, a 25-dot Chalkley microscope eyepiece graticule [34] has been used to quantify tumor angiogenesis (*see* Subheading 3). This method is not only objective, since no decision is required as to whether adjacent stained structures are separate, but is rapid (2–3 min per section) and reproducible and gives independent prognostic information in breast [59, 90, 91] and bladder [92] cancers. Thus, it is currently the preferred method contained in a multicenter discussion paper [32].

The final consideration in quantifying angiogenesis is the difference in the value used for stratification into different study groups. This alone will result in different conclusions being drawn from the same data set. Studies have used the highest, the mean, the median [85], tertiles [59], mean count in node-negative patients with recurrence [81] or variable cutoffs given as a function of tumor area [6, 35], or microscope magnification [88]. The use of the median and tertile groups avoids strong assumptions about the relationship between tumor vascularity and other variables including survival and is therefore useful clinically. However, there is some loss of information making it optimal to use continuous data where possible.

5   Other Measures of Tumor Vascularity

5.1   Other Measures of Angiogenesis

The microvessels highlighted by immunohistochemistry in a tissue section are the conclusion of a dynamic multistep process. The evolving neovasculature is the result of a complex interplay between extracellular matrix remodeling, endothelial cell migration and proliferation, capillary differentiation, and anastomosis [93, 94]. Although it may soon be possible to measure these continuous processes in vivo, human tissue measurement of molecules involved in these events might be surrogate end points of angiogenesis. Thus partly due to many of the inherent and methodological difficulties of vascular counts, these alternative strategies for quantifying tumor angiogenesis have also been pursued.

5.2 Angiogenic Factors and Receptors

Angiogenesis is the result of the net change in the balance of angiogenic stimulators and inhibitors (i.e., gain of promoters and/or loss of inhibitors). There are now numerous reports documenting upregulation of several angiogenic factors and their receptors at the mRNA and protein level using a variety of techniques in a range of histological tumor types including breast [95–102].

The prototypical factor is vascular endothelial growth factor (VEGF). Bevacizumab, a monoclonal antibody against VEGF, and several small-molecule inhibitors of vascular endothelial growth factor receptors (VEGFRs) have been approved for clinical use in several types of cancer [103, 104]. Bevacizumab (Avastin) is currently approved by the FDA for treatment of metastatic colorectal carcinoma, metastatic renal cell carcinoma, non-small cell carcinoma of the lung, and glioblastoma [105]. Previous approval for the treatment of metastatic HER2-negative breast cancer was withdrawn in 2011 due to the lack of demonstrated clinical benefit [106]. Small-molecule inhibitors of VEGFRs currently in clinical use include sorafenib, used in the treatment of advanced renal cell carcinoma, unresectable hepatocellular carcinoma, and recurrent or metastatic differentiated thyroid carcinoma refractory to radioactive iodine therapy, and sunitinib, used to treat metastatic renal cell carcinoma, gastrointestinal stromal tumor, and pancreatic neuroendocrine tumors [103, 107, 108].

A significant relationship between tumor VEGF levels and microvessel density has been shown in breast tumors [109–111]. Furthermore, some studies demonstrated that tumor VEGF expression levels gave prognostic information in breast carcinomas [112–117] that the power of which may be improved by combining it in a ratio with soluble flt-1 [118]. Studies have also demonstrated the association of VEGF with prognosis in other cancer types, including soft tissue sarcomas [119], prostate carcinoma [120], colorectal carcinoma [121], pancreatic adenocarcinoma [122], and glioblastoma [123].

However, this is a complex pathway with four direct members VEGF-A, VEGF-B, VEGF-C, and VEGF-D with the VEGF-A and VEGF-C genes generating additional isoforms through alternative splicing. These variably bind to three receptors, VEGF receptor (VEGFR)-1, VEGFR-2, and VEGFR-3, with the latter largely restricted to lymphatic endothelium in normal tissues. VEGF-C and

VEGF-D are expressed in areas of lymphatic sprouting during embryonic development, and some studies have shown a correlation between VEGF-C levels [124, 125] or VEGFR-3 [124] and lymph node metastases. Other family ligand members and co-receptors such as placenta growth factor and neuropilins, respectively, further complicate the pathway. However, not all studies have shown associations with clinicopathological factors [124–126]. It is also unknown for human breast tumors what is the dominant factor or isoforms at different stages of neoplastic progression. Some studies of VEGF-D expression in breast, colon, and endometrium have suggested that this may also be an independent prognostic marker [127–129]. It has been further suggested that VEGF-D may be particularly important in inflammatory breast cancer [130].

However, in our experience, the assessment of VEGF in histological sections by immunohistochemistry is not straightforward due to the presence of widespread nonspecific staining.

Thymidine phosphorylase (TP) is another angiogenic factor that appears to be important in human cancer. This migratory rather than mitogenic angiogenic factor is expressed in in situ [131], and invasive breast cancer [132] has also been associated in some studies with microvessel density [132, 133] and patient survival [134–137].

Different tumors use different angiogenic factors during the various phases of their development (breast carcinomas co-express VEGF and TP [138] whereas they are reciprocally expressed in bladder cancers [139, 140]), and it is likely that determining specific profiles for individual tumor types might assume greater importance in quantitative tumor angiogenesis. Nevertheless, some sera investigations may be compromised by the high VEGF levels present in platelets and not reflect tumor-derived VEGF, suggesting plasma measurement may be more accurate [141].

5.3 Endothelial Cell Proliferation

It is now possible to measure endothelial cell proliferation in tissue sections using double immunohistochemistry with antibodies to endothelial markers to discriminate endothelial cells from other tissue elements in conjunction with antibodies to proliferation markers. We have used combinations of CD31 or CD34 (for endothelium) and BrdU and Ki67 (MIB-1) (as proliferation markers) with good results, but some of the newer cell cycle markers that can also be used on archival tissue, such as minichromosome maintenance proteins 2 and 5, may also be of use [142, 143]. This technique allows simultaneous assessment of tumor and endothelial cell proliferation. High vascular proliferation index has been associated with aggressive features and poorer survival in breast cancer [51] and may aid the stratification of patients for novel therapies. Studies of lymphatic endothelial cell proliferation using lymphatic markers and proliferation markers can also be undertaken.

5.4 Vessel Maturation Index

A late event in the establishment of a tumor blood supply, which accompanies downregulation of endothelial cell proliferation, is pericyte recruitment and secretion of a basement membrane. This basement membrane is irregular and is composed of abnormal ratios of fibronectin, laminin, and collagen depending on the maturation state of the capillary. Although many studies have documented the heterogeneity, few studies have assessed the significance. Nevertheless, some studies have examined the ratio of endothelial cells with a pericyte [144] or basement membrane [25] cover as a surrogate of vessel maturation. There is great variation between tumor types [144, 145], but studies in breast carcinomas have shown that the vessel maturation index (VMI) gives a different measure to that of microvessel density. There is continual remodeling of vessels in normal breast, a subset of patients can be identified who have an elevated risk of node recurrence, and the vessel maturation index potentially gives more functional information [25].

5.5 Cell Adhesion Molecules

Increasing evidence suggests that many of the endothelial cell adhesion molecules (CAMs) of the immunoglobulin, cadherin, selectin, and integrin superfamilies, which have physiological roles in immune trafficking and tumor metastasis, also play a major role in angiogenesis [146]. In particular, differential VE-cadherin expression, thought to be controlled by VEGF and Notch signaling, has recently been shown to play an important role in the proper formation of vessels during sprouting angiogenesis. Under high VEGF conditions, such as in tumors, VE-cadherin expression is dysregulated and differential endothelial cell movement is impaired resulting in abnormally formed vessels [147].

Some clinical studies indicated that melanoma patients with upregulated CAMs on endothelium have a significantly worse prognosis and these studies validated the interest in CAMs and their cognate ligands in tumor angiogenesis [148, 149]. Indeed soluble CAMs are readily identified in sera of cancer-bearing patients, although their relationship to tumor angiogenesis is yet unknown [150, 151]. Similarly integrins, including b3av, also appear to be upregulated in human breast carcinomas compared to normal or benign breast and might also be a potential surrogate marker for angiogenesis [152, 153]. Therefore cell adhesion molecules are potential targets for anticancer therapies. These targeted agents include intetumumab, a monoclonal antibody to $\alpha v\beta 1$, $\alpha v\beta 3$, $\alpha v\beta 5$, and $\alpha v\beta 6$ integrins, and cilengitide, an inhibitor of integrins $\alpha v\beta 3$ and $\alpha v\beta 5$. These agents have been evaluated in phase II clinical trials in a variety of cancers including prostate cancer [154], non-small cell lung carcinoma [155], melanoma [156, 157], and glioblastoma [158]. However, recently, a phase III clinical trial of cilengitide in addition to standard treatment in glioblastoma reported no survival benefit [159].

5.6 Proteolytic Enzymes

Several studies have demonstrated that proteolytic enzymes including the plasminogen activators and the matrix metalloproteinases that are important in tumor cell invasion and migration are also important in angiogenesis [160–165]. Although no correlation was observed between microvessel density and both uPA and PAI-1 [166], the poor prognosis in tumors [167–173] associated with elevated levels of the uPA system are likely to be partly due to the angiogenic activity of these tumors. Thus measurement of proteases, particularly components of the urokinase system, might give some indication of the angiogenic activity of a tumor.

5.7 Hypoxic Markers

Once a tumor vasculature has been established, there is still continued remodeling of vessels. The remodeling process is likely to be related to an exaggerated stress response and therefore is profoundly influenced by the tumor microenvironment. Hypoxia has been frequently reported despite the increased microvessel density in tumors. The hypoxia may be due to a reduction in blood flow from structural differences in blood vessels [30, 174] or from the influence of permeability factors such as VEGF-A [175] or from shunting of blood across the tumor vascular bed as reported in several tumor types [176]. In any of these scenarios, tissue hypoxia results in stabilization of the hypoxic inducible factors that mediate transcription of angiogenic pathways thereby enhancing tumor angiogenesis. A pivotal pathway is the regulation by hypoxia of VEGF through the transcription factor hypoxia-inducible factor (HIF).

Hypoxia-inducible factor 1α (and 2α) binds to the aryl-hydrocarbon nuclear translocator (ARNT) (HIF-1β), which then binds a specific DNA hypoxia response element increasing mRNA transcription. In normoxia the HIF-α units are unstable and are rapidly degraded by the proteasome pathway through ubiquitin E3 ligase complex, the recognition component of the von Hippel-Lindau (VHL). This is regulated through enzymatic hydroxylation of either two critical prolyl residues within the oxygen-dependent degradation domains of HIF-α subunits by prolyl hydroxylase-1, hydroxylase-2, and hydroxylase-3 (PHDs) and dioxygen as a co-substrate which results in one oxygen incorporated into the prolyl residue of HIF-α, the other into 2-oxoglutarate to yield succinate and carbon dioxide. However, in hypoxic conditions, since no oxygen is available for hydroxylation, the HIF-α subunit is stabilized and translocated to the nucleus where it binds to HIF-β. The complex then recruits co-activators and binds a specific DNA hypoxia response element (HRE) resulting in increased mRNA transcription (reviewed in [177]). A further level of control is mediated by factor inhibitor of HIF (FIH) that interferes with coactivator binding.

Many HIF target genes are beneficial to the tumor including those involved in iron metabolism (transferrin and ceruloplasmin), angiogenesis (VEGF, VEGFR-1, TP, Ang2), glucose metabolism (glucose transporters), proliferation (IGF II), endothelial adhesion, and pH regulation (carbonic anhydrase IX). Antibodies to both HIFs [178] and carbonic anhydrase IX [179] have been shown to be surrogates of hypoxia [180] and associated with a poor prognosis [181, 182]. Thus measurement of the factors involved in the regulation of the hypoxic response may be clinically important in patient management [183, 184].

Indeed, HIF-1α immunohistochemistry has been reported to correlate with microvascular density [123], VEGF-C expression and lymphatic microvascular density [185, 186], and overall and disease-free survival in breast cancer patients [186].

Nevertheless, these techniques require a degree of quantitation and are currently unsuitable for a general diagnostic pathology laboratory. The presence of a fibrotic focus as a surrogate marker of hypoxia is being examined. This is defined as a scar-like area, consisting of fibroblasts and collagen fibers in the center of an invasive ductal carcinoma of the breast. It was first proposed in 1996 by Hasebe et al. as an indicator of tumor aggressiveness [187–190], and the presence of a fibrotic focus is associated with a higher microvessel density [191] and more latterly a marker of intratumoral hypoxia. Thus a fibrotic focus may be a useful surrogate marker of hypoxia-driven ongoing angiogenesis [192].

5.7.1 Assessment of Lymphangiogenesis

In recent years, there has been considerable interest in the role of the lymphatics in solid tumors. Previously thought to act as passive conduits in the dissemination of malignancy, it is now realized that lymphatic vessels actively modulate tumor progression [193].

The lymphatic vascular system consists of a hierarchical arrangement of vessels. The smallest lymphatic channels through which interstitial fluid and cells enter are known as the initial lymphatics. These drain into pre-collecting lymphatics which in turn drain into collecting lymphatics. The collecting lymphatics then transport the lymphatic fluid to the lymph nodes [193].

Similar to blood vessels, the lymphatics can also be highlighted with immunohistochemical markers, the most commonly used being the monoclonal antibody D2-40, which targets podoplanin, and LYVE-1, which targets endothelial hyaluronan receptor-1. D2-40 is expressed by endothelial cells of all lymphatic vessel subtypes, while LYVE-1 is expressed in small lymphatic vessels but not by collecting lymphatics [193]. LYVE-1 polyclonal antibodies have been reported to be more sensitive than LYVE-1 monoclonal antibodies in detection of lymphatic vessels in breast cancer tissue [194]. Both D2-40 and LYVE-1 require a degree of operator input in using computerized image analysis systems as D2-40 also stains myoepithelial cells and certain tumors [195], while LYVE-1 also stains the endothelial cells lining hepatic sinusoids [196], pulmonary capillaries, and pulmonary arteries [195], and its expression is affected by inflammatory processes [197].

Assessment of lymphatic vessels is similar to that of blood vessels. Hot spots are identified by low-power magnification scanning of sections immunohistochemically stained with D2-40 or another lymphatic marker. Lymphatic vessels in these hot spots are then counted over several high-power fields and the counts averaged to generate a lymphatic microvascular density score. High lymphatic microvascular density has been associated with poorer survival in breast cancer, non-small cell carcinoma of the lung, and pancreatic adenocarcinoma [198–200] and the presence of lymph node metastases in microinvasive ductal carcinoma in situ [201] and colorectal carcinoma [202]. However several other studies did not observe correlation of lymphatic microvascular density with clinical outcome [203–205].

6   Notes

1. The tumor block should be selected by examining hematoxylin and eosin stained slides.

2. If using TMAs the periphery of the tumor should be selected since this is the area where angiogenesis is most active; additionally it is likely that at least three tissue cores should be examined.

3. 8 μm cryostat sections can also be used but the area of tumor assessed is less representative.

4. If sections continually float off after antigen retrieval, drying at 56 °C overnight or using distilled water when cutting sections will increase tissue adherence.

5. Conjugated polymers or APAAP methodology is preferred in tissues such as liver and kidney that contain high endogenous biotin.

6. PBS works well for chain polymer-conjugated or StreptABC methodologies, and Tris-buffered saline (TBS) works well for alkaline phosphatase anti-alkaline phosphatase (APAAP).

7. Chalkley graticule can be obtained from Graticules Ltd, Morely Road, Botany Trading Estate, Tonbridge Wells, Kent, TN9 1ZN, UK. NB. The size of the graticule required will depend on the eyepiece diameter of the microscope lens.

8. Studies using a magnification of ×200–400 and field areas of between 0.74 and 0.15 mm^2 have derived prognostic information.

9. One block is justified by the prevailing evidence that shows a high concordance in vessel number between different blocks [78, 206, 207].

10. Stromal cell immunoreactivity may interfere with microvessel counts if using CD34 as a vascular marker.

7   Tissue Microarrays (TMAs) and Their Assessment of Staining

Tissue microarrays (TMAs) have become an important tool for the high-throughput assessment of candidate markers. They have also helped standardize testing across samples where it was previously difficult to ensure reproducibility when using cohorts of individual slides. They have also enabled more accurate correlation analyses since the same part (core) of the tumor is being examined across markers in contrast to the use of whole tissue sections when different parts of so-called positive tumors were being compared. Interestingly for several markers such as estrogen receptor in breast tumors, the use of a single core is representative of the whole tumor.

An important consideration when using TMAs for quantifying vessel density is that vessels may not be evenly distributed throughout the tumor, and as such, sampling may not include vascular hot spots which typically occur at the tumor periphery. Nonetheless, good correlations of microvascular density between TMAs and whole sections have been reported [70, 208].

Most TMAs use the core system pioneered by Kononen et al. [209] where cores of tissue from donor blocks are arrayed at high density into a recipient block using a dedicated instrument. These may hold many hundreds of tumor cores and give 150 serial sections. When constructing TMAs we usually take four cores of tumor and two cores of normal all arrayed in separate recipient blocks to give six TMAs that are representative of normal and tumor taking into account any heterogeneity that may be apparent for some novel markers. More recently this technology has been extended to cutting edge matrix assembly arrays that maximize array density [210].

There is no single correct method for the scoring of immunohistochemistry in tissue sections. Although some authorities do not use intensity in their system, we use a combination of both intensity and proportion of cell staining when assessing tissue microarrays. However, the range of intensity scores is dependent on the dynamic range of the antibody. Thus, for most epitopes, we use the standard 0 (no staining), 1 (weak staining), 2 (moderate staining), and 3 (strong staining), but for the HIFs, we use 0, 1, and 2 since these demonstrate a narrower range of staining in our hands. For the proportion of cells staining, we use the broad categories of 0, 1–10 %, 11–50 %, 51–80 %, and 81–100 % that are reproducible in our hands. For analysis we have used the product of intensity and proportion of cells staining with variable cutoff depending on the distribution of cases. We have tended to use medians and tertiles to avoid assumptions on what level of staining is significant.

The analysis of data derived from TMAs has proved challenging. Akin to the bioinformatic analysis of cDNA microarrays, similar algorithms of interrogation are required. We have used several methods including multivariate logistic regression and the elastic net method [211], but others are also available [212, 213].

8   Summary

Continuing research into angiogenesis using quantitative data will not only broaden our understanding of the angiogenic process but will have several potential clinical applications beyond its use for prognosis [38]. It might help in stratifying patients for cytotoxic therapy [214], aid monitoring, and prediction of their response [215], and, with the advent of anti-angiogenesis and vascular targeting, treatment could be stratified and altered based on these angiogenic measurements. The next few years will provide the data as to the reliability of quantitation of angiogenesis in tissue sections. During this time it is also probable that basic research will describe several candidate molecules that might become objective, sensitive, and specific enough to supersede the presently used assays.

References

1. Folkman J (1971) Tumour angiogenesis: therapeutic implications. N Engl J Med 285:82–86

2. Brem S, Cotran R, Folkman J (1972) Tumor angiogenesis: a quantitative method for histological grading. J Natl Cancer Inst 48:347–356

3. Mlynek M, van Beunigen D, Leder L-D, Streffer C (1985) Measurement of the grade of vascularisation in histological tumour tissue sections. Br J Cancer 52:945–948

4. Svrivastava A, Laidler P, Davies R, Horgan K, Hughes L (1988) The prognostic significance of tumor vascularity in intermediate-thickness (0.76–4.0 mm thick) skin melanoma. Am J Pathol 133:419–423

5. Porschen R, Classen S, Piontek M, Borchard F (1994) Vascularization of carcinomas of the esophagus and its correlation with tumor proliferation. Cancer Res 54(2):587–591

6. Weidner N, Semple JP, Welch WR, Folkman J (1991) Tumor angiogenesis and metastasis—correlation in invasive breast carcinoma. N Engl J Med 324(1):1–8

7. Fox SB (1997) Tumour angiogenesis and prognosis. Histopathology 30(3):294–301

8. Fox S, Harris A (2004) The biology of breast tumor angiogenesis. In: Harris J, Lippman ME, Morrow M, Osborne CK (eds) Diseases of the breast, 3rd edn. Lippincott Williams & Wilkins, Philadelphia, pp 441–458

9. Folkman J (1990) What is the evidence that tumours are angiogenesis dependent. J Natl Cancer Inst 82:4–6

10. Pezzella F, Pastorin OU, Tagliabue E, Andreola S, Sozzi G, Gasparini G, Menard S, Gatter K, Harris A, Fox S et al (1996) Non-small-cell lung carcinoma tumor growth without morphological evidence of neo-angiogenesis. Am J Pathol 151(5):1417–1423

11. Pezzella F (2000) Evidence for novel non-angiogenic pathway in breast-cancer metastasis. Breast Cancer Progression Working Party. Lancet 355(9217):1787–1788

12. Holash J, Maisonpierre PC, Compton D, Boland P, Alexander CR, Zagzag D, Yancopoulos GD, Wiegand SJ (1999) Vessel cooption, regression, and growth in tumors mediated by angiopoietins and VEGF. Science 284(5422):1994–1998

13. Holash J, Wiegand SJ, Yancopoulos GD (1999) New model of tumor angiogenesis: dynamic balance between vessel regression and growth mediated by angiopoietins and VEGF. Oncogene 18(38):5356–5362

14. Vermeulen PB, Colpaert C, Salgado R, Royers R, Hellemans H, Van Den Heuvel E, Goovaerts G, Dirix LY, Van Marck E (2001) Liver metastases from colorectal adenocarcinomas grow in three patterns with different angiogenesis and desmoplasia. J Pathol 195(3):336–342

15. Shirakawa K, Wakasugi H, Heike Y, Watanabe I, Yamada S, Saito K, Konishi F (2002) Vasculogenic mimicry and pseudo-comedo formation in breast cancer. Int J Cancer 99(6):821–828

16. Asahara T, Masuda H, Takahashi T, Kalka C, Pastore C, Silver M, Kearne M, Magner M, Isner JM (1999) Bone marrow origin of endothelial progenitor cells responsible for postnatal vasculogenesis in physiological and pathological neovascularization. Circ Res 85(3):221–228

17. Gunsilius E, Duba HC, Petzer AL, Kahler CM, Grunewald K, Stockhammer G, Gabl C, Dirnhofer S, Clausen J, Gastl G (2000) Evidence from a leukaemia model for maintenance of vascular endothelium by bone-marrow-derived endothelial cells. Lancet 355(9216):1688–1691

18. Rafii S (2000) Circulating endothelial precursors: mystery, reality, and promise [comment]. J Clin Invest 105(1):17–19

19. Asahara T, Takahashi T, Masuda H, Kalka C, Chen D, Iwaguro H, Inai Y, Silver M, Isner JM (1999) VEGF contributes to postnatal neovascularization by mobilizing bone marrow-derived endothelial progenitor cells. EMBO J 18(14):3964–3972

20. Lyden D, Hattori K, Dias S, Costa C, Blaikie P, Butros L, Chadburn A, Heissig B, Marks W, Witte L et al (2001) Impaired recruitment of bone-marrow-derived endothelial and hematopoietic precursor cells blocks tumor angiogenesis and growth. Nat Med 7(11):1194–1201

21. Patan S, Munn LL, Jain RK (1996) Intussusceptive microvascular growth in a human colon adenocarcinoma xenograft: a novel mechanism of tumor angiogenesis. Microvasc Res 51(2):260–272

22. Patan S (2000) Vasculogenesis and angiogenesis as mechanisms of vascular network formation, growth and remodeling. J Neurooncol 50(1–2):1–15

23. Fox S, Gatter K, Bicknell R, Going J, Stanton P, Cooke T, Harris A (1993) Relationship of endothelial cell proliferation to tumor vascularity in human breast cancer. Cancer Res 53:9161–9163

24. Kakolyris S, Giatromanolaki A, Koukourakis M, Leigh IM, Georgoulias V, Kanavaros P, Sivridis E, Gatter KC, Harris AL (1999) Assessment of vascular maturation in non-small cell lung cancer using a novel basement membrane component, LH39: correlation with p53 and angiogenic factor expression. Cancer Res 59(21):5602–5607

25. Kakolyris S, Fox SB, Koukourakis M, Giatromanolaki A, Brown N, Leek RD, Taylor M, Leigh IM, Gatter KC, Harris AL (2000) Relationship of vascular maturation in breast cancer blood vessels to vascular density and metastasis, assessed by expression of a novel basement membrane component, LH39. Br J Cancer 82(4):844–851

26. Sood AK, Seftor EA, Fletcher MS, Gardner LM, Heidger PM, Buller RE, Seftor RE, Hendrix MJ (2001) Molecular determinants of ovarian cancer plasticity. Am J Pathol 158(4):1279–1288

27. Cao Z, Bao M, Miele L, Sarkar FH, Wang Z, Zhou Q (2013) Tumour vasculogenic mimicry is associated with poor prognosis of human cancer patients: a systemic review and meta-analysis. Eur J Cancer 49(18):3914–3923

28. Folberg R, Hendrix MJ, Maniotis AJ (2000) Vasculogenic mimicry and tumor angiogenesis. Am J Pathol 156(2):361–381

29. McDonald DM, Munn L, Jain RK (2000) Vasculogenic mimicry: how convincing, how novel, and how significant? Am J Pathol 156(2):383–388

30. Warren B (1979) The vascular morphology of tumors. In: Peterson H (ed) Tumor blood circulation. CRC Press, Boca Raton, FL, pp 1–47

31. Chang YS, di Tomaso E, McDonald DM, Jones R, Jain RK, Munn LL (2000) Mosaic blood vessels in tumors: frequency of cancer cells in contact with flowing blood. Proc Natl Acad Sci U S A 97(26):14608–14613

32. Vermeulen PB, Gasparini G, Fox SB, Colpaert C, Marson LP, Gion M, Belien JA, de Waal RM, Van Marck E, Magnani E et al (2002) Second international consensus on the methodology and criteria of evaluation of angiogenesis quantification in solid human tumours. Eur J Cancer 38(12):1564–1579

33. Barbareschi M, Weidner N, Gasparini G, Morelli L, Forti S, Eccher C, Fina P, Caffo O, Leonardi E, Mauri F et al (1995) Microvessel quantitation in breast carcinomas. Appl Immunochem 3(2):75–84

34. Chalkley H (1943) Method for the quantitative morphological analysis of tissues. J Natl Cancer Inst 4:47–53

35. Weidner N, Folkman J, Pozza F, Bevilacqua P, Allred EN, Moore DH, Meli S, Gasparini G (1992) Tumor angiogenesis: a new significant and independent prognostic indicator in early-stage breast carcinoma. J Natl Cancer Inst 84(24):1875–1887

36. Fox SB, Leek RD, Bliss J, Mansi JL, Gusterson B, Gatter KC, Harris AL (1997) Association of tumor angiogenesis with bone marrow micrometastases in breast cancer patients. J Natl Cancer Inst 89(14):1044–1049

37. Burrows FJ, Thorpe PE (1994) Vascular targeting—a new approach to the therapy of solid tumors. Pharmacol Ther 64(1):155–174

38. Fox S, Harris A (1997) Markers of tumor angiogenesis: clinical applications in prognosis and anti-angiogenic therapy. Invest New Drugs 15:15–28

39. Dales JP, Garcia S, Andrac L, Carpentier S, Ramuz O, Lavaut MN, Allasia C, Bonnier P, Charpin C (2004) Prognostic significance of angiogenesis evaluated by CD105 expression compared to CD31 in 905 breast carcinomas: correlation with long-term patient outcome. Int J Oncol 24(5):1197–1204

40. Yao Y, Kubota T, Takeuchi H, Sato K (2005) Prognostic significance of microvessel density determined by an anti-CD105/endoglin monoclonal antibody in astrocytic tumors: comparison with an anti-CD31 monoclonal antibody. Neuropathology 25(3):201–206

41. Nassiri F, Cusimano MD, Scheithauer BW, Rotondo F, Fazio A, Yousef GM, Syro LV, Kovacs K, Lloyd RV (2011) Endoglin (CD105): a review of its role in angiogenesis and tumor diagnosis, progression and therapy. Anticancer Res 31(6):2283–2290

42. Miyata Y, Sagara Y, Watanabe S, Asai A, Matsuo T, Ohba K, Hayashi T, Sakai H (2013) CD105 is a more appropriate marker for evaluating angiogenesis in urothelial cancer of the upper urinary tract than CD31 or CD34. Virchows Arch 463(5):673–679

43. Minhajat R, Mori D, Yamasaki F, Sugita Y, Satoh T, Tokunaga O (2006) Organ-specific endoglin (CD105) expression in the angiogenesis of human cancers. Pathol Int 56(12):717–723

44. Sica G, Lama G, Anile C, Geloso MC, La Torre G, De Bonis P, Maira G, Lauriola L, Jhanwar-Uniyal M, Mangiola A (2011) Assessment of angiogenesis by CD105 and nestin expression in peritumor tissue of glioblastoma. Int J Oncol 38(1):41–49

45. Saroufim A, Messai Y, Hasmim M, Rioux N, Iacovelli R, Verhoest G, Bensalah K, Patard JJ, Albiges L, Azzarone B et al (2014) Tumoral CD105 is a novel independent prognostic marker for prognosis in clear-cell renal cell carcinoma. Br J Cancer 110(7):1778–1784

46. Rosen LS, Gordon MS, Robert F, Matei DE (2014) Endoglin for targeted cancer treatment. Curr Oncol Rep 16(2):365

47. Matsuda Y, Hagio M, Ishiwata T (2013) Nestin: a novel angiogenesis marker and possible target for tumor angiogenesis. World J Gastroenterol 19(1):42–48

48. Teranishi N, Naito Z, Ishiwata T, Tanaka N, Furukawa K, Seya T, Shinji S, Tajiri T (2007) Identification of neovasculature using nestin in colorectal cancer. Int J Oncol 30(3):593–603

49. Ishiwata T, Matsuda Y, Naito Z (2011) Nestin in gastrointestinal and other cancers: effects on cells and tumor angiogenesis. World J Gastroenterol 17(4):409–418

50. Yamahatsu K, Matsuda Y, Ishiwata T, Uchida E, Naito Z (2012) Nestin as a novel therapeutic target for pancreatic cancer via tumor angiogenesis. Int J Oncol 40(5):1345–1357

51. Kruger K, Stefansson IM, Collett K, Arnes JB, Aas T, Akslen LA (2013) Microvessel proliferation by co-expression of endothelial nestin and Ki-67 is associated with a basal-like phenotype and aggressive features in breast cancer. Breast 22(3):282–288

52. Smolle J, Soyer HP, Hofmann-Wellenhof R, Smolle-Juettner FM, Kerl H (1989) Vascular architecture of melanocytic skin tumors. Pathol Res Pract 185:740–745

53. Cockerell CJ, Sonnier G, Kelly L, Patel S (1994) Comparative analysis of neovascularisation in primary cutaneous melanoma and Spitz nevus. Am J Dermatopathol 16:9–13

54. Folberg R, Rummelt V, Ginderdeuren R-V, Hwang T, Woolson R, Pe'er J, Gruman L (1993) The prognostic value of tumor blood vessel morphology in primary uveal melanoma. Ophthalmology 100:1389–1398

55. Pezzella F, Dibacco A, Andreola S, Nicholson AG, Pastorino U, Harris AL (1996) Angiogenesis in primary lung-cancer and lung secondaries. Eur J Cancer 32A:2494–2500

56. Balsat C, Signolle N, Goffin F, Delbecque K, Plancoulaine B, Sauthier P, Samouelian V, Beliard A, Munaut C, Foidart JM et al (2014) Improved computer-assisted analysis of the global lymphatic network in human cervical tissues. Mod Pathol 27(6):887–898

57. Wakui S (1992) Epidermal growth factor receptor at endothelial cell and pericyte interdigitation in human granulation tissue. Microvasc Res 44(3):255–262

58. Visscher D, Smilanetz S, Drozdowicz S, Wykes S (1993) Prognostic significance of image morphometric microvessel enumeration in breast carcinoma. Anal Quant Cytol 15:88–92

59. Fox SB, Leek RD, Weekes MP, Whitehouse RM, Gatter KC, Harris AL (1995) Quantitation and prognostic value of breast cancer angiogenesis: comparison of microvessel density, Chalkley count, and computer image analysis. J Pathol 177(3):275–283

60. Simpson J, Ahn C, Battifora H, Esteban J (1994) Vascular surface area as a prognostic indicator in invasive breast carcinoma. Lab Invest 70:22A

61. Brawer MK, Deering RE, Brown M, Preston SD, Bigler SA (1994) Predictors of pathologic stage in prostatic carcinoma. The role of neovascularity. Cancer 73(3):678–687

62. Furusato M, Wakui S, Sasaki H, Ito K, Ushigome S (1994) Tumour angiogenesis in latent prostatic carcinoma. Br J Cancer 70(6):1244–1246

63. Bigler S, Deering R, Brawer M (1993) Comparisons of microscopic vascularity in benign and malignant prostate tissue. Hum Pathol 24:220–226

64. Williams JK, Carlson GW, Cohen C, Derose PB, Hunter S, Jurkiewicz MJ (1994) Tumor angiogenesis as a prognostic factor in oral cavity tumors. Am J Surg 168(5):373–380

65. Wesseling P, van der Laak JA, Link M, Teepen HL, Ruiter DJ (1998) Quantitative analysis of microvascular changes in diffuse astrocytic neoplasms with increasing grade of malignancy. Hum Pathol 29(4):352–358

66. Charpin C, Devictor B, Bergeret D, Andrac L, Boulat J, Horschowski N, Lavaut MN, Piana L (1995) CD31 quantitative immunocytochemical assays in breast carcinomas. Correlation with current prognostic factors. Am J Clin Pathol 103(4):443–448

67. Van der Laak J, Westphal J, Schalkwijk L, Pahplazt M, Ruiter D, de Waal R, de Wilde P (1998) An improved procedure to quantify tumour vascularity using true colour image analysis: comparison with the manual hotspot procedure in a human melanoma xenograft model. J Pathol 184:136–143

68. Karslioglu Y, Yigit N, Onguru O (2014) Chalkley method in the angiogenesis research and its automation via computer simulation. Pathol Res Pract 210(3):161–168

69. Hansen TF, Nielsen BS, Jakobsen A, Sorensen FB (2013) Visualising and quantifying angiogenesis in metastatic colorectal cancer: a comparison of methods and their predictive value for chemotherapy response. Cell Oncol 36(4):341–350

70. Mohammed ZM, Orange C, McMillan DC, Mallon E, Doughty JC, Edwards J, Going JJ (2013) Comparison of visual and automated assessment of microvessel density and their impact on outcome in primary operable invasive ductal breast cancer. Hum Pathol 44(8):1688–1695

71. Asioli S, Eusebi V, Gaetano L, Losi L, Bussolati G (2008) The pre-lymphatic pathway, the roots of the lymphatic system in breast tissue: a 3D study. Virchows Arch 453(4):401–406

72. Di Ieva A (2010) Angioarchitectural morphometrics of brain tumors: are there any potential histopathological biomarkers? Microvasc Res 80(3):522–533

73. Zudaire E, Gambardella L, Kurcz C, Vermeren S (2011) A computational tool for quantitative analysis of vascular networks. PLoS One 6(11):e27385

74. Mikalsen LT, Dhakal HP, Bruland OS, Naume B, Borgen E, Nesland JM, Olsen DR (2013) The clinical impact of mean vessel size and solidity in breast carcinoma patients. PLoS One 8(10):e75954

75. McCarthy SA, Kuzu I, Gatter KC, Bicknell R (1991) Heterogeneity of the endothelial cell and its role in organ preference of tumour metastasis. Trends Pharmacol Sci 12(12):462–467

76. Carnochan P, Briggs JC, Westbury G, Davies AJ (1991) The vascularity of cutaneous melanoma: a quantitative histological study of lesions 0.85–1.25 mm in thickness. Br J Cancer 64(1):102–107

77. Vesalainen S, Lipponen P, Talja M, Alhava E, Syrjanen K (1994) Tumor vascularity and basement membrane structure as prognostic factors in T1-2M0 prostatic adenocarcinoma. Anticancer Res 14:709–714

78. Van Hoef ME, Knox WF, Dhesi SS, Howell A, Schor AM (1993) Assessment of tumour vascularity as a prognostic factor in lymph node negative invasive breast cancer. Eur J Cancer 29A(8):1141–1145

79. Hall NR, Fish DE, Hunt N, Goldin RD, Guillou PJ, Monson JR (1992) Is the relationship between angiogenesis and metastasis in breast cancer real? Surg Oncol 1(3):223–229

80. Ottinetti A, Sapino A (1988) Morphometric evaluation of microvessels surrounding hyperplastic and neoplastic mammary lesions. Breast Cancer Res Treat 11:241–248

81. Bosari S, Lee AK, DeLellis RA, Wiley BD, Heatley GJ, Silverman ML (1992) Microvessel quantitation and prognosis in invasive breast carcinoma. Hum Pathol 23(7):755–761

82. Bundred N, Bowcott M, Walls J, Faragher E, Knox F (1994) Angiogenesis in breast cancer predicts node metastasis and survival. Br J Surg 81:768 (Abstract)

83. Li VW, Folkerth RD, Watanabe H, Yu C, Rupnick M, Barnes P, Scott RM, Black PM, Sallan SE, Folkman J (1994) Microvessel count and cerebrospinal fluid basic fibroblast growth factor in children with brain tumours. Lancet 344(8915):82–86

84. Parums D, Cordell J, Micklem K, Heryet A, Gatter K, Mason D (1990) JC70: a new monoclonal antibody that detects vascular endothelium associated antigen on routinely

processed tissue sections. J Clin Pathol 43:752–757

85. Horak ER, Harris AL, Stuart N, Bicknell R (1993) Angiogenesis in breast cancer. Regulation, prognostic aspects, and implications for novel treatment strategies. Ann N Y Acad Sci 698(71):71–84

86. Sightler H, Borowsky A, Dupont W, Page D, Jensen R (1994) Evaluation of tumor angiogenesis as a prognostic marker in breast cancer. Lab Invest 70:22A (abstract)

87. Barnhill RL, Fandrey K, Levy MA, Mihm MJ, Hyman B (1992) Angiogenesis and tumor progression of melanoma. Quantification of vascularity in melanocytic nevi and cutaneous malignant melanoma. Lab Invest 67(3): 331–337

88. Sahin A, Sneige N, Singletary E, Ayala A (1992) Tumor angiogenesis detected by Factor-VIII immunostaining in node-negative breast carcinoma (NNBC): a possible predictor of distant metastasis. Mod Pathol 5:17A (abstract)

89. Axelsson K, Ljung BM, Moore DH II, Thor AD, Chew KL, Edgerton SM, Smith HS, Mayall BH (1995) Tumor angiogenesis as a prognostic assay for invasive ductal breast carcinoma [see comments]. J Natl Cancer Inst 87(13):997–1008

90. Fox SB, Leek RD, Smith K, Hollyer J, Greenall M, Harris AL (1994) Tumor angiogenesis in node-negative breast carcinomas—relationship with epidermal growth factor receptor, estrogen receptor, and survival. Breast Cancer Res Treat 29(1):109–116

91. Hansen S, Grabau DA, Sorensen FB, Bak M, Vach W, Rose C (2000) The prognostic value of angiogenesis by Chalkley counting in a confirmatory study design on 836 breast cancer patients. Clin Cancer Res 6(1):139–146

92. Dickinson AJ, Fox SB, Persad RA, Hollyer J, Sibley GN, Harris AL (1994) Quantification of angiogenesis as an independent predictor of prognosis in invasive bladder carcinomas. Br J Urol 74(6):762–766

93. Paweletz N, Knierim M (1989) Tumor-related angiogenesis. Crit Rev Oncol Hematol 9(3):197–242

94. Blood CH, Zetter BR (1990) Tumor interactions with the vasculature: angiogenesis and tumor metastasis. Biochim Biophys Acta 1032(1):89–118

95. Brown LF, Berse B, Jackman RW, Tognazzi K, Guidi AJ, Dvorak HF, Senger DR, Connolly JL, Schnitt SJ (1995) Expression of vascular permeability factor (vascular endothelial growth factor) and its receptors in breast cancer. Hum Pathol 26(1):86–91

96. Moghaddam A, Bicknell R (1992) Expression of platelet-derived endothelial cell growth factor in Escherichia coli and confirmation of its thymidine phosphorylase activity. Biochemistry 31(48):12141–12146

97. Anandappa SY, Winstanley JH, Leinster S, Green B, Rudland PS, Barraclough R (1994) Comparative expression of fibroblast growth factor mRNAs in benign and malignant breast disease. Br J Cancer 69(4):772–776

98. Relf M, LeJeune S, Scott PA, Fox S, Smith K, Leek R, Moghaddam A, Whitehouse R, Bicknell R, Harris AL (1997) Expression of the angiogenic factors vascular endothelial cell growth factor, acidic and basic fibroblast growth factor, tumor growth factor beta-1, platelet-derived endothelial cell growth factor, placenta growth factor, and pleiotrophin in human primary breast cancer and its relation to angiogenesis. Cancer Res 57(5):963–969

99. Garver RJ, Radford DM, Donis KH, Wick MR, Milner PG (1994) Midkine and pleiotrophin expression in normal and malignant breast tissue. Cancer 74(5):1584–1590

100. Smith K, Fox SB, Whitehouse R, Taylor M, Greenall M, Clarke J, Harris AL (1999) Upregulation of basic fibroblast growth factor in breast carcinoma and its relationship to vascular density, oestrogen receptor, epidermal growth factor receptor and survival. Ann Oncol 10(6):707–713

101. Wong SY, Purdie AT, Han P (1992) Thrombospondin and other possible related matrix proteins in malignant and benign breast disease. An immunohistochemical study. Am J Pathol 140(6):1473–1482

102. Visscher DW, DeMattia F, Ottosen S, Sarkar FH, Crissman JD (1995) Biologic and clinical significance of basic fibroblast growth factor immunostaining in breast carcinoma. Mod Pathol 8:665–670

103. Ivy SP, Wick JY, Kaufman BM (2009) An overview of small-molecule inhibitors of VEGFR signaling. Nat Rev Clin Oncol 6(10):569–579

104. Mittal K, Ebos J, Rini B (2014) Angiogenesis and the tumor microenvironment: vascular endothelial growth factor and beyond. Semin Oncol 41(2):235–251

105. FDA approval for bevacizumab. http://www.cancer.gov/cancertopics/druginfo/fda-bevacizumab

106. Proposal to withdraw approval for the breast cancer indication for Avastin (bevacizumab). http://www.fda.gov/downloads/NewsEvents/Newsroom/UCM280546.pdf

107. FDA approval for sunitinib malate. http://www.cancer.gov/cancertopics/druginfo/fda-sunitinib-malate

108. FDA approval for sorafenib tosylate. http://www.cancer.gov/cancertopics/druginfo/fda-sorafenib-tosylate

109. Toi M, Kondo S, Suzuki H, Yamamoto Y, Inada K, Imazawa T, Taniguchi T, Tominaga T (1996) Quantitative analysis of vascular endothelial growth factor in primary breast cancer. Cancer 77(6):1101–1106

110. Lantzsch T, Hefler L, Krause U, Kehl A, Goepel C, Koelbl H, Dunst J, Lampe D (2002) The correlation between immunohistochemically-detected markers of angiogenesis and serum vascular endothelial growth factor in patients with breast cancer. Anticancer Res 22(3):1925–1928

111. Valkovic T, Dobrila F, Melato M, Sasso F, Rizzardi C, Jonjic N (2002) Correlation between vascular endothelial growth factor, angiogenesis, and tumor-associated macrophages in invasive ductal breast carcinoma. Virchows Arch 440(6):583–588

112. Linderholm B, Tavelin B, Grankvist K, Henriksson R (1998) Vascular endothelial growth factor is of high prognostic value in node-negative breast carcinoma. J Clin Oncol 16(9):3121–3128

113. Gasparini G, Toi M, Gion M, Verderio P, Dittadi R, Hanatani M, Matsubara I, Vinante O, Bonoldi E, Boracchi P et al (1997) Prognostic-significance of vascular endothelial growth-factor protein in node-negative breast-carcinoma. J Natl Cancer Inst 89(2):139–147

114. Obermair A, Bancher-Todesca D, Bilgi S, Kaider A, Kohlberger P, Mullauer-Ertl S, Leodolter S, Gitsch G (1997) Correlation of vascular endothelial growth factor expression and microvessel density in cervical intraepithelial neoplasia. J Natl Cancer Inst 89(16):1212–1217

115. Manders P, Beex LV, Tjan-Heijnen VC, Geurts-Moespot J, Van Tienoven TH, Foekens JA, Sweep CG (2002) The prognostic value of vascular endothelial growth factor in 574 node-negative breast cancer patients who did not receive adjuvant systemic therapy. Br J Cancer 87(7):772–778

116. Eppenberger U, Kueng W, Schlaeppi JM, Roesel JL, Benz C, Mueller H, Matter A, Zuber M, Luescher K, Litschgi M et al (1998) Markers of tumor angiogenesis and proteolysis independently define high- and low-risk subsets of node-negative breast cancer patients. J Clin Oncol 16(9):3129–3136

117. Coradini D, Boracchi P, Daidone MG, Pellizzaro C, Miodini P, Ammatuna M, Tomasic G, Biganzoli E (2001) Contribution of vascular endothelial growth factor to the Nottingham prognostic index in node-negative breast cancer. Br J Cancer 85(6):795–797

118. Toi M, Bando H, Ogawa T, Muta M, Hornig C, Weich HA (2002) Significance of vascular endothelial growth factor (VEGF)/soluble VEGF receptor-1 relationship in breast cancer. Int J Cancer 98(1):14–18

119. Kilvaer TK, Smeland E, Valkov A, Sorbye SW, Bremnes RM, Busund LT, Donnem T (2014) The VEGF- and PDGF-family of angiogenic markers have prognostic impact in soft tissue sarcomas arising in the extremities and trunk. BMC Clin Pathol 14(1):5

120. El-Gohary YM, Silverman JF, Olson PR, Liu YL, Cohen JK, Miller R, Saad RS (2007) Endoglin (CD105) and vascular endothelial growth factor as prognostic markers in prostatic adenocarcinoma. Am J Clin Pathol 127(4):572–579

121. Saad RS, Liu YL, Nathan G, Celebrezze J, Medich D, Silverman JF (2004) Endoglin (CD105) and vascular endothelial growth factor as prognostic markers in colorectal cancer. Mod Pathol 17(2):197–203

122. Georgiadou D, Sergentanis TN, Sakellariou S, Filippakis GM, Zagouri F, Vlachodimitropoulos D, Psaltopoulou T, Lazaris AC, Patsouris E, Zografos GC (2014) VEGF and Id-1 in pancreatic adenocarcinoma: prognostic significance and impact on angiogenesis. Eur J Surg Oncol 40(10):1331–1337

123. Clara CA, Marie SK, de Almeida JR, Wakamatsu A, Oba-Shinjo SM, Uno M, Neville M, Rosemberg S (2014) Angiogenesis and expression of PDGF-C, VEGF, CD105 and HIF-1alpha in human glioblastoma. Neuropathology 34(4):343–352

124. Gunningham S, Currie M, Cheng H, Scott P, Robinson B, Harris A, Fox S (2000) The short form of the alternatively spliced flt-4 but not its ligand VEGF-C is related to lymph node metastasis in human breast cancers. Clin Cancer Res 6:4278–4286

125. Kinoshita J, Kitamura K, Kabashima A, Saeki H, Tanaka S, Sugimachi K (2001) Clinical significance of vascular endothelial growth factor-C (VEGF-C) in breast cancer. Breast Cancer Res Treat 66(2):159–164

126. Gunningham SP, Currie MJ, Han C, Robinson BA, Scott PA, Harris AL, Fox SB (2001) VEGF-B expression in human primary breast cancers is associated with lymph node metastasis but not angiogenesis. J Pathol 193(3):325–332

127. Onogawa S, Kitadai Y, Tanaka S, Kuwai T, Kimura S, Chayama K (2004) Expression of VEGF-C and VEGF-D at the invasive edge correlates with lymph node metastasis and prognosis of patients with colorectal carcinoma. Cancer Sci 95(1):32–39

128. Yokoyama Y, Charnock-Jones DS, Licence D, Yanaihara A, Hastings JM, Holland CM, Emoto M, Sakamoto A, Sakamoto T, Maruyama H et al (2003) Expression of vascular endothelial growth factor (VEGF)-D and its receptor, VEGF receptor 3, as a prognostic factor in endometrial carcinoma. Clin Cancer Res 9(4):1361–1369

129. Nakamura Y, Yasuoka H, Tsujimoto M, Yang Q, Imabun S, Nakahara M, Nakao K, Nakamura M, Mori I, Kakudo K (2003) Prognostic significance of vascular endothelial growth factor D in breast carcinoma with long-term follow-up. Clin Cancer Res 9(2):716–721

130. Kurebayashi J, Otsuki T, Kunisue H, Mikami Y, Tanaka K, Yamamoto S, Sonoo H (1999) Expression of vascular endothelial growth factor (VEGF) family members in breast cancer. Jpn J Cancer Res 90(9):977–981

131. Engels K, Fox SB, Whitehouse RM, Gatter KC, Harris AL (1997) Up-regulation of thymidine phosphorylase expression is associated with a discrete pattern of angiogenesis in ductal carcinomas in situ of the breast. J Pathol 182(4):414–420

132. Fox SB, Westwood M, Moghaddam A, Comley M, Turley H, Whitehouse RM, Bicknell R, Gatter KC, Harris AL (1996) The angiogenic factor platelet-derived endothelial cell growth factor/thymidine phosphorylase is up-regulated in breast cancer epithelium and endothelium. Br J Cancer 73(3): 275–280

133. Toi M, Hoshina S, Taniguchi T, Yamamoto Y, Ishitsuka H, Tominaga T (1995) Expression of platelet derived endothelial cell growth factor/thymidine phosphorylase in human breast cancer. Int J Cancer 64:79–82

134. Toi M, Ueno T, Matsumoto H, Saji H, Funata N, Koike M, Tominaga T (1999) Significance of thymidine phosphorylase as a marker of protumor monocytes in breast cancer. Clin Cancer Res 5(5):1131–1137

135. Yang Q, Barbareschi M, Mori I, Mauri F, Muscara M, Nakamura M, Nakamura Y, Yoshimura G, Sakurai T, Caffo O et al (2002) Prognostic value of thymidine phosphorylase expression in breast carcinoma. Int J Cancer 97(4):512–517

136. Kanzaki A, Takebayashi Y, Bando H, Eliason JF, Watanabe Si S, Miyashita H, Fukumoto M, Toi M, Uchida T (2002) Expression of uridine and thymidine phosphorylase genes in human breast carcinoma. Int J Cancer 97(5):631–635

137. Nagaoka H, Iino Y, Takei H, Morishita Y (1998) Platelet-derived endothelial cell growth factor/thymidine phosphorylase expression in macrophages correlates with tumor angiogenesis and prognosis in invasive breast cancer. Int J Oncol 13(3):449–454

138. Toi M, Yamamoto Y, Inada K, Hoshina S, Suzuki H, Kondo S, Tominaga T (1995) Vascular endothelial growth factor and platelet-derived endothelial growth factor are frequently co-expressed in highly vascularized breast cancer. Clin Cancer Res 1:961–964

139. O'Brien T, Fox S, Dickinson A, Turley H, Westwood M, Moghaddam A, Gatter K, Bicknell R, Harris A (1996) Expression of the angiogenic factor thymidine phosphorylase/platelet derived endothelial cell growth factor in primary bladder cancers. Cancer Res 56:4799–4804

140. O'Brien TS, Smith K, Cranston D, Fuggle S, Bicknell R, Harris AL (1995) Urinary basic fibroblast growth factor in patients with bladder cancer and benign prostatic hypertrophy. Br J Urol 76(3):311–314

141. Adams J, Carder PJ, Downey S, Forbes MA, MacLennan K, Allgar V, Kaufman S, Hallam S, Bicknell R, Walker JJ et al (2000) Vascular endothelial growth factor (VEGF) in breast cancer: comparison of plasma, serum, and tissue VEGF and microvessel density and effects of tamoxifen. Cancer Res 60(11):2898–2905

142. Freeman A, Morris LS, Mills AD, Stoeber K, Laskey RA, Williams GH, Coleman N (1999) Minichromosome maintenance proteins as biological markers of dysplasia and malignancy. Clin Cancer Res 5(8):2121–2132

143. Stoeber K, Swinn R, Prevost AT, de Clive-Lowe P, Halsall I, Dilworth SM, Marr J, Turner WH, Bullock N, Doble A et al (2002) Diagnosis of genito-urinary tract cancer by detection of minichromosome maintenance 5 protein in urine sediments. J Natl Cancer Inst 94(14):1071–1079

144. Eberhard A, Kahlert S, Goede V, Hemmerlein B, Plate KH, Augustin HG (2000) Heterogeneity of angiogenesis and blood vessel maturation in human tumors: implications for antiangiogenic tumor therapies. Cancer Res 60(5):1388–1393

145. Barresi V, Branca G, Caffo M, Caltabiano R, Ieni A, Vitarelli E, Lanzafame S, Tuccari G (2014) Immuno-expression of endoglin and smooth muscle actin in the vessels of brain metastases. Is there a rational for anti-angio-

genic therapy? Int J Mol Sci 15(4): 5663–5679

146. Francavilla C, Maddaluno L, Cavallaro U (2009) The functional role of cell adhesion molecules in tumor angiogenesis. Semin Cancer Biol 19(5):298–309

147. Bentley K, Franco CA, Philippides A, Blanco R, Dierkes M, Gebala V, Stanchi F, Jones M, Aspalter IM, Cagna G et al (2014) The role of differential VE-cadherin dynamics in cell rearrangement during angiogenesis. Nat Cell Biol 16(4):309–321

148. Schadendorf D, Heidel J, Gawlik C, Suter L, Czarnetzki BM (1995) Association with clinical outcome of expression of VLA-4 in primary cutaneous malignant melanoma as well as P-selectin and E-selectin on intratumoral vessels. J Natl Cancer Inst 87:366–371

149. Kageshita T, Hamby CV, Hirai S, Kimura T, Ono T, Ferrone S (2000) Alpha(v)beta3 expression on blood vessels and melanoma cells in primary lesions: differential association with tumor progression and clinical prognosis. Cancer Immunol Immunother 49(6):314–318

150. Kageshita T, Yoshii A, Kimura T, Kuriya N, Ono T, Tsujisaki M, Imai K, Ferrone S (1993) Clinical relevance of ICAM-1 expression in primary lesions and serum of patients with malignant melanoma. Cancer Res 53:4927–4932

151. Banks RE, Gearing AJ, Hemingway IK, Norfolk DR, Perren TJ, Selby PJ (1993) Circulating intercellular adhesion molecule-1 (ICAM-1), E-selectin and vascular cell adhesion molecule-1 (VCAM-1) in human malignancies. Br J Cancer 68(1):122–124

152. Brooks PC, Stromblad S, Klemke R, Visscher D, Sarkar FH, Cheresh DA (1995) Antiintegrin $\beta_3\alpha_v$ blocks human breast cancer growth and angiogenesis in human skin. J Clin Invest 96:1815–1822

153. Gasparini G, Brooks PC, Biganzoli E, Vermeulen PB, Bonoldi E, Dirix LY, Ranieri G, Miceli R, Cheresh DA (1998) Vascular integrin alpha(v)beta3: a new prognostic indicator in breast cancer. Clin Cancer Res 4(11):2625–2634

154. Heidenreich A, Rawal SK, Szkarlat K, Bogdanova N, Dirix L, Stenzl A, Welslau M, Wang G, Dawkins F, de Boer CJ et al (2013) A randomized, double-blind, multicenter, phase 2 study of a human monoclonal antibody to human alphanu integrins (intetumumab) in combination with docetaxel and prednisone for the first-line treatment of patients with metastatic castration-resistant prostate cancer. Ann Oncol 24(2):329–336

155. Manegold C, Vansteenkiste J, Cardenal F, Schuette W, Woll PJ, Ulsperger E, Kerber A, Eckmayr J, von Pawel J (2013) Randomized phase II study of three doses of the integrin inhibitor cilengitide versus docetaxel as second-line treatment for patients with advanced non-small-cell lung cancer. Invest New Drugs 31(1):175–182

156. O'Day S, Pavlick A, Loquai C, Lawson D, Gutzmer R, Richards J, Schadendorf D, Thompson JA, Gonzalez R, Trefzer U et al (2011) A randomised, phase II study of intetumumab, an anti-alphav-integrin mAb, alone and with dacarbazine in stage IV melanoma. Br J Cancer 105(3):346–352

157. Kim KB, Prieto V, Joseph RW, Diwan AH, Gallick GE, Papadopoulos NE, Bedikian AY, Camacho LH, Hwu P, Ng CS et al (2012) A randomized phase II study of cilengitide (EMD 121974) in patients with metastatic melanoma. Melanoma Res 22(4):294–301

158. Scaringi C, Minniti G, Caporello P, Enrici RM (2012) Integrin inhibitor cilengitide for the treatment of glioblastoma: a brief overview of current clinical results. Anticancer Res 32(10):4213–4223

159. Stupp R, Hegi ME, Gorlia T, Perry JR, Erridge S, Reardons DA, Markivskyy A, Wick W, Hong YK, Weller M (2013) Standard chemotherapy +/− cilengitide in newly diagnosed glioblastoma (GBM): updated results and subgroup analyses of the international randomized phase III CENTRIC trial (EORTC trial #26071-22072/Canadian Brain Tumor Consortium). In: European Multidisciplinary Cancer Congress 2013, vol 49. Eur J Cancer, Amsterdam, p S775

160. Pepper MS (2001) Role of the matrix metalloproteinase and plasminogen activator-plasmin systems in angiogenesis. Arterioscler Thromb Vasc Biol 21(7):1104–1117

161. John A, Tuszynski G (2001) The role of matrix metalloproteinases in tumor angiogenesis and tumor metastasis. Pathol Oncol Res 7(1):14–23

162. Haas TL, Madri JA (1999) Extracellular matrix-driven matrix metalloproteinase production in endothelial cells: implications for angiogenesis. Trends Cardiovasc Med 9(3–4):70–77

163. Lochter A, Bissell MJ (1999) An odyssey from breast to bone: multi-step control of mammary metastases and osteolysis by matrix metalloproteinases. APMIS 107(1):128–136

164. Parfyonova YV, Plekhanova OS, Tkachuk VA (2002) Plasminogen activators in vascular remodeling and angiogenesis. Biochemistry 67(1):119–134

165. Nielsen BS, Sehested M, Kjeldsen L, Borregaard N, Rygaard J, Dano K (1997) Expression of matrix metalloprotease-9 in vascular pericytes in human breast cancer. Lab Invest 77(4):345–355

166. Fox S, Taylor M, Grondahl-Hansen J, Kakolyris S, Gatter K, Harris A (2001) Plasminogen activator inhibitor-1 as a measure of vascular remodelling in breast cancer. J Pathol 195:236–243

167. Grøndahl-Hansen J, Christensen IJ, Rosenquist C, Brunner N, Mouridsen HT, Danø K, Blichert-Toft M (1993) High levels of urokinase-type plasminogen activator and its inhibitor PAI-1 in cytosolic extracts of breast carcinomas are associated with poor prognosis. Cancer Res 53(11):2513–2521

168. Grøndahl-Hansen J, Peters HA, van Putten WL, Look MP, Pappot H, Rønne E, Danø K, Klijn JGM, Brunner N, Foekens JA (1995) Prognostic significance of the receptor for urokinase plasminogen activator in breast cancer. Clin Cancer Res 1:1079–1087

169. Grøndahl-Hansen J, Hilsenbeck SG, Christensen IJ, Clark GM, Osborne CK, Brünner N (1997) Prognostic significance of PAI-1 and uPA in cytosolic extracts obtained from node-positive breast cancer patients. Breast Cancer Res Treat 43(2):153–163

170. Janicke F, Pache L, Schmitt M, Ulm K, Thomssen C, Prechtl A, Graeff H (1994) Both the cytosols and detergent extracts of breast cancer tissues are suited to evaluate the prognostic impact of the urokinase-type plasminogen activator and its inhibitor, plasminogen activator inhibitor type 1. Cancer Res 54(10):2527–2530

171. Foekens JA, Look MP, Peters HA, van Putten WL, Portengen H, Klijn JG (1995) Urokinase-type plasminogen activator and its inhibitor PAI-1: predictors of poor response to tamoxifen therapy in recurrent breast cancer. J Natl Cancer Inst 87(10):751–756

172. Duffy MJ (2002) Urokinase plasminogen activator and its inhibitor, PAI-1, as prognostic markers in breast cancer: from pilot to level 1 evidence studies. Clin Chem 48(8):1194–1197

173. Harbeck N, Schmitt M, Kates RE, Kiechle M, Zemzoum I, Janicke F, Thomssen C (2002) Clinical utility of urokinase-type plasminogen activator and plasminogen activator inhibitor-1 determination in primary breast cancer tissue for individualized therapy concepts. Clin Breast Cancer 3(3):196–200

174. Warren B, Greenblatt M, Kommineni V (1972) Tumor angiogenesis: ultrastructure of endothelial cells in mitosis. Br J Exp Pathol 53:216–224

175. Dvorak HF, Nagy JA, Feng D, Brown LF, Dvorak AM (1999) Vascular permeability factor/vascular endothelial growth factor and the significance of microvascular hyperpermeability in angiogenesis. Curr Top Microbiol Immunol 237:97–132

176. Vaupel P, Kallinowski F, Okunieff P (1989) Blood flow, oxygen and nutrient supply, and metabolic microenvironment of human tumors: a review. Cancer Res 49(23):6449–6465

177. Harris AL (2002) Hypoxia—a key regulatory factor in tumour growth. Nat Rev Cancer 2(1):38–47

178. Talks KL, Turley H, Gatter KC, Maxwell PH, Pugh CW, Ratcliffe PJ, Harris AL (2000) The expression and distribution of the hypoxia-inducible factors HIF-1alpha and HIF-2alpha in normal human tissues, cancers, and tumor-associated macrophages. Am J Pathol 157(2):411–421

179. Wykoff CC, Beasley NJ, Watson PH, Turner KJ, Pastorek J, Sibtain A, Wilson GD, Turley H, Talks KL, Maxwell PH et al (2000) Hypoxia-inducible expression of tumor-associated carbonic anhydrases. Cancer Res 60(24):7075–7083

180. Loncaster JA, Harris AL, Davidson SE, Logue JP, Hunter RD, Wycoff CC, Pastorek J, Ratcliffe PJ, Stratford IJ, West CM (2001) Carbonic anhydrase (CA IX) expression, a potential new intrinsic marker of hypoxia: correlations with tumor oxygen measurements and prognosis in locally advanced carcinoma of the cervix. Cancer Res 61(17):6394–6399

181. Bos R, van der Groep P, Greijer AE, Shvarts A, Meijer S, Pinedo HM, Semenza GL, van Diest PJ, van der Wall E (2003) Levels of hypoxia-inducible factor-1alpha independently predict prognosis in patients with lymph node negative breast carcinoma. Cancer 97(6):1573–1581

182. Swinson DE, Jones JL, Richardson D, Wykoff C, Turley H, Pastorek J, Taub N, Harris AL, O'Byrne KJ (2003) Carbonic anhydrase IX expression, a novel surrogate marker of tumor hypoxia, is associated with a poor prognosis in non-small-cell lung cancer. J Clin Oncol 21(3):473–482

183. Qin C, Wilson C, Blancher C, Taylor M, Safe S, Harris AL (2001) Association of ARNT splice variants with estrogen receptor-negative breast cancer, poor induction of vascular endothelial growth factor under hypoxia, and poor prognosis. Clin Cancer Res 7(4):818–823

184. Schindl M, Schoppmann SF, Samonigg H, Hausmaninger H, Kwasny W, Gnant M,

Jakesz R, Kubista E, Birner P, Oberhuber G et al (2002) Overexpression of hypoxia-inducible factor 1alpha is associated with an unfavorable prognosis in lymph node-positive breast cancer. Clin Cancer Res 8(6):1831–1837

185. Okada K, Osaki M, Araki K, Ishiguro K, Ito H, Ohgi S (2005) Expression of hypoxia-inducible factor (HIF-1alpha), VEGF-C and VEGF-D in non-invasive and invasive breast ductal carcinomas. Anticancer Res 25(4):3003–3009

186. Schoppmann SF, Fenzl A, Schindl M, Bachleitner-Hofmann T, Nagy K, Gnant M, Horvat R, Jakesz R, Birner P (2006) Hypoxia inducible factor-1alpha correlates with VEGF-C expression and lymphangiogenesis in breast cancer. Breast Cancer Res Treat 99(2):135–141

187. Hasebe T, Sasaki S, Imoto S, Mukai K, Yokose T, Ochiai A (2002) Prognostic significance of fibrotic focus in invasive ductal carcinoma of the breast: a prospective observational study. Mod Pathol 15(5):502–516

188. Hasebe T, Tsuda H, Hirohashi S, Shimosato Y, Iwai M, Imoto S, Mukai K (1996) Fibrotic focus in invasive ductal carcinoma: an indicator of high tumor aggressiveness. Jpn J Cancer Res 87(4):385–394

189. Van den Eynden GG, Colpaert CG, Couvelard A, Pezzella F, Dirix LY, Vermeulen PB, Van Marck EA, Hasebe T (2007) A fibrotic focus is a prognostic factor and a surrogate marker for hypoxia and (lymph)angiogenesis in breast cancer: review of the literature and proposal on the criteria of evaluation. Histopathology 51(4):440–451

190. Mujtaba SS, Ni YB, Tsang JY, Chan SK, Yamaguchi R, Tanaka M, Tan PH, Tse GM (2013) Fibrotic focus in breast carcinomas: relationship with prognostic parameters and biomarkers. Ann Surg Oncol 20(9):2842–2849

191. Jitsuiki Y, Hasebe T, Tsuda H, Imoto S, Tsubono Y, Sasaki S, Mukai K (1999) Optimizing microvessel counts according to tumor zone in invasive ductal carcinoma of the breast. Mod Pathol 12(5):492–498

192. Colpaert CG, Vermeulen PB, Fox SB, Harris AL, Dirix LY, Van Marck EA (2003) The presence of a fibrotic focus in invasive breast carcinoma correlates with the expression of carbonic anhydrase IX and is a marker of hypoxia and poor prognosis. Breast Cancer Res Treat 81(2):137–147

193. Stacker SA, Williams SP, Karnezis T, Shayan R, Fox SB, Achen MG (2014) Lymphangiogenesis and lymphatic vessel remodelling in cancer. Nat Rev Cancer 14(3):159–172

194. Kato T, Prevo R, Steers G, Roberts H, Leek RD, Kimura T, Kameoka S, Nishikawa T, Kobayashi M, Jackson DG et al (2005) A quantitative analysis of lymphatic vessels in human breast cancer, based on LYVE-1 immunoreactivity. Br J Cancer 93(10):1168–1174

195. Baluk P, McDonald DM (2008) Markers for microscopic imaging of lymphangiogenesis and angiogenesis. Ann N Y Acad Sci 1131:1–12

196. Mouta Carreira C, Nasser SM, di Tomaso E, Padera TP, Boucher Y, Tomarev SI, Jain RK (2001) LYVE-1 is not restricted to the lymph vessels: expression in normal liver blood sinusoids and down-regulation in human liver cancer and cirrhosis. Cancer Res 61(22):8079–8084

197. Johnson LA, Prevo R, Clasper S, Jackson DG (2007) Inflammation-induced uptake and degradation of the lymphatic endothelial hyaluronan receptor LYVE-1. J Biol Chem 282(46):33671–33680

198. Wang J, Guo Y, Wang B, Bi J, Li K, Liang X, Chu H, Jiang H (2012) Lymphatic microvessel density and vascular endothelial growth factor-C and -D as prognostic factors in breast cancer: a systematic review and meta-analysis of the literature. Mol Biol Rep 39(12):11153–11165

199. Wang J, Li K, Wang B, Bi J (2012) Lymphatic microvessel density as a prognostic factor in non-small cell lung carcinoma: a meta-analysis of the literature. Mol Biol Rep 39(5):5331–5338

200. Kurahara H, Takao S, Shinchi H, Maemura K, Mataki Y, Sakoda M, Hayashi T, Kuwahata T, Minami K, Ueno S et al (2010) Significance of lymphangiogenesis in primary tumor and draining lymph nodes during lymphatic metastasis of pancreatic head cancer. J Surg Oncol 102(7):809–815

201. Lee SK, Cho EY, Kim WW, Kim SH, Hur SM, Kim S, Choe JH, Kim JH, Kim JS, Lee JE et al (2010) The prediction of lymph node metastasis in ductal carcinoma in situ with microinvasion by assessing lymphangiogenesis. J Surg Oncol 102(3):225–229

202. Saad RS, Kordunsky L, Liu YL, Denning KL, Kandil HA, Silverman JF (2006) Lymphatic microvessel density as prognostic marker in colorectal cancer. Mod Pathol 19(10):1317–1323

203. Minardi D, d'Anzeo G, Lucarini G, Filosa A, Zizzi A, Simonetti O, Polito M Jr, Offidani AM, Di Primio R, Montironi R et al (2011) D2-40 immunoreactivity in penile squamous cell carcinoma: a marker of aggressiveness. Hum Pathol 42(11):1596–1602

204. Rudno-Rudzinska J, Kielan W, Grzebieniak Z, Dziegiel P, Donizy P, Mazur G, Knakiewicz M, Frejlich E, Halon A (2013) High density of peritumoral lymphatic vessels measured by D2-40/podoplanin and LYVE-1 expression in gastric cancer patients: an excellent prognostic indicator or a false friend? Gastric Cancer 16(4):513–520

205. Zorgetto VA, Silveira GG, Oliveira-Costa JP, Soave DF, Soares FA, Ribeiro-Silva A (2013) The relationship between lymphatic vascular density and vascular endothelial growth factor A (VEGF-A) expression with clinical-pathological features and survival in pancreatic adenocarcinomas. Diagn Pathol 8:170

206. de Jong JS, van Diest PJ, Baak JP (1995) Heterogeneity and reproducibility of microvessel counts in breast cancer. Lab Invest 73(6):922–926

207. Martin L, Holcombe C, Green B, Leinster SJ, Winstanley J (1997) Is a histological section representative of whole tumour vascularity in breast cancer? Br J Cancer 76(1):40–43

208. Chen B, Fang WK, Wu ZY, Xu XE, Wu JY, Fu JH, Yao XD, Huang JH, Chen JX, Shen JH et al (2014) The prognostic implications of microvascular density and lymphatic vessel density in esophageal squamous cell carcinoma: comparative analysis between the traditional whole sections and the tissue microarray. Acta Histochem 116(4):646–653

209. Kononen J, Bubendorf L, Kallioniemi A, Barlund M, Schraml P, Leighton S, Torhorst J, Mihatsch MJ, Sauter G, Kallioniemi OP (1998) Tissue microarrays for high-throughput molecular profiling of tumor specimens. Nat Med 4(7):844–847

210. LeBaron MJ, Crismon HR, Utama FE, Neilson LM, Sultan AS, Johnson KJ, Andersson EC, Rui H (2005) Ultrahigh density microarrays of solid samples. Nat Methods 2(7):511–513

211. Generali D, Buffa FM, Berruti A, Brizzi MP, Campo L, Bonardi S, Bersiga A, Allevi G, Milani M, Aguggini S et al (2009) Phosphorylated ERα, HIF-1α, and MAPK signaling as predictors of primary endocrine treatment response and resistance in patients with breast cancer. J Clin Oncol 27(2): 227–234

212. Liu X, Minin V, Huang Y, Seligson DB, Horvath S (2004) Statistical methods for analyzing tissue microarray data. J Biopharm Stat 14(3):671–685

213. Zhang DH, Salto-Tellez M, Chiu LL, Shen L, Koay ES (2003) Tissue microarray study for classification of breast tumors. Life Sci 73(25):3189–3199

214. Protopapa E, Delides GS, Revesz L (1993) Vascular density and the response of breast carcinomas to mastectomy and adjuvant chemotherapy. Eur J Cancer 29A(8):1141–1145

215. Fox S, Engels K, Comley M, Whitehouse R, Turley H, Gatter K, Harris A (1997) Relationship of elevated tumour thymidine phosphorylase in node positive breast carcinomas to the effects of adjuvant CMF. Ann Oncol 8:271–275

Chapter 2

Immunohistochemical Methods for Measuring Tissue Lymphangiogenesis

Daniel J. Royston, Steven Clasper, and David G. Jackson

Abstract

The field of lymphatic research has benefited enormously from the discovery of "marker" proteins that permit not only the identification and quantitation of lymphatic vessels in tissue sections for tumor pathology but also the isolation of primary lymphatic endothelial cells for basic research. This chapter focuses on the use of these markers for the immunohistochemical analysis of lymphangiogenesis in both frozen and paraffin-embedded tissue sections and discusses current protocols including newer versions employing biotin tyramide amplification and their associated problems.

Key words Lymphatic, Lymphangiogenesis, LYVE-1, Podoplanin, Immunofluorescence, Peroxidase, Chalkley grid

1 Introduction

The measurement of lymphangiogenesis is of great significance in understanding the role of this process during many pathological conditions. Clearly the most prominent example has been cancer, specifically the metastatic spread of tumors through lymphatic vessels; other examples include lymphedema, wound healing, and inflammation [1, 2]. The measurement of lymphangiogenesis within a tissue is usually approached by assessing the density of lymphatic vessels, although in the case of tumor vessels, it should be borne in mind that isolated measurements of this kind can mask artifactual increases that are caused by tissue compression and fortuitous growth next to preexisting lymphatic networks as well as genuine proliferation [3].

Traditionally the microscopic identification of lymphatic vessels has relied on the skilled analysis of morphology, absence of red blood cells within the lumen, and negative staining for blood vascular markers. However, the ready availability of lymphatic marker proteins has simplified this task [3–5]. Here we describe the use of two of these markers, LYVE-1 [6, 7] and podoplanin [8], to

Stewart G. Martin and Peter W. Hewett (eds.), *Angiogenesis Protocols*, Methods in Molecular Biology, vol. 1430,
DOI 10.1007/978-1-4939-3628-1_2, © Springer Science+Business Media New York 2016

identify lymphatic vessels in frozen or paraffin sections of the human tissue. The level of lymphangiogenesis itself is determined by measurement of lymphatic vessel density (LVD) either throughout the tissue specimen or within vessel "hot spots" using the Chalkley point graticule method (*see* refs. 9, 10), combined with estimation based on markers of nuclear division [3]. In addition, we outline the use of a particularly sensitive immunostaining technique (Dako CSA kit) that can be used to detect extremely low levels of these or other novel target antigens expressed by lymphatic vessels [11].

2 Materials

2.1 General

1. Slide staining tray with lightproof lid.
2. Coplin jars.
3. PBS.
4. Hydrophobic pen (Abcam).
5. Normal goat serum.
6. Fetal calf serum.
7. Polyclonal rabbit antihuman LYVE-1 (Abcam, R&D Systems, Reliatech).
8. Mouse monoclonal antihuman podoplanin D2-40 (Signet Laboratories).
9. Mouse monoclonal antihuman Ki-67 (BD Pharmingen).
10. D2-40 (Dako, Abcam, Thermo Scientific).
11. Clear nail varnish.
12. Chalkley eyepiece graticule.

2.2 Frozen Sections

1. Acetone.
2. Paraformaldehyde. (Dissolve 8 g of paraformaldehyde in 90 ml water on a heated stirring block in a fume hood. Add a few drops of 10 M NaOH to help the powder dissolve. Return the solution to neutral pH using 1 M HCl and an indicator paper. Store at −20 °C in small aliquots. To use, melt the solution in a heated water bath then add an equal volume of 2× PBS.)

2.3 Paraffin Sections

1. Microwave-safe dish and slide rack.
2. Citroclear® (HD Supplies).
3. Ethanol.
4. 10 mM sodium citrate pH 6.0/Dako target retrieval solution.

2.4 Immunohisto-chemistry

1. Bovine serum albumin.
2. EnVision kits (HRP anti-mouse, HRP anti-rabbit, and G|2 doublestain) (Dako).

3. Catalyzed Signal Amplification (CSA) System (Dako K1500).

4. Aquamount (BDH).

2.5 Immuno-fluorescence

1. Goat anti-rabbit IgG Alexafluor 488 conjugated (Molecular Probes).

2. Goat anti-mouse IgG Alexafluor 568 conjugated (Molecular Probes).

3. Streptavidin, Alexafluor 488/568 conjugated (Molecular probes).

4. Vectashield with Dapi (Vector Laboratories).

3   Methods

3.1 Pretreatment of Frozen Sections

3.1.1 Fixing

1. Allow slides of sections of approximately 10 μm thickness to equilibrate to room temperature.

2. Label slides with a pencil noting specimen and primary antibody(s) to be used.

3. (a) If the podoplanin antibody D2-40 is to be used, then cover section with 4 % paraformaldehyde for 10 min. Rinse slides carefully over a sink using a wash bottle of PBS (*see* **Note 1**).

   (b) If D2-40 is not being used, then place slides in a coplin jar containing 100 % acetone for 2 min to fix and then remove and air-dry.

4. Place slides in a jar of PBS for approximately 5 min to allow embedding compound around the sections to dissolve.

3.1.2 Blocking

1. Carefully dry the slides using tissue or paper towel and draw around each specimen with a hydrophobic pen to retain the small antibody volumes on the section.

2. Place slides in a staining tray and block nonspecific antibody-binding sites by applying approximately 200 μl of 5 % v/v goat serum in PBS to each section and incubating for 20 min (*see* **Note 2**).

3. Rinse slides carefully over a sink using a wash bottle of PBS (*see* **Note 1**) and place in a jar of PBS for 5 min.

4. Proceed to the appropriate section for either immunohisto-chemistry or immunofluorescence.

3.2 Pretreatment of Paraffin Sections

3.2.1 Dewaxing

1. Label slides with a pencil noting the specimen and primary antibody(s) to be used.

2. Place slides in a coplin jar containing Citroclear® for approximately 5 min. Remove and drain.

3. Place slides in a second coplin jar containing Citroclear® for approximately 5 min. Remove and drain (*see* **Note 3**).

4. Place slides in a coplin jar containing 100 % ethanol for approximately 5 min. Remove and drain.

5. Place slides in a second coplin jar containing 100 % ethanol for approximately 5 min. Remove and drain.

6. Place slides in a coplin jar containing 50 % v/v ethanol for approximately 5 min. Remove and drain.

7. Place slides in a coplin jar containing water for approximately 5 min. Remove and drain.

3.2.2 Antigen Retrieval

1. Preheat a covered microwave-safe dish containing enough 10 mM citrate buffer (pH 6.0)/Dako antigen retrieval buffer to cover a rack of slides to 100 °C in a microwave.

2. Place the rack of slides into the heated buffer, re-cover, and simmer for 10 min.

3. Place sections in a jar of PBS for approximately 5 min to cool.

3.2.3 Blocking

1. Carefully dry the slides using tissue or paper towel and draw around each specimen with a hydrophobic pen to retain the small antibody volumes on the section.

2. Place slides in a staining tray and block nonspecific antibody-binding sites by applying approximately 200 µl of 5 % v/v goat serum in PBS to each section and incubating for 20 min (*see* **Note 2**).

3. Rinse slides carefully over a sink using a wash bottle of PBS [1] and place in a jar of PBS for 5 min.

4. Proceed to the appropriate section for either immunohisto-chemistry or immunofluorescence.

3.3 Immunohisto-chemistry

3.3.1 Dako EnVision Staining

Primary Antibody

1. Carefully dry slides and place in the staining tray.

2. Apply peroxidase block from the Dako EnVision kit (blocks endogenous peroxidase activity that is present in certain tissues) to cover each section. Incubate for approximately 5 min.

3. Rinse slides carefully over a sink using a wash bottle of PBS and place in a jar of PBS for 5 min.

4. Carefully dry slides and place in the staining tray.

5. Apply primary antibody (anti-LYVE-1 or D2-40) at 5 µg/ml in 1 % w/v BSA in PBS. Use approximately 200 µl per section. Incubate for 30 min [4].

6. Rinse slides carefully over a sink using a wash bottle of PBS and place in a jar of PBS for 5 min.

Secondary Antibody
and Mounting

1. Carefully dry slides and place in the staining tray.

2. Apply enough of the appropriate Dako EnVision secondary HRP conjugate to cover each section (anti-rabbit for LYVE-1, anti-mouse for D2-40). Incubate for 30 min.

3. Rinse slides carefully over a sink using a wash bottle of PBS and place in a jar of PBS for 5 min.

4. Mix the EnVision peroxidase substrate as per the instructions and apply to each section. Incubate for 5–10 min.

5. Wash slides using distilled water from a wash bottle.

6. Place slides in a jar of hematoxylin solution for approximately 2 min.

7. Wash slides with normal tap water.

8. Carefully dry slides and place in the staining tray.

9. Apply a few drops of Aquamount medium to each section and place a coverslip on top.

10. Invert each slide and press down firmly on a flat pile of paper towels to evenly spread the mounting medium.

11. Seal each slide by painting around the edge of the coverslip with clear nail varnish.

3.3.2 Dako Catalyzed Signal Amplification (CSA) Staining

The Dako CSA System offers a more sensitive immunohistochemical (IHC) staining procedure for the detection of extremely low levels of marker antigen. The method involves the peroxidase-catalyzed oxidation and deposition of biotinylated tyramide, followed by a secondary reaction with streptavidin peroxidase.

1. Carefully dry slides and place in staining tray.

2. Apply peroxidase block from Dako CSA kit (blocks endogenous peroxidase activity that is present in certain tissues). Incubate for approximately 5 min.

3. Place in a fresh buffer (0.05 M Tris–HCl pH 7.6 containing 0.3 M NaCl and 0.1 % w/v Tween 20).

4. Apply protein block solution from Dako CSA kit. Incubate for 5 min, but do not rinse off protein block solution.

5. Tap off excess protein block solution and apply primary antibody. Use approximately 200 µl per section and incubate for 15 min.

6. Rinse gently with buffer and place in up to three fresh buffer baths for 3–5 min.

7. Apply Link antibody (anti-mouse or anti-rabbit) from Dako CSA kit. Incubate for 15 min.

8. Repeat *step 6*.

9. Apply streptavidin-biotin complex (prepared 30 min in advance as per CSA instructions) and incubate for 15 min.

10. Repeat *step 6*.

11. Apply amplification reagent from Dako CSA kit and incubate for 15 min.

12. Repeat *step 6*.

13. Apply streptavidin peroxidase from Dako CSA kit and incubate for 15 min.

14. Repeat *step 6*.

15. Apply substrate-chromogen (prepared in advance as per CSA instructions) and incubate for 5 min. Rinse gently for 5 min in distilled water.

16. Place slides in a jar of hematoxylin solution for approximately 2 min.

17. Wash slides with normal tap water.

18. Carefully dry slides and place in the staining tray.

19. Apply a few drops of Aquamount medium to each section and place a coverslip on top.

20. Invert each slide and press down firmly on a flat pile of paper towels to evenly spread the mounting medium.

21. Seal each slide by painting around the edge of the coverslip with clear nail varnish.

3.4 Immunofluorescence

3.4.1 Standard Immunofluorescence Staining

Primary Antibodies

1. Carefully dry slides and place in the staining tray.

2. Apply primary antibodies (anti-LYVE-1 and D2-40) at 10 μg/ml each in 5 % v/v FCS/PBS. Use approximately 200 μl per section. Incubate for 30 min (*see* **Note 4**).

3. Rinse slides carefully over a sink using a wash bottle of PBS and place in a jar of PBS for 5 min.

Secondary Antibodies and Mounting

1. Carefully dry slides and place in the staining tray.

2. Dilute both fluorescently labeled secondary antibodies together 1 in 500 in 5 % v/v FCS/PBS and apply approximately 200 μl to each section. Incubate for 30 min, ensuring that the light-proof lid is in place for this step.

3. Rinse slides carefully over a sink using a wash bottle of PBS and place in a jar of PBS for 5 min.

4. Carefully dry slides and place in the staining tray.

5. Apply a few drops of Vectashield to each section and place a coverslip on top.

6. Invert each slide and press down firmly on a flat pile of paper towels to evenly spread the mounting medium.

7. Seal each slide by painting around the edge of the coverslip with nail varnish.

8. Store slides at 4 °C in a lightproof box.

1. Perform *steps 1–12* as for Dako CSA kit immunohistochemistry.

2. Add fluorescently labeled streptavidin 1 in 200 in 5 % v/v FCS/PBS and apply approximately 200 μl to each section. Incubate for 30 min, ensuring that the lightproof lid is in place for this step.

3. Rinse slides carefully over a sink using a wash bottle of PBS and place in a jar of PBS for 5 min.

4. Carefully dry slides and place in the staining tray.

5. Apply a few drops of Vectashield to each section and place a coverslip on top.

6. Invert each slide and press down firmly on a flat pile of paper towels to evenly spread the mounting medium.

7. Seal each slide by painting around the edge of the coverslip with nail varnish.

8. Store slides at 4 °C in a lightproof box.

Note: For double immunofluorescent labeling of slides using the Dako CSA kit, a second compatible antibody of a different species that does not require significant amplification can be added after *step 3*. The protocol outlined in Subheading 3.4 for conventional immunofluorescence staining should then be followed.

3.5 Identification of Lymphatic Vessels

3.5.1 Detection of Lymphatic Vessels

Lymphatic vessels should be clearly visible as stained structures within the tissue. However there may be a wide variation in size and morphology. Initial lymphatics may appear as small structures (diameter range 10–50 μm), while larger capillaries may appear as elongated structures, sometimes with a collapsed lumen (diameter range 100–200 μm). Tumor-associated lymphatic vessels may appear as small basketlike structures within the tumor mass or as a continuous endothelium surrounding a tumor embolus [12]. Figure 1 shows typical lymphatic morphologies in normal and tumor tissue detected by immunohistochemistry with antiserum against LYVE-1. Additionally the contrast in morphology between a lymphatic and a blood vessel can be seen.

3.5.2 Confirmation of Lymphatic Identity

Positive staining of a structure for a single lymphatic marker protein should not be regarded as definitive proof of lymphatic identity. For sections stained with antibodies to LYVE-1 and podoplanin by immunofluorescence, it is a simple matter to check the co-expression of marker proteins. For specimens stained by immunohistochemistry, we recommend the staining of consecutive serial sections with each antibody to confirm identification. Two-color staining is also possible using immunohistochemistry, for example, by combining peroxidase and alkaline phosphatase-conjugated antibodies. However, it is this author's opinion that the procedure is more satisfactory for mutually exclusive staining of distinct cell types or for cases where the two markers localize to different

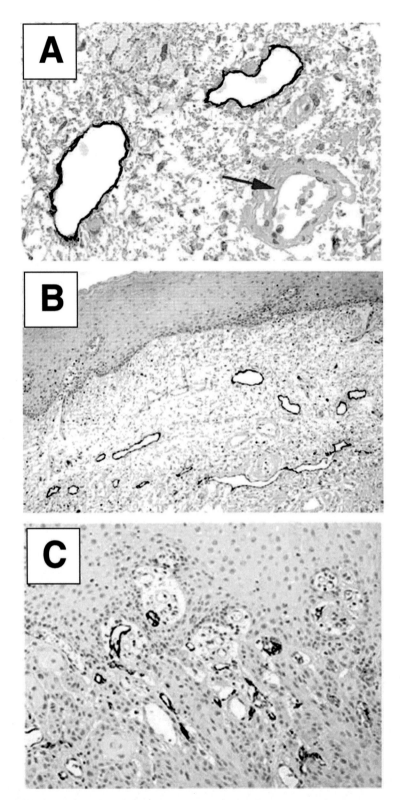

Fig. 1 Immunohistochemical staining of lymphatic vessels in soft tissue. Patent lymphatic vessels of the normal human tongue show strong staining of LYVE-1, while the erythrocyte-containing blood vessel (*red arrow*) shows no staining and a typical thickened vessel wall (×32 objective) (**a**). Abundant LYVE-1 positive lymphatic vessels can be seen beneath the epithelium (×10 objective) (**b**). LYVE-1 positive intratumoral lymphatic vessels within squamous carcinoma of the tongue (×20 objective) (**c**). Adapted from Beasley et al. [12]

regions of the same cell as described later. Regardless of whether immunohistochemistry or immunofluorescence staining is chosen, the results should still be interpreted with caution (*see* **Note 5**). Discrimination between lymphatic vessels and blood vessels is rarely a problem when using these markers together (*see* **Note 6**).

3.5.3 Vessel Hot Spots

In normal tissue, e.g., the dermis, lymphatic vessels may be evenly distributed. However, in tumors these vessels may be concentrated within discrete areas termed "hot spots" induced by agents enriched in the local microenvironment such as lymphangiogenic growth factors (VEGF-C, VEGF-D, PDGF, etc.). Hot spots rather than randomly chosen areas are frequently targeted for tumor vessel counts in the assessment of lymphangiogenesis, although the validity of this practice has been disputed (*see*, e.g., ref. 13).

3.6 Measurement of Mean Lymphatic Vessel Density

This method relies upon the ability of the observer to distinguish discrete lymphatic vessels and count them within a known area, thus giving an actual measurement of lymphatic vessel density within the plane of the section.

3.6.1 Calculation of Area of View

Consult the microscope manufacturer's handbook to obtain the field of view distance in mm for each objective lens to be used. Use this number to calculate the area of view with the formula:

$$\text{Area}(\text{mm}^2) = \pi(\text{field of view} / 2)^2$$

3.6.2 Counting of Lymphatic Vessels

1. Using a low power objective, a field of view containing lymphatic vessels (e.g., hot spot) should be identified, and a suitable objective lens should be chosen to magnify this area. This size of objective should be constant throughout all of the samples.

2. Each discrete individual stained lymphatic structure (see above), irrelevant of size, is counted as a vessel, and the total within the area of view is recorded.

3. A different region within the same section is then chosen and the vessel number again recorded. This is repeated at least three times and the mean vessel number is calculated.

4. The process is repeated for each section, making sure to keep the objective lens constant.

3.6.3 Calculation of Lymphatic Vessel Density

1. The vessel density is calculated for each section:

$$\text{Lymphatic vessel density} \ (\text{mm}^{-2}) = \text{mean vessel number} / \text{area of view} \ (\text{mm}^2)$$

2. The application of a suitable statistical method can be used to compare mean vessel densities between tissues and hence levels of lymphangiogenesis (*see* **Note 7**).

3.7 Chalkley Counting

This method does not rely on the observer's assessment of individual vessels, but instead effectively measures the area covered by vessels [9]. A Chalkley eyepiece graticule is required to fit the microscope. This is a rotatable graticule marked with 25 randomly placed spots. Due to the requirement to be able to see the graticule markings against the background of the illuminated slide, this method is only suitable for sections stained by immunohistochemistry, not immunofluorescence.

3.7.1 Determining the Chalkley Count

1. Using a low power objective, a field of view containing lymphatic vessels (i.e., hot spot) should be identified, and a suitable objective lens should be chosen to magnify this area. This size of objective should be constant throughout all of the samples.

2. The graticule is carefully rotated until the maximum number of dots overlaps lymphatic vessels (note that these do not need to be separate vessels), and the number of these dots (maximum score 25) is recorded. *See* Fig. 2 for an example of this.

3. The procedure is repeated a minimum of three times using different regions within the same section.

4. The mean number of dots overlapping lymphatic vessels on a section is the Chalkley count.

5. The application of a suitable statistical method can be used to compare Chalkley counts between tissues (applying cutoffs where appropriate) and hence levels of lymphangiogenesis (*see* **Note 7**).

3.8 Lymphatic Endothelial Cell Proliferation

The basis of this method is that by co-staining with both a lymphatic marker and a proliferation marker, it allows identification of actively dividing lymphatic endothelial cells and hence an accurate measurement of ongoing lymphangiogenesis at the time the tissue was taken. As this method requires two-color staining, it is perhaps most suitable for immunofluorescence; however two-color immunohistochemistry may be used as the lymphatic marker, and proliferation marker proteins are located on the cell surface and nucleus, respectively. Extreme care must be taken in the identification of lymphatic vessels as only a single lymphatic marker antibody is used.

3.8.1 For Frozen Sections

1. Frozen sections must be fixed and permeabilized to allow access to the nuclear compartment. Follow the instructions described in Subheading 3.1.1 for fixing with paraformaldehyde.

2. Before blocking, rinse the section and incubate for 5 min with 0.5 % (w/v) Triton X-100 in PBS.

3. Rinse with PBS and proceed to blocking.

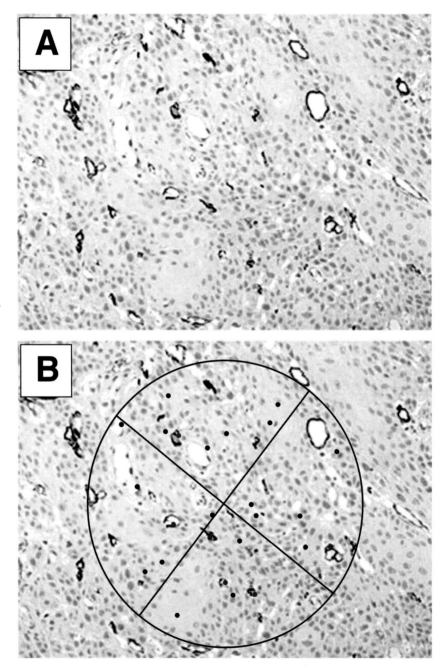

Fig. 2 The Chalkley counting method for estimating lymphatic vessel density. Intratumoral lymphatic vessels of tongue squamous carcinoma stained for expression of LYVE-1 (**a**) are overlayed with a representation of a Chalkley grid (**b**). Dots which overlap lymphatic vessels are highlighted in *red* (×20 objective). Adapted from Beasley et al. [12]

3.8.2 Immunofluorescence

The protocol for frozen or paraffin sections should be followed with the exchange of the podoplanin D2-40 antibody with a mouse monoclonal raised against the proliferation marker Ki-67.

3.8.3 Immunohistochemistry

The Dako EnVision G|2 Doublestain system should be used with anti LYVE-1 and anti Ki-67 according to the manufacturer's instructions.

3.8.4 Evaluation of Staining

1. Proliferating lymphatic vessels will contain cells which are stained positively for both LYVE-1 and Ki-67 expression (*see* Fig. 3). The observer should be aware that tumor tissue and areas of inflammation or wound healing are likely to contain an abundance of proliferating non-lymphatic cells.

2. Mean proliferating lymphatic vessel density may be calculated in a way analogous to the calculation of mean lymphatic density as described in Subheading 3.6 when only vessels containing proliferating cells are scored.

3. Immunohistochemistry may be combined with a Chalkley graticule to give a Chalkley count for proliferating vessels.

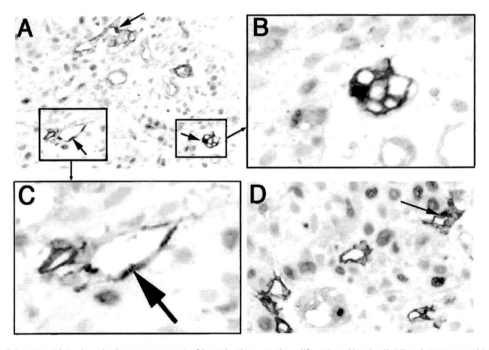

Fig. 3 Immunohistochemical measurement of lymphatic vessel proliferation. Newly dividing intratumoral lymphatic vessels are detected by double staining for the Ki-67 proliferation antigen (*brown* nuclear staining) and LYVE-1 (*pink* membrane staining). Panels (**a–d**) show examples of dividing LYVE-1/Ki-67 double-positive small lymph vessels (*black arrows*) surrounded by large numbers of LYVE-1-negative/Ki-67-positive squamous carcinoma cells. Panel (**a**) ×20 objective, panels (**b**) and (**c**) ×100 objective, and panel (**d**) ×40 objective. Adapted from Beasley et al. [12]

4. Using either method, a measurement of the ratio of proliferating to nonproliferating vessels gives an insight into the degree of lymphangiogenesis in the tissue.

3.9 Measuring Lymphangiogenesis in Mouse Tissues

While the analysis of lymphangiogenesis within human tissue samples is clearly important in a diagnostic and prognostic role, it is appreciated that experimental models are widely used both during the study of adult disease and embryonic development. The techniques above are equally valid using mouse tissues providing the correct antibodies are used (*see* **Note 8**). However, the lack of a Dako EnVision kit for rat antibodies limits the immunohistochemistry that can be performed using the particular protocols described above. As an alternative, we suggest use of the Vector Laboratories ImmPRESS™ HRP Anti-Rat IgG, Mouse adsorbed (Peroxidase) Polymer detection kit (Catalogue no. MP-7444).

4 Notes

1. Use a wash bottle with a wide bore spout. Hold the slide almost vertically and aim the jet at a point above the section, allowing the PBS to gently flow down over the section.

2. It is beneficial from this stage onward to ensure that the sections do not dry out. Therefore the slides should be dried in small batches before adding the next solution.

3. When processing large numbers of slides, it may be easier to place the dewaxing solutions in individual small glass tanks and transfer the slides in a rack from tank to tank. Dewaxing solutions may be stored and used for several cycles before replacement with fresh solutions.

4. The use of a negative control antibody is necessary to confirm the validity of the staining. This should either be an isotype-matched antibody or a preimmune serum from the relevant species.

5. LYVE-1, while being a widely used marker protein for lymphatic endothelium, is expressed by other cell types including certain macrophages, liver sinusoidal endothelium, and certain lung alveolar cells. In light of this, single cells staining positive for LYVE-1 should be identified with care, while the use of LYVE-1 as a marker for hepatic lymphatic endothelium is not recommended. It is also worth noting that the expression of LYVE-1 on lymphatic endothelium is downregulated during inflammation. This may clearly lead to an underestimation of lymphatic vessel density during inflammatory conditions. Similarly, podoplanin is expressed by several different cell types (e.g., epithelia and fibroblasts) in addition to lymphatic endothelia and is present in several tumor types, particularly at the

invasive front. Care is therefore required in the interpretation of tumor lymphangiogenesis using this marker. Should problems be encountered with the specificity of LYVE-1 and podoplanin expression in the chosen tissue, antibodies to other marker proteins may be tried. These include the nuclear transcription factor Prox1 and VEGFR3, although again it should be borne in mind that these proteins are also not exclusively expressed by lymphatic endothelium, the former being expressed by hepatocytes and the latter by macrophages and blood vessels associated with tumors and wound healing.

6. Identification of blood vessels in human tissues can be confirmed by positive staining with the antibody PAL-E.

7. It is recommended that all slides are evaluated either single- or double-blindedly by two independent observers to prevent bias.

8. Polyclonal antisera to mouse LYVE-1 are commercially available (R&D Systems, Reliatech, etc.) as is a hamster monoclonal antibody to podoplanin. Most rabbit antisera to human Prox1 appear to cross-react with the mouse protein. The rat antibody MECA32 works well as a mouse blood endothelial-specific marker.

References

1. Alitalo K, Tammela T, Petrova TV (2005) Lymphangiogenesis in development and human disease. Nature 438:946–953

2. Stacker S, Hughes RA, Williams RA, Achen MG (2006) Current strategies for modulating lymphangiogenesis signalling pathways in human disease. Curr Med Chem 13:783–792

3. Van der Auwera I, Cao Y, Tille JC et al (2006) First International consensus on the methodology of lymphangiogenesis quantification in solid human tumours. Br J Cancer 95:1611–1625

4. Jackson DG (2001) New molecular markers for the study of tumour lymphangiogenesis. Anticancer Res 21:4279–4283

5. Sleeman JP, Krishnan J, Kirkin V, Baumann P (2001) Markers for the lymphatic endothelium: in search of the holy grail? Microsc Res Tech 55:61–69

6. Banerji S, Ni J, Wang SX et al (1999) LYVE-1, a new homologue of the CD44 glycoprotein, is a lymph-specific receptor for hyaluronan. J Cell Biol 144:789–801

7. Jackson DG (2004) Biology of the lymphatic marker LYVE-1 and applications in research into lymphatic trafficking and lymphangiogenesis. APMIS 112:526–538

8. Breiteneder-Geleff S, Soleiman A, Kowalski H et al (1999) Angiosarcomas express mixed endothelial phenotypes of blood and lymphatic capillaries: podoplanin as a specific marker for lymphatic endothelium. Am J Pathol 154:385–394

9. Fox SB, Leek RD, Weekes MP, Whitehouse RM, Gatter KC, Harris AL (1995) Quantitation and prognostic value of breast cancer angiogenesis: comparison of microvessel density, Chalkley count, and computer image analysis. J Pathol 177:275–283

10. Vermeulen PB, Gasparini G, Fox SB et al (1996) Quantification of angiogenesis in solid human tumours: an international consensus on the methodology and criteria of evaluation. Eur J Cancer 32:2474–2484

11. Clasper S, Royston D, Baban D, Cao Y, Ewers S, Butz S, Vestweber D, Jackson DG (2008) A novel gene expression profile in lymphatics associated with tumor growth and nodal metastasis. Cancer Res 68:7293–7303

12. Beasley NJ, Prevo R, Banerji S et al (2002) Intratumoral lymphangiogenesis and lymph node metastasis in head and neck cancer. Cancer Res 62:1315–1320

13. Shields JD, Borsetti M, Rigby H et al (2004) Lymphatic density and metastatic spread in human malignant melanoma. Br J Cancer 90:693–700

Chapter 3

Immunohistochemical Assessment of Leukocyte Involvement in Angiogenesis

Narmeen S. Ahmad, Stewart G. Martin, and Sarah J. Storr

Abstract

Angiogenesis is a hallmark of cancer and is important for tumor growth, development, and metastasis. Leukocytes, including neutrophils, eosinophils, basophils, lymphocytes, and monocytes, are found invading many solid tumors, and this inflammation is often associated with tumorigenesis. Tumor-associated macrophages have been shown to be involved in tumor migration and metastasis and are modulators of tumor vascularization. Tumor-associated macrophages are a source of angiogenic factors, and pro-inflammatory cytokines involved in angiogenesis, lymphangiogenesis, and metastasis. Here we describe a method of quantifying the number of macrophages and their class within tumor tissue which can be compared with tumor blood and lymphatic microvessel density as a measure of angiogenesis and lymphangiogenesis. Although not described in depth, application of the methodology is described for other leukocyte populations, such as tumor-infiltrating lymphocytes.

Key words Angiogenesis, Tumor-associated macrophages, CD68, CD163

1 Introduction

Angiogenesis and lymphangiogenesis are essential for tumor growth and metastasis, and leukocytes, including neutrophils, eosinophils, basophils, lymphocytes, and monocytes, are linked with these processes [1]. Macrophages are produced by the differentiation of monocytes and have been characterized as key regulators of tumor vascularization through the release of numerous growth factors, proteolytic enzymes, cytokines, chemokines, and inflammatory mediators that can regulate tumor growth, angiogenesis, lymphangiogenesis, invasion, and metastasis [2–5]. Macrophages can release a number of well-characterized growth factors such as vascular endothelial growth factor (VEGF) [6], basic fibroblast growth factor (bFGF), epidermal growth factor (EGF), and transforming growth factor-α (TGF-α). VEGF secreted by macrophages is a potent angiogenic factor that enhances endothelial cell proliferation and has been observed in tumors with high

Stewart G. Martin and Peter W. Hewett (eds.), *Angiogenesis Protocols*, Methods in Molecular Biology, vol. 1430, DOI 10.1007/978-1-4939-3628-1_3, © Springer Science+Business Media New York 2016

microvessel density [7–9]. In addition, the inflammatory cytokines, tumor necrosis factor (TNF)-α, interleukin (IL)-6, and IL-8, are produced by macrophages and have been shown to be associated with elevated microvessel density [10–13]. The quantitation of macrophages and leukocytes within tissue is becoming increasingly important due to their importance in various clinical pathologies [4, 14, 15], with others interested in tumor-infiltrating lymphocytes and their assessment by multispectral imaging approaches [16]. The focus of this chapter is the quantification and classification of macrophages and the type of macrophage. Macrophages can be polarized into either classically activated macrophages (type 1) or alternatively activated macrophages (type 2). Type 1 macrophages are pro-inflammatory and can activate type 1 T-helper cells (Th1), whereas type 2 express anti-inflammatory cytokines and can activate type 2 T-helper cells (Th2). Type 2 macrophages can function in tissue remodeling, wound healing, and tumor angiogenesis and are regarded as tumor-associated macrophages [17, 18].

Here we describe the use of two markers to identify macrophages within human tissue: the first marker being CD68, which is expressed by monocytes, macrophages, myeloid cells, dendritic cells, and fibroblasts, among others [19, 20], which has been used to stain cells of the monocyte lineage in numerous tissue types [3, 4], and the second marker being CD163 which is also expressed on cells of the monocyte lineage and functions as a hemoglobin/haptoglobin scavenger receptor which appears to play a role in the anti-inflammatory function of type 2 macrophages [15, 21–23]. The use of these two markers allows discrimination of macrophage polarization in tissue by determining levels of total macrophages and type 2 macrophages. Quantification of macrophages is achieved using the Chalkley overlap morphometric technique. These methodologies should be used in combination with methods to determine tumor microvessel density (blood and lymphatic), which are described in other chapters to allow any relationships between the two to be investigated, and, by default, involvement in angiogenesis and lymphangiogenesis, respectively. Although parallel section staining is described using chromogens, double immunohistochemistry can be performed using fluorescence. Although we describe methodology for type 1 and type 2 macrophages, a range of markers can be used for other leukocyte populations such as CD3 and CD8 for tumor-infiltrating lymphocytes.

2   Materials

1. Pencil or pen (xylene and ethanol resistant).
2. Hot plate.
3. Coplin jars.

4. Microwave oven.

5. Sequenza trays and Shandon (Sequenza) cover plates.

6. Glass coverslips.

7. Chalkley graticule: A 25-dot Chalkley graticule microscope eyepiece to quantify macrophages within tissue.

8. Industrial methylated spirit (IMS)/ethanol.

9. Xylene.

10. Antigen retrieval buffer: 10 mM sodium citrate buffer pH6.0. Dissolve 2.94 g sodium citrate tribasic dihydrate in 1 L dH$_2$O.

11. TBS: Tris-buffered saline, pH 7.6.

12. Peroxidase blocking solution: Add 10 % (v/v) H$_2$O$_2$ in methanol.

13. Blocking solution: Add 2 % (v/v) horse serum in TBS. Prepare on the day of use.

14. Antibodies to CD68 (e.g., clone KP1; Abcam) and CD163 (e.g., clone 10D6, Thermo Scientific).

15. Avidin-biotin complex (ABC) staining kit (e.g. Vectastain Elite ABC kit; *see* **Note 1**) and appropriate biotinylated secondary antibodies. ABC should be prepared 30 min prior to use by adding 1:100 reagent A and 1:100 reagent B to TBS.

16. Chromogen: Add diaminobenzidine (DAB) at 2 % (v/v) in substrate buffer (Dako). Prepare prior to use (*see* **Note 2**).

17. Gills formula hematoxylin.

18. DPX mounting agent.

19. Tissue sections: Cut consecutive 4 μm sections of formalin-fixed paraffin-embedded (FFPE) tissue on glass slides.

3   Methods

Here we describe the use of CD68 and CD163 to detect macrophages in consecutively cut breast cancer FFPE tissue. Additional consecutive sections can be cut to determine angiogenesis or lymphangiogenesis if direct correlations between markers are required. It should be noted that other tissue types may require optimization of antibody dilution, antigen retrieval, and/or general immunohistochemical procedure.

3.1   Dewaxing

1. Label slides with a pencil or pen (xylene and ethanol resistant), noting specimen details and primary antibody to be used (*see* **Note 3**).

2. Place 4 μm formalin-fixed paraffin-embedded sections on a 60 °C hot plate for 10 min.

3. Allow to cool and place in slide rack.

4. Sections should be dewaxed and dehydrated by 5 min incubation in each of the following series of solutions in coplin jars: (1) xylene, (2) xylene, (3) ethanol/IMS, (4) ethanol/IMS, and (5) ethanol/IMS.

5. Slides should be placed in water for 5 min (*see* **Note 4**).

3.2 Antigen Retrieval

1. Place the sections in microwave-safe dish containing enough antigen retrieval buffer to fully cover the slides (*see* **Note 5**).

2. Heat the slides in a microwave at 750 W for 10 min, followed by 10 min at 450 W.

3. Slides should slowly be cooled under running tap water.

3.3 Blocking

1. Working in large bath of tap water, load slides onto the Shandon cover plates, and then place in Sequenza trays (*see* **Note 6**).

2. Add 100 μL of peroxidase blocking solution per slide and incubate for 10 min.

3. Wash the slides three times for 5 min with TBS (each wash should be 5 min).

4. Add 100 μL of blocking solution to each slide and incubate for 30 min (*see* **Note 7**).

3.4 Primary and Secondary Antibodies

1. Apply 100 μL primary antibody (1:100 anti-CD68 or 1:35 anti-CD163 in blocking solution) per slide, and incubate at room temperature for 1 h (*see* **Note 8**).

2. Wash the slides three times for 5 min with TBS.

3. Dilute secondary antibody (1:50) in blocking solution and add 100 μL per slide. Cover and leave at room temperature for 1 h.

4. Wash the slides three times for 5 min with TBS.

3.5 Staining

1. Add 100 μL of ABC per slide and incubate for 30 min.

2. Wash the slides three times for 5 min with TBS.

3. Add 100 μL diluted chromogen to each slide and incubate for 5 min (*see* **Note 2**).

4. Wash slides three times for 5 min with water.

5. Add 100 μL hematoxylin per slide and incubate for 1 min.

6. Rinse the slides immediately with water, remove from Shandon cover plates, and place slides into a slide rack.

7. Slides should be placed for 5 min in each of the following series of solutions: (1) water, (2) ethanol/IMS, (3) ethanol/IMS, (4) ethanol/IMS, (5) xylene, and (6) xylene.

8. Apply one or two drops of DPX to glass coverslips.

9. Place the slides tissue face down onto the coverslips and allow the weight of the slide to spread the DPX mounting agent.

10. Allow slides to dry overnight

3.6 Selection of Hot Spots and Chalkley Overlap Morphometric Technique

The Chalkley overlap morphometric technique has been successfully implemented to measure microvessel density in an objective manner and is also suitable for quantifying macrophages [24, 25]. Other methods can be used such as computer-aided image analysis or counting macrophage in fields of view, which may be more suitable if fluorescent staining is to be employed as is not possible the Chalkley graticule with immunofluorescence [26, 27].

1. Following the staining of two consecutive sections with anti-CD68 and anti-CD163 antibodies, the tumor areas should be scanned at low magnification (40×) and divided subjectively into two areas: the intratumoral (IT) area represents 2/3 of the tumor area, and the peripheral area (PP) represents 1/3 of the outer tumor area (*see* Fig. 1).

2. The tissue is scanned to identify possible "hot spot" areas at 100× magnification.

3. The eyepiece graticule is orientated on the tissue so that the maximum number of dots overlap stained macrophages at 200× magnification; the macrophages should then be counted (Fig. 2a, b) (*see* **Note 9**).

4. This should be done three times in different regions within the same intratumoral and peripheral tumor areas.

5. The macrophage count is the mean number of macrophages counted in the three "hot spots" for each area.

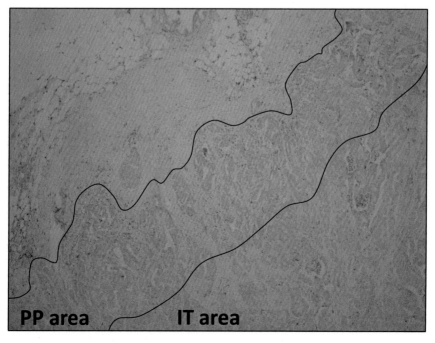

Fig. 1 Representative photomicrograph of breast cancer tissue stained with anti-CD68 antibody with a representation of how tumor sections are divided into two areas to quantify macrophages; intratumoral (IT) and peripheral tumor (PP) areas are shown (×40 magnification)

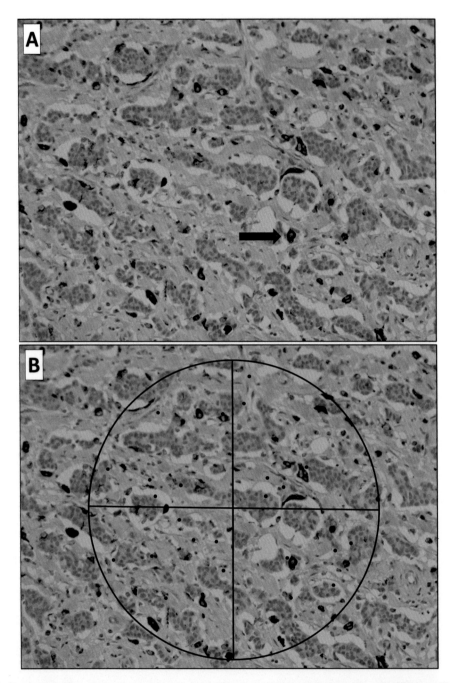

Fig. 2 Representative photomicrograph of breast cancer tissue stained with anti-CD68 antibody: (**a**) Macrophages have been stained with anti-CD68 antibody (*arrow*). (**b**) Illustration showing how the Chalkley graticule is used for counting macrophages (×200 magnification). A 25-dot Chalkley graticule is orientated over each "hot spot" so that the highest number of dots on the graticule overlaps stained macrophages

4   Notes

1. Many ABC staining kits are commercially available; however we have successfully used the Vectastain Elite ABC kit, Vector Laboratories, USA with this method.

2. Once prepared DAB solution should be kept in the dark and used within 6 h.

3. In each run a positive and negative control must be included to confirm that the experiment has been successful. For positive control, lymph node sections can be used for CD68 and CD163 staining. For negative controls, sections are stained following the same procedure but omitting primary antibody.

4. Once the tissue has been deparaffinized and rehydrated, it must not be allowed to dry as this may interfere with staining.

5. Optimization of antigen retrieval buffer and microwave time may be required for different tissue types and should be determined empirically. In addition, antigen retrieval should be optimized for all new antibodies.

6. Fill the Sequenza reservoir half full with TBS to rinse the slides, and watch it flow through to ensure there are no air bubbles (as this will impair staining). If the TBS runs through a Sequenza plate either very quickly or very slowly, this indicates an air bubble, in which case the slide must be reloaded into the plate. Humidity trays can be used instead of the Sequenza system; however the use of humidity trays will require more antibody and reagents.

7. Make up sufficient buffer to also dilute the primary and secondary antibodies.

8. The concentration of primary antibody should be optimized to ensure ideal staining; in addition, the specificity of primary antibody should be confirmed which can be done using techniques such as Western blotting or by blocking binding of the antibody with a neutralizing peptide prior to immunohistochemistry.

9. To confirm the scoring of the first assessor, a percentage of slides should be evaluated by an independent assessor, blinded to the first assessor's data and clinicopathological criteria of the patients.

Acknowledgments

The authors wish to acknowledge a grant from Breast Cancer Campaign (2011NovSP025) which funded this area of their research.

References

1. Ruegg C (2006) Leukocytes, inflammation, and angiogenesis in cancer: fatal attractions. J Leukoc Biol 80:682–684

2. Shih YY, Hsu YH, Duong TQ et al (2011) Longitudinal study of tumor-associated macrophages during tumor expansion using MRI. NMR Biomed 24:1353–1360

3. Ammar A, Mohammed RA, Salmi M et al (2011) Lymphatic expression of CLEVER-1 in breast cancer and its relationship with lymph node metastasis. Anal Cell Pathol (Amst) 34:67–78

4. Storr SJ, Safuan S, Mitra A et al (2012) Objective assessment of blood and lymphatic vessel invasion and association with macrophage infiltration in cutaneous melanoma. Mod Pathol 25:493–504

5. Lewis CE, Pollard JW (2006) Distinct role of macrophages in different tumor microenvironments. Cancer Res 66:605–612

6. Lewis JS, Landers RJ, Underwood JC et al (2000) Expression of vascular endothelial growth factor by macrophages is up-regulated in poorly vascularized areas of breast carcinomas. J Pathol 192:150–158

7. Barbera-Guillem E, Nyhus JK, Wolford CC et al (2002) Vascular endothelial growth factor secretion by tumor-infiltrating macrophages essentially supports tumor angiogenesis, and IgG immune complexes potentiate the process. Cancer Res 62:7042–7049

8. Ferrara N, Gerber HP, LeCouter J (2003) The biology of VEGF and its receptors. Nat Med 9:669–676

9. Li C, Shintani S, Terakado N et al (2005) Microvessel density and expression of vascular endothelial growth factor, basic fibroblast growth factor, and platelet-derived endothelial growth factor in oral squamous cell carcinomas. Int J Oral Maxillofac Surg 34:559–565

10. Cohen T, Nahari D, Cerem LW et al (1996) Interleukin 6 induces the expression of vascular endothelial growth factor. J Biol Chem 271:736–741

11. Fujimoto J, Aoki I, Khatun S et al (2002) Clinical implications of expression of interleukin-8 related to myometrial invasion with angiogenesis in uterine endometrial cancers. Ann Oncol 13:430–434

12. Leibovich SJ, Polverini PJ, Shepard HM et al (1987) Macrophage-induced angiogenesis is mediated by tumour necrosis factor-alpha. Nature 329:630–632

13. Wei LH, Kuo ML, Chen CA et al (2001) Interleukin-6 in cervical cancer: the relationship with vascular endothelial growth factor. Gynecol Oncol 82:49–56

14. Medrek C, Ponten F, Jirstrom K et al (2012) The presence of tumor associated macrophages in tumor stroma as a prognostic marker for breast cancer patients. BMC Cancer 12:306

15. Shabo I, Stal O, Olsson H et al (2008) Breast cancer expression of CD163, a macrophage scavenger receptor, is related to early distant recurrence and reduced patient survival. Int J Cancer 123:780–786

16. Webb JR, Milne K, Watson P et al (2014) Tumor-infiltrating lymphocytes expressing the tissue resident memory marker CD103 are associated with increased survival in high-grade serous ovarian cancer. Clin Cancer Res 20:434–444

17. Allavena P, Sica A, Solinas G et al (2008) The inflammatory micro-environment in tumor progression: the role of tumor-associated macrophages. Crit Rev Oncol Hematol 66:1–9

18. Gordon S (2003) Alternative activation of macrophages. Nat Rev Immunol 3:23–35

19. Ramprasad MP, Terpstra V, Kondratenko N et al (1996) Cell surface expression of mouse macrosialin and human CD68 and their role as macrophage receptors for oxidized low density lipoprotein. Proc Natl Acad Sci U S A 93:14833–14838

20. Kunisch E, Fuhrmann R, Roth A et al (2004) Macrophage specificity of three anti-CD68 monoclonal antibodies (KP1, EBM11, and PGM1) widely used for immunohistochemistry and flow cytometry. Ann Rheum Dis 63:774–784

21. Schaer DJ, Schaer CA, Buehler PW et al (2006) CD163 is the macrophage scavenger receptor for native and chemically modified hemoglobins in the absence of haptoglobin. Blood 107:373–380

22. Edin S, Wikberg ML, Dahlin AM et al (2012) The distribution of macrophages with a M1 or M2 phenotype in relation to prognosis and the molecular characteristics of colorectal cancer. PLoS One 7, e47045

23. Buechler C, Ritter M, Orso E et al (2000) Regulation of scavenger receptor CD163 expression in human monocytes and macrophages by pro- and antiinflammatory stimuli. J Leukoc Biol 67:97–103

24. Leek RD, Lewis CE, Whitehouse R et al (1996) Association of macrophage infiltration with angiogenesis and prognosis in invasive breast carcinoma. Cancer Res 56:4625–4629

25. Vermeulen PB, Gasparini G, Fox SB et al (2002) Second international consensus on the

methodology and criteria of evaluation of angiogenesis quantification in solid human tumours. Eur J Cancer 38:1564–1579

26. Tan KL, Scott DW, Hong F et al (2012) Tumor-associated macrophages predict inferior outcomes in classic Hodgkin lymphoma: a correlative study from the E2496 Intergroup trial. Blood 120:3280–3287

27. Berbic M, Schulke L, Markham R et al (2009) Macrophage expression in endometrium of women with and without endometriosis. Hum Reprod 24:325–332

Part II

Cell Isolation Techniques

Chapter 4

Isolation and Culture of Human Endothelial Cells from Micro- and Macro-vessels

Peter W. Hewett

Abstract

The endothelium from different vascular beds exhibits a high degree of phenotypic heterogeneity. Endothelial cells (EC) can be harvested easily from large vessels by mechanical removal or collagenase digestion. In particular, the human umbilical vein has been used due to its wide availability, and the study of ECs derived from it has undoubtedly greatly advanced our knowledge of vascular biology. However, the majority of the body's endothelium (>95 %) forms the microvasculature, and it is these cells providing the interface between the blood and tissues that play a critical role in the development of new blood vessels. This has led to the establishment of techniques for the isolation of microvascular ECs (MEC) from different tissues to provide more physiologically relevant in vitro models of angiogenesis and EC function.

In this chapter the use of superparamagnetic beads (Dynabeads) coated with anti-PECAM-1 (CD31) antibodies (PECA-beads) to culture MECs from human adipose tissue is described along with the standard methods used to characterize them. Adipose tissue is an ideal source of MECs as it is composed mainly of adipocytes with a very rich microvasculature and is easy to disaggregate. Furthermore, it can be obtained in large quantities during plastic surgery procedures. Adipose obtained at reduction mammoplasty or abdominoplasty is first dissected free of the connective tissue, minced finely, and subjected to collagenase type II digestion. The adipocytes are removed by centrifugation to obtain a microvessel rich pellet, which is further disaggregated with trypsin/EDTA solution. Following filtration to remove fragments of the connective tissue, the pellet is incubated with PECA-beads and microvessel fragments/ECs and washed and harvested using a magnet. In addition, the adaptation of this basic technique for the isolation of the human lung and stomach MECs is also described along with common methods for the preparation of large vessel endothelial cells.

Key words Endothelial cells, Microvascular, Adipose, Dynabeads, PECAM-1/CD31, von Willebrand Factor, E-selectin

1 Introduction

Human ECs derived from large vessels including the aorta and umbilical [1] and saphenous veins have proven an abundant and convenient tool for the investigation of many aspects of endothelial biology. However, the endothelium demonstrates a high degree of functional, morphological, biochemical, and molecular diversity

Stewart G. Martin and Peter W. Hewett (eds.), *Angiogenesis Protocols*, Methods in Molecular Biology, vol. 1430,
DOI 10.1007/978-1-4939-3628-1_4, © Springer Science+Business Media New York 2016

between organs and within the different vascular beds of a given organ [2–6]. This phenotypic heterogeneity has highlighted the need for reliable techniques for MEC isolation and culture from a variety of tissues in order to establish more realistic in vitro models.

Many techniques have been developed to enrich ECs from tissue homogenates, either directly or after a period in culture [6]. Most methods use as their starting point tissue homogenization and digestion and are often hampered by low EC yield and problems of contaminating cell populations that readily adapt to culture. Some tissues are inherently better suited to MEC isolation such as the brain and adipose, which have high microvascular densities, and can be disaggregated easily [2, 6–9]. Wagner and Matthews [7] were the first to utilize adipose tissue from the rat epididymal fat pad for the isolation of microvascular endothelium. The major advantage of this tissue is the difference in buoyant densities between the adipocytes and stromal component allowing their separation by centrifugation.

The development of superparamagnetic beads (Dynabeads™) coupled to endothelial-specific ligands represented a major advance in the purification of MECs from mixed-cell populations. The original technique described by Jackson and colleagues [10] employed the lectin *Ulex europaeus* agglutinin-1 (*UEA*-1), which binds specifically to α-fucosyl residues of EC glycoproteins. However, *UEA*-1 also binds to some epithelial and mesothelial cells [1, 11]. We refined this technique by coupling antibodies raised against platelet EC adhesion molecule-1 (PECAM-1/CD31) [12], a pan-endothelial marker [3, 4], to Dynabeads (PECA-beads), and have used these to prepare MECs from various human tissues [13, 14]. Similarly, other endothelial markers, including CD34, have been used to isolate MECs. However, the majority of EC markers, including PECAM-1 and CD34 cross-react with subpopulations of hematopoietic cells which share common developmental origins. However, hematopoietic cells do not usually represent a significant problem for MEC isolation due to their limited viability in culture. Mesothelial cells exhibit similar morphology and share common markers with ECs and so represent a potentially difficult contaminant of EC cultures isolated from tissues surrounded by a serosal layer, such as the omentum and lung. [6, 11]. However, the absence of constitutive PECAM-1 expression in these cells means that the use of PECA-bead selection should eliminate mesothelial cell contamination of endothelial isolates [11].

In this chapter we describe in detail the use of PECA-beads to isolate MECs from human adipose tissue and methods for the routine culture of these cells. In addition, the adaptation of this purification technique for the culture of MECs from the human lung and stomach [13, 14] is provided alongside routine methods for the preparation of large vessel endothelial cells from the umbilical vein and aorta.

2   Materials

All solutions should be warmed to 37 °C prior to use:

1. *10 % BSA solution:* Dissolve 10 g of bovine serum albumin (BSA) in 100 ml of calcium-magnesium-free Dulbecco's phosphate-buffered saline (PBS/A), 0.22 μm filter sterilize and store at 4 °C.

2. *Antibiotic/antimycotic solution:* Dilute 100× antibiotic/antimycotic solution (Sigma) in PBS/A to give final concentration of 100 U/ml penicillin, 0.1 mg/ml streptomycin, and 0.25 μg/ml amphotericin B. Aliquot and store the 100× stock solution at −20 °C.

3. *Collagenase solution:* Dissolve type II collagenase (Sigma) at 2000 U/ml in Hank's balanced salt solution (HBSS) containing 0.5 % (w/v) BSA, 0.22 μm filter sterilize, aliquot and store at −20 °C.

4. *Trypsin/EDTA solution:* Dilute 10× stock solution of porcine trypsin (2.5 %) in PBS/A, and add 1 mM (0.372 g/L) ethylenediaminetetraacetic acid (EDTA), 0.22 μm filter sterilize, aliquot and store at −20 °C.

5. *Gelatin solution:* Dilute stock (2 %) porcine gelatin solution in PBS/A to give a 0.2 % (v/v) solution and store at 4 °C. To coat tissue culture dishes, add 0.2 % gelatin solution and incubate for 1 h at 37 °C or overnight at 4 °C. Remove the gelatin solution from the flasks immediately prior to plating the cells.

6. *Growth medium:* Many different growth media have been described for maintaining EC in culture (*see* **Note 1**). Supplement M199 (with Earle's salts) with 14 ml/L of 1M *N*-[2-hydroxyethyl] piperazine-*N'*-[2-hydroxy-propane] sulfonic acid (HEPES) solution, 20 ml/L of 7.5 % sodium hydrogen carbonate solution, and 20 ml/L 200 mM l-glutamine solution and mix 1:1 with Ham F12 nutrient mix. To 680 ml of medium M199/Ham F12 solution, add 20 ml of penicillin (100 U/ml)/streptomycin (100 μg/ml) solution, 1500 U/L of heparin, 300 ml of iron-supplemented calf serum (CS) (*see* **Note 2**), 1 μg/ml hydrocortisone, 5 ng/ml basic fibroblast growth factor (bFGF/FGF-2), and 20 ng/ml epidermal growth factor (EGF) (PeproTech EC Ltd, London, UK) (*see* **Note 3**). Store growth medium at 4 °C.

7. *Cryopreservation medium:* Growth medium containing 10 % (v/v) tissue culture grade dimethyl sulfoxide.

8. *Dispase solution:* Dissolve 2 U/ml dispase in medium M199 containing 20 % CS, 0.22 μm filter sterilize, aliquot and store at −20 °C. *Note:* This is only required for the isolation of human lung MEC.

2.2 Preparation of PECA-Beads

Mix 0.1–0.2 mg of mouse anti-PECAM-1 monoclonal antibody (e.g., clone 9G11 R&D Systems) in sterile PBS/A containing 0.1 % BSA (PBS/A+0.1 % BSA) per 10 mg of Dynabeads-M450 (Thermofisher Dynal AS—*see* **Note 4**) pre-coated with pan anti-mouse IgG$_2$ (*see* **Note 5**). Incubate on a rotary stirrer for 16 h at 4 °C. Remove free antibody by washing four times for 10 min and then overnight in PBS/A+0.1 % BSA. PECA-beads maintain their activity for more than 6 months if sterile and stored at 4 °C. However, it is necessary to wash the beads with PBS/A + 0.1 % BSA to remove any free antibody prior to use.

2.3 Equipment for EC Isolation

A class II laminar flow cabinet is essential for all procedures involving the handling of tissue and cultured cells to maintain sterility and protect the operator.

Scalpels, scissors, and forceps are required for the isolation procedures and should be sterilized by autoclaving at 121 °C for 30 min.

100 μm nylon filters: Cover the top of a polypropylene funnel ~10 cm with 100 μm nylon mesh and sterilize by autoclaving (*see* **Note 6**).

Magnet: A suitable magnet is required for the magnetic cell selection system employed which accepts 15 ml tubes (*see* **Note 7**).

Disposable Sterile Plastics

1. 25 and 75 cm^2 tissue culture flasks.

2. Large plastic dishes (e.g., bioassay dishes, Nunc, Naperville, IL, USA).

3. 30 ml universal tubes.

4. 50 ml centrifuge tubes.

5. Multiwell glass chamber slides.

6. 20 ml Luer-Lock syringes.

7. Luer-Lock three-way stopcocks.

2.4 Antibodies for EC Characterization

There are many commercial antibodies available against endothelial markers:

1. Monoclonal antibodies against human PECAM-1, E-selectin, and vWF (e.g., clone F8/86; Dako, High Wycombe, Bucks, UK).

2. Fluorescein isothiocyanate (FITC)-conjugated anti-mouse secondary antibodies.

3. *Nuclear counterstain:* Dissolve Hoechst 33342 (Sigma) in PBS/A 10 μg/ml to give a 10 μg/ml solution, aliquot and store at –20 °C.

3 Methods

3.1 Isolation and Culture of Human Adipose MECs

3.1.1 Collection of Tissue

A suitable large sterile container is required for the collection of adipose tissue obtained during breast or abdominal reductive surgery (*see* **Note 8**). The fat can be processed immediately or stored for up to 48 h at 4 °C.

3.1.2 Isolation of Adipose MECs

1. Working under sterile conditions in a class II cabinet, place the tissue on a large sterile dish (e.g., bioassay dish, Nunc) and wash with 2 % antibiotic/antimycotic solution. Avoiding areas of dense (white) connective tissue (that are often prevalent in breast tissue) and visible blood vessels, scrape the fat free from the connective tissue with two scalpel blades.

2. Chop the fat up finely and aliquot 10–20 g into sterile 50 ml centrifuge tubes. Add 10 ml of PBS/A and 5–10 ml of the type II collagenase solution. Shake the tubes vigorously to further break up the fat and incubate with end-over-end mixing on a rotary stirrer at 37 °C for approximately 1 h. Following digestion the fat should have broken down and no spicules should be evident.

3. Centrifuge the digests at $500 \times g$ for 5 min, discard the fatty (top) layer, and retain the cell pellet with some of the lower (aqueous) layer. Add PBS/A, and recentrifuge $500 \times g$ for 5 min.

4. Resuspend the cell pellet in 10 % BSA solution and centrifuge ($200 \times g$, 10 min). Discard the supernatant and repeat the centrifugation in 10 % BSA solution. Wash the pellet with 50 ml of PBS/A. Viewed under the light microscope, the tissue digest should contain obvious microvessel fragments in addition to single cells and debris.

5. Resuspend the pellet obtained in 5 ml of trypsin/EDTA solution and incubate for 10–15 min with occasional agitation at 37 °C. Add 20 ml HBSS containing 5 % CS (HBSS+5 %CS) and mix thoroughly to neutralize the trypsin. We have found it advantageous to break up the microvessel fragments and cell clumps further with trypsin/EDTA as this reduces the number of contaminating cells co-isolated with the ECs during PECA-bead purification.

6. Filter the suspension through 100 μm nylon mesh to remove fragments of sticky connective tissue. Centrifuge the filtrate at $700 \times g$, for 5 min, and resuspend the resulting pellet in ~1–2 ml of ice-cold HBSS+5 %CS.

7. Add approximately 50 μl of PECA beads and incubate for ~20 min at 4 °C with occasional agitation (*see* **Note 9**). Add HBSS+5 %CS to a final volume of ~12 ml, mix thoroughly, and

select the microvessel fragments/ECs using a suitable magnet (*see* **Note 7**) for 3 min. Repeat the cell selection process a further three to five times by washing the magnetically separated material in ~12 ml of HBSS+5 %CS and reselecting the microvessel fragments with the magnet.

8. Suspend the magnetically separated cells in growth medium (*see* **Note 10**) and seed at high density onto 0.2 % gelatin-coated 25 cm² tissue culture flasks and incubate at 37 °C in a humidified atmosphere of 5 % CO₂.

3.1.3 Adipose MECs in Culture

Following the PECA-bead selection procedure, small microvessel fragments and single cells coated with Dynabeads should be evident under light microscopy (*see* **Note 11**). After 24 h the cells adhere to the flasks and start to grow out from the microvessel fragments present to form distinct colonies. Human mammary microvessel EC (HuMMEC) isolated using this technique grow to confluence forming contact-inhibited cobblestone-like monolayers within 10–14 days depending on the initial seeding density. We have successfully cultured these cells to passage 8 without observable change in their morphology, but routinely use them in experiments between passages 3–6.

3.1.4 Subculture of Adipose MEC

Maintain the MECs at 37 °C, 5 % CO₂ changing the medium every 3–4 days. When confluent EC can be passaged using trypsin/EDTA solution, onto 0.2 % gelatin-coated dishes at a split ratio of 1:4 as follows:

1. Discard the old medium and wash the cell monolayer twice with 5–10 ml of PBS/A. Add a few ml of trypsin/EDTA solution, wash it over the cell monolayer, remove the surplus leaving the cells just covered, and incubate at 37 °C for 1–2 min. Monitor the cells regularly under the microscope until they round up and detach (*see* **Note 12**). Striking the flask sharply helps to dislodge the cells and break up cell aggregates.

2. Add sufficient growth medium to achieve a split ratio of ~1:4 and plate the cells onto gelatin-coated flasks.

3.1.5 Microcarrier Beads

Microcarrier beads can be used to continuously culture EC without the need to use trypsin. Following hydration and sterilization according to the manufacturer's instructions, add the gelatin-coated Cytodex 3™ microcarrier beads (Sigma) to the medium and allow the EC to crawl onto and attach to the microcarriers. Agitate the flasks occasionally to facilitate seeding carriers over a period of 2–3 days. Remove the beads and place into a fresh gelatin-coated flasks containing growth medium and allow the cells to attach to the flask agitating occasionally to ensure good distribution of the cells. Once sufficient cells have attached, the beads may be removed and placed into a fresh flask and the process is repeated.

3.1.6 Cryopreserv ation of EC	Following trypsinization (*see* Subheading 3.1.4), suspend ECs at ~2×10^6/ml in the cryopreservation medium and dispense into suitable cryovials. Cool the vials to –80 °C at 1 °C/min and store under liquid nitrogen.
3.1.7 Maintaining the Purity of MEC Cultures	It may be necessary to reselect the ECs with PECA-beads and/or perform minor manual "weeding" to maintain the purity of cultures. Reselection with PECA-beads can be performed as described above (*see* Subheading 3.1.2, **step 7**) following removal of the cells from flasks using trypsin/EDTA solution (*see* Subheading 3.1.4). Provided that there are clear morphological differences between contaminating cell populations and the ECs (*see* **Note 13**), it is relatively straightforward, although time consuming, to physically remove them. Manual weeding should be performed with the stage of a phase contrast microscope within a class II cabinet to ensure sterile conditions. A needle or Pasteur pipette is used to carefully remove contaminating cells from around EC colonies. The medium is discarded and the adherent cells washed with several changes of sterile PBS/A to remove the dislodged contaminating cells.
3.2 Characteriz ation of ECs	Cobblestone morphology is very typical of ECs derived from many tissues, and they are usually readily distinguished from the typical fibroblastoid contaminating cell populations. However, a more elongated morphology has been reported for human MEC derived from some tissues, and similar elongated phenotypes forming "swirling" monolayers are often observed following stimulation of ECs with growth factors. EC will form "capillary-like" tube networks within few hours of plating on matrices such as growth factor-reduced Matrigel™. This phenomenon also occurs in many types of MECs if cultures are left for several days at confluence (Fig. 1). However, the formation of "capillary-like" structures is not an exclusive property of ECs in culture.
3.2.1 EC Morphology in Culture	
3.2.2 Key EC Markers	There are many criteria in which EC identification may be based which have been reviewed extensively (*see* refs. 2–4, 6). Many endothelial markers/properties are not unique to ECs and several may be required to confirm endothelial identity. ECs isolated from different vascular beds may also display phenotypic heterogeneity, and lack of a particular marker may not preclude the endothelial origin of isolates [2]. It is often useful to demonstrate the absence of markers characteristic of potential contaminating cell populations such as smooth muscle α-actin-positive stress fibers and the intermediate filament protein, desmin, which are expressed by smooth muscle cells and pericytes [15]. A number of good endothelial markers have been identified, including endothelial cell adhesion molecule (ICAM-2/CD102), endothelial cell-selective adhesion molecule (ESAM), and vascular

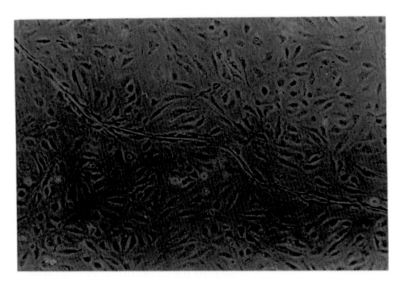

Fig. 1 Photomicrograph of MEC isolated from the human mammary adipose (at passage 2) demonstrating typical cobblestone morphology and tube formation on the surface of the post-confluent EC monolayer

endothelial cadherin (VE cadherin) [4]. Here we focus on von Willebrand factor (vWF), PECAM-1 [12], and E-selectin (endo-thelial-leukocyte adhesion molecule-1/CD62E) [16] that we believe to be useful for the rapid identification of ECs.

vWF is only expressed at significant levels in ECs, platelets, megakaryocytes, and the syncytiotrophoblast of the placenta. In ECs it is stored in the rod-shaped Weibel-Palade bodies, which produce characteristic punctate perinuclear staining. These organ-elles are present in large vessel EC but have been reported to be scarce or absent in the capillary endothelium of various species [1–4, 6]. However, typical granular perinuclear staining for vWF has been reported in cultured human kidney, dermis [10], synovium, lung [13], stomach [14], decidua, heart, adipose [9, 13], and brain [8] MECs.

PECAM-1 is constitutively expressed on the surface of ECs (>10^6 molecules/cell) and to a lesser extent in platelets, granulo-cytes, and a subpopulation of CD8+ lymphocytes [6, 12]. PECAM-1 staining of ECs in vitro is characterized by typical intense membrane fluorescence at points of cell-cell contact (*see* Fig. 2).

E-selectin: Strong expression of E-selectin following stimula-tion with pro-inflammatory cytokines appears to be a unique char-acteristic of ECs [13]. It is not expressed constitutively by the majority of ECs, but stimulation with tumor necrosis factor-α (TNF-α) or interleukin-1β (IL-1β) leads to intense E-selectin stain-ing of the EC plasma membrane reaching a maximum after 4–8 h (*see* Fig. 3).

Fig. 2 Immunofluorescent staining of platelet endothelial cell adhesion molecule-1 (PECAM-1/CD31) in the human mammary adipose MEC

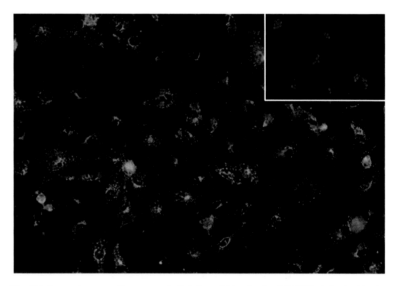

Fig. 3 Intense immunofluorescent staining of E-selectin (CD62E) detected in the human mammary microvessel MEC following 6 h exposure to 10 ng/ml tumor necrosis factor-α (TNF-α), which is absent in unstimulated control cells (*inset*)

3.2.3 Immunocytofluo rescent Characterization of ECs

Outlined below is a simple protocol for the immunofluorescent detection of EC markers:

1. *Preparation of ECs on glass slides.* Multiwell glass chamber slides are extremely useful for this purpose as multiple tests can be performed on the same slide conserving both reagents and cells. ECs are cultured on chamber slides that have been pretreated

for 1 h with 5 μg/cm² bovine fibronectin (Sigma) in PBS/A or 0.2 % gelatin. When sufficient cells are present, discard the medium and wash twice with PBS/A prior to fixation. Different fixatives can be employed depending on the activity of the antibody used. Acetone fixation is suitable for most antibodies: Place the slides (*see* **Note 14**) in cold acetone (−20 °C, 10 min), air dry and store frozen at −80 °C. Alternatively, fix cells in 3.7 % formaldehyde solution for 30 min at room temperature. Formaldehyde does not permeabilize the plasma membrane, and further treatment with 0.1 % Nonidet P-40 or Triton X-100 is required to detect cytoplasmic/nuclear antigens.

2. *Immunocytoflourescent staining.* Warm up slides to room temperature and wash with PBS/A (2×5 min). Block slides for 20 min with 10 % normal serum from the species in which the secondary antibody was raised to prevent nonspecific binding of the secondary antibody.

3. Incubate slides with a predetermined or the manufacturer's recommended concentration of primary antibody in PBS/A for 60 min at room temperature.

4. Wash slides with PBS/A (3×5 min) and incubate with the appropriate FITC-labeled secondary antibody at 1:50 dilution in PBS/A for 30 min to 1 h at room temperature and protect direct light.

5. Counterstain cells with Hoechst 33342 (10 μg/ml) in PBS/A for 10 min to facilitate assessment of EC purity.

6. Wash slides in PBS/A (3×5 min), mount in 50 % (v/v) glycerol in PBS/A. Stained slides can be stored for several months in the dark at 4 °C.

Controls: To avoid false positives generated by nonspecific binding of secondary antibodies, it is essential to include negative controls of cells treated as described above, but with an isotype-matched control antibody and/or PBS/A substituted for the primary antibody. It is also useful to include as controls of other cell types such as fibroblasts, smooth muscle cells, and previously characterized ECs to act as negative and positive control cells, respectively.

E-selectin: Cells are incubated with 1–10 ng/ml TNF-α or IL-1β in growth medium for 4–6 h prior to fixation to induce E-selectin expression. Unstimulated cells should be used as controls.

3.2.4 *Other Properties of Human Adipose MECs*

These cells possess typical EC characteristics including scavenger receptors for acetylated low-density lipoprotein, expression of the transforming growth factor-β co-receptor, endoglin (CD105), and high levels of angiotensin-converting enzyme activity. All the EC types that we have cultured also express the vascular endothelial cell growth factor (VEGF) receptors Flt-1, Flt-4, and KDR/Flk-1

[17] and proliferate and express tissue factor in response to VEGF. Similarly, the angiopoietin receptor, Tie-2/Tek and Tie-1, are also expressed by these cells.

3.3 Isolation of MECs from Other Vascular Beds

We have adapted the basic method for the selection of adipose MECs to isolate ECs from other tissues. Here we outline briefly the modifications that have been made for the isolation of the human lung [12] and stomach MECs [18].

3.3.1 MECs from the Human Lung

Although it has a high microvascular density, the lung is composed of many diverse cell types that readily adapt to culture and is generally more difficult to obtain than adipose tissue. We have successfully isolated EC from the normal lung from transplant donors and diseased tissue from transplant recipients [13]. To ensure that MECs are harvested, a thin strip of tissue at the periphery of the lung is used. As the amount of tissue available is usually limited, and we have found that the yield of cells following direct Dynabead selection is low, it is better to allow the cells to proliferate in culture for a few days and then perform the magnetic purification with PECA-beads before they became overgrown with contaminating cells:

1. Cut small peripheral sections of the lung (3–5 cm long ~1 cm from the periphery) and wash in antibiotic/antimycotic solution.

2. Dissect the underlying tissue from the pleura and chop it up very finely between scalpel blades or by using a tissue chopper.

3. Wash the minced tissue above sterile 20 μm nylon mesh (prepared as described under Subheading 2.3, *see* **Note 6**) to filter out blood cells and fine debris.

4. Incubate the retained material overnight in a dispase solution on a rotary stirrer at 37 °C overnight.

5. Pellet the digest, resuspended in ~5 ml of trypsin/EDTA solution, and incubate at 37 °C for 15 min.

6. Add growth medium and remove fragments of undigested tissue by filtration through 100 μm nylon mesh.

7. Pellet and resuspend the cells in growth medium and plate onto gelatin-coated dishes.

8. Monitor the cultures daily, trypsinize, and select the ECs using PECA-beads before they became overgrown by contaminating cells (*see* Subheading 3.1.7).

3.3.2 Human Stomach MECs

MECs can be cultured from stomach mucosa obtained from biopsies or organ donors [14]:

1. Expose the stomach mucosa washed with antibioitic/antimycotic solution.

2. Dissect the mucosa from the underlying muscle, chop into 2–3 mm pieces, and incubate in 1 mM EDTA in HBSS at 37 °C in a shaking water bath for 30 min.

3. Transfer the pieces of mucosa to collagenase type II solution for 60 min, and then trypsin/EDTA solution for 15 min, in a shaking water bath at 37 °C.

4. Using a blunt dissecting tool, scrape the mucosa and submucosa from the white fibrous tissue.

5. Suspend the mucosal tissue in HBSS+20 % CS and wash through 100 μm nylon mesh.

6. Centrifuge the filtrate ($700 \times g$, 5 min) and resuspend the pellet in ~12 ml of HBSS+5 %CS. Proceed with PECA-bead selection (*see* Subheading 3.1.2, **step** 7).

3.4 Human Large Vessel EC Isolation

3.4.1 Preparation of HUVEC

The umbilical vein provides an abundant source of large vessel EC, which have been used extensively due to the availability of this postnatally redundant tissue and ease with which pure HUVEC preparations can be obtained. The procedure given below has been adapted from the method originally reported by Jaffe and colleagues [1]:

1. Collect umbilical cords in a suitable covered container and store at 4 °C (*see* **Note 15**).

2. Working over a large dish or tray to contain any spills, wash off any blood from the surface of the cord with antibiotic/antimycotic solution and inspect the cord very carefully for the presence of clamp marks and damaged areas, which must be avoided to prevent smooth muscle cell contamination. Cut a length (>20 cm) of cord free of clamp marks using cord scissors (*see* **Note 16**).

3. Cannulate the umbilical vein using a three-way disposable Luer-Lock stopcock and secure using ligature. The umbilical vein with its larger and thinner walls is relatively easy to distinguish from the two narrow umbilical arteries surrounded by the smooth muscle.

4. Attach a syringe to the stopcock and wash the cord through gently with ~30 ml of PBS/A until all the residual blood is removed.

5. Perfuse the cord with ~5–10 ml of 0.1 % (v/v) type I collagenase (e.g., CLS collagenase, Worthington) solution in HBSS (prepared as described for type II collagenase—*see* Subheading 2.1) carefully displacing any residual air, and clamp off the free end.

6. Place the cord on a large plastic tray (e.g., Nunc bioassay dish), cover and place in a humidified incubator at 37 °C for ~15 min.

7. Gently massage the cord and collect the collagenase solution from the vein into a 30 ml universal tube. Wash through with ~20 ml of HBSS+5 % CS. Centrifuge at $200 \times g$ for 10 min and

resuspend the pellet in growth medium (*see* **Note 1**) and plate cells onto gelatin-coated 25 cm² flasks. Under the phase contrast microscope sheets of ECs in addition to many single cells, red blood cells and debris should be apparent.

8. Allow cells to attach to the flask for at least 4 h, or overnight, and then wash the cells twice with PBS/A to remove non-adherent cells and debris. Add fresh growth medium and incubate the flask at 37 °C, in a humidified 5 % CO_2 incubator changing the medium every 2–3 days until the cells reach confluence. Passage the cells at a one-to-three ratio as described above (*see* Subheading 3.1.4). HUVEC dedifferentiate quite rapidly in culture and we do not routinely use them beyond passage 4.

3.4.2 Human Aortic ECs

A relatively simple method for the isolation of human aortic ECs (HAEC) is outlined below as described previously [18]. It is also possible to physically remove the EC by gently scraping of the luminal surface of the aorta rather than using collagenase digestion (*see* **Note 17**). This method can be directly applied to isolate EC from the aortas of other large mammals:

1. Wash sections of human thoracic aorta obtained at postmortem are thoroughly in antibiotic/antimycotic solution.

2. Cut the vessel open longitudinally to expose the surface of the luminal endothelium and lay flat on a sterile petri dish. Add collagenase solution (*see* **Note 17**) over the exposed luminal surface, cover with lid, and incubate at 37 °C for ~15 min in a humidified incubator.

3. Wash the surface of the aorta with ~5 ml of HBSS+5 % CS, drain and collect into a 30 ml universal tube. Centrifuge at $200 \times g$ for 10 min, resuspend the cell pellet in ~5 ml of growth medium and plate onto gelatin-coated T25 flasks, and incubate in a 37 °C, in a humidified 5 % CO_2 incubator.

4. After 12–24 h, discard the medium, wash twice with PBS/A (10 ml), and add fresh medium and subculture the cells as described above passaging with trypsin at confluence (*see* Subheading 3.1.4).

4 Notes

1. Many different growth media have been described for the culture of ECs. Some MECs have very specific requirements such as the presence of human serum, while large vessel ECs tend to be far less fastidious in their growth requirements. This medium M199/Ham F12 nutrient mix-based recipe is relatively inexpensive and works well for a range of ECs, but researchers may wish to optimize their media further. MCDB 131 [19] con-

taining the supplements described (Subheading 2.1, **Item 6**) represents an excellent alternative and can be used with lower concentrations of serum. There are now also several commercial sources of optimized EC media that are based on MCDB 131. For example EGM™-1/2 and EGM™-MV-2 from Lonza (Lonza, San Diego, CA, USA) and although more expensive, they are very convenient and provide reliable support.

2. Iron-supplemented bovine calf serum (CS) provides an economical alternative to FBS and in our hands supports the proliferation and survival of the EC well (*see* **Note 1**). However, it is necessary to batch test all bovine sera to ensure optimal growth of ECs—some MECs such as those isolated from decidualized endometrium are reported to require the presence of human serum or specific growth factors such as VEGF [20].

3. We use recombinant FGF-2 and EGF routinely, but VEGF (5–10 ng/ml) can also be added and provides excellent support for the majority of cultured EC, but is more expensive. These growth factors may also be substituted with EC growth supplement (ECGS) derived from bovine brain, which is rich in acidic and basic FGF and can be obtained from various suppliers.

4. There are alternatives to the use of Dynabeads, for example, the Miltenyi Biotec MACS system. These use composite iron oxide and polysaccharide beads (50 nm diameter) and require a MACS separation system comprising a disposable filtration column which is placed in a magnetic field. They offer the advantage that they are small and so do not interfere with cell attachment and subsequent use of the cells in procedures such as flow cytometry.

5. Precoated Dynabeads (and CELLection™ beads—*see* **Note 10**) carrying various secondary antibodies (e.g., pan anti-mouse) are available from Dynal and are very convenient. However, anti-immunoglobulin-coated beads can be prepared as follows: Incubate the secondary antibody (150 µg/ml) in 0.17 M sodium tetraborate buffer (pH 9.5; 0.22 µm sterile filtered) with tosyl-activated Dynabeads-M450 for 24 h on a rotary stirrer at room temperature. Wash the beads 4× for 10 min and then overnight in PBS/A+0.1 % BSA on a rotary stirrer at 4 °C before proceeding to coat them with the primary antibody as described (*see* Subheading 2.2). Although we have found selection of MECs to be more reliable using anti-PECAM-1 antibodies [13], *UEA*-1 lectin can be used directly conjugated to tosyl-activated Dynabeads as described above.

6. There are now several commercial suppliers of sterile disposable cell strainers that provide a very convenient alternative.

7. The MPC-1 magnet that we have used for many years is no longer manufactured, but suitable alternatives are available from Invitrogen Dynal AS in the DynaMag range which will accept 15 ml to 50 ml tubes.

8. Omental adipose tissue obtained through general abdominal surgery can also be used. However, care should be taken to remove the fat from the omental membranes that are covered with a layer of mesothelium prior to dissection. Using PECA-bead selection, we have not found mesothelial cell contamination of MEC cultures to be a problem.

9. The cell PECA-bead suspension is incubated at 4 °C during the purification steps to minimize nonspecific phagocytosis of Dynabeads.

10. We do not routinely remove Dynabeads following cell selection. However, it may be necessary to remove the Dynabeads if you are using the cells to perform techniques such as flow cytometry. CELLection™ Dynabeads (Dynal) coated with the anti-PECAM-1 antibody can be used to select the MECs. In this system antibodies are conjugated to the Dynabeads via a DNA linker that can be cleaved with DNase-1 to release the beads from the cells following selection.

11. Dynabeads are internalized within ~24 h of selection and are diluted to negligible numbers/cell by the first passage, through cell division. Consistent with the original observations of Jackson and colleagues [10] using *UEA*-I-coated Dynabeads, we have not observed any adverse effects on the adherence, proliferation, or morphology of EC following PECA-bead selection [12].

12. To maintain cell viability, it is important to rapidly remove the endothelial cells from the flasks as they are very sensitive to trypsin exposure. The use of a trypsin inhibitor to neutralize the tryptic activity immediately following detachment from the flask has been reported to prolong the viability of EC cultures. Mung bean trypsin inhibitor (Sigma) can be used for this purpose, although we have not assessed its effect on endothelial viability.

13. The major contaminating cell population observed in unselected adipose MEC cultures demonstrates a distinct fibroblastic morphology.

14. The plastic wells must be removed from the multiwell chamber slides as the acetone will rapidly dissolve the plastic. Depending on the type used, it may be possible to retain the gasket to make it easier to keep reagents on individual wells separate during the staining procedure.

15. In our hands storage of the umbilical cords at 4 °C for up to 48 h after delivery does not appear to adversely affect the yield or viability of the isolated cells.

16. Specialized umbilical cord scissors can be obtained cheaply from a number of suppliers and make the task of cutting lengths of cord a lot easier and safer than using scalpels.

17. The luminal surface of the aorta can be simply scraped very gently to remove the aortic EC. However, we have observed greater contamination of cultures with smooth muscle cells from the underlying intima using this approach compared with collagenase digestion. Following isolation of EC, smooth muscle cells can be explanted from the aortic segments.

References

1. Jaffe EA, Nachman RL, Becker CG, Minidi CR (1973) Culture of human endothelial cells derived from umbilical veins: identification by morphological and immunological criteria. J Clin Invest 52:2745–2756

2. Kumar S, West DC, Ager M (1987) Heterogeneity in endothelial cells from large vessels and microvessels. Differentiation 36:57–70

3. Kuzu I, Bicknell R, Harris AM, Jones M, Gatter KG, Mason DY (1992) Heterogeneity of vascular endothelial cells with relevance to diagnosis of vascular tumors. J Clin Pathol 45:143–148

4. Garlanda C, Dejana E (1997) Heterogeneity of endothelial cells: specific markers. Arterioscler Thromb Vasc Biol 17:1193–1202

5. Trepel M, Arap W, Pasqualini R (2002) In vivo phage display and vascular heterogeneity: implications for targeted medicine. Curr Opin Chem Biol 6:399–404

6. Hewett PW, Murray JC (1993) Human microvessel endothelial cells: isolation, culture and characterization. In Vitro Cell Dev Biol 29A:823–830

7. Wagner RC, Matthews MA (1975) The isolation and culture of capillary endothelium from epididymal fat. Microvasc Res 10:286–297

8. Dorovini-Zis K, Prameya R, Bowman PD (1991) Culture and characterization of microvessel endothelial cells derived from human brain. Lab Invest 64:425–436

9. Hewett PW, Murray JC, Price EA, Watts ME, Woodcock M (1993) Isolation and characterization of microvessel endothelial cells from human mammary adipose tissue. In Vitro Cell Dev Biol 29A:325–331

10. Jackson CJ, Garbett PK, Nissen B, Schrieber L (1990) Binding of human endothelium to Ulex europaeus -1 coated dynabeads: application to the isolation of microvascular endothelium. J. Cell Sci 96:257–262

11. Hewett PW, Murray JC (1994) Human omental mesothelial cells: a simple method for isolation and discrimination from endothelial cells. In Vitro Cell Dev Biol 30A:145–147

12. Newman PJ, Berndt MC, Gorski J, White GC II, Lyman S, Paddock C, Muller WA (1990) PECAM-1 (CD31) cloning and relation to adhesion molecules of the immunoglobulin gene superfamily. Science 247:1219–1222

13. Hewett PW, Murray JC (1993) Immunomagnetic purification of human microvessel endothelial cells using Dynabeads coated with monoclonal antibodies to PECAM-1. Eur J Cell Biol 62:451–454

14. Hull MA, Hewett PW, Brough JL, Hawkey CJ (1996) Isolation and culture of human gastric endothelial cells. Gastroenterology 111:1230–1240

15. Diaz-Flores L, Gutiérrez R, Varela H, Rancel N, Valladares F (1991) Microvascular pericytes: A review of their morphological and functional characteristics. Histol Histopathol 6:269–286

16. Bevilacqua MP, Pober JS, Mendrich DL, Cotran RS, Grimbone MA (1987) Identification of an inducible endothelial-leukocyte adhesion molecule. Proc Natl Acad Sci U S A 84:9238–9242

17. Ferrara N, Gerber HP, LeCoulter J (2003) The biology of VEGF and its receptors. Nat Med 6:669–676

18. Antonov AS, Nikolaeva MA, Klueva TS, Romanov YA, Babaev VR, Bystrevskaya VB, Perov NA, Repin VS, Smirnov VN (1986) Primary culture of endothelial cells from atherosclerotic human aorta. Atherosclerosis 59:1–19

19. Knedler A, Ham RG (1987) Optimized medium for clonal growth of human microvascular endothelial cells with minimal serum. In Vitro Cell Dev Biol 23:481–491

20. Grimwood J, Bicknell R, Rees MCP (1995) The Isolation characterisation and culture of human decidual endothelium. Hum Reprod 10:101–108

Chapter 5

Isolation, Identification, and Culture of Human Lymphatic Endothelial Cells

Zerina Lokmic

Abstract

A protocol describing the isolation of foreskin lymphatic endothelial cells (LECs) and lymphatic malformation lymphatic endothelial cells (LM LECs) is presented herein. To isolate LECs and LM LECs, tissues are mechanically disrupted to make a single-cell suspension, which is then enzymatically digested in dispase and collagenase type II. LECs and LM LECs, in the resulting single-cell suspension, are then sequentially labeled with antibodies recognizing fibroblast and endothelial cell surface antigens CD34 and CD31 and separated from the remaining components in the cell suspension by capture with magnetic beads. Viable LECs and LM LECs are then seeded and expanded on fibronectin-coated flasks. LEC and LM LEC purity is determined immunohistochemically using cell surface markers CD31, CD34, podoplanin, VEGFR-3 and nuclear marker PROX-1. Cells whose purity is >98 % are used for experiments between passage 4 and 6.

Key words Human lymphatic endothelial cells, Isolation, Foreskin, Lymphatic malformation, Magnetic beads, Cell culture

1 Introduction

As a part of a healthy lymphatic system, lymphatic endothelial cells (LECs) play a role in reabsorption of lymph, an excess tissue fluid containing macromolecules, white and occasionally red blood cells, and facilitate the passage of lymph to general blood circulation. LECs are found on the luminal surface of initial and colleting lymphatic capillaries, lymphatic veins, and lymphatic valves which are found throughout the lymphatic system. As part of the initial lymphatic capillary which has discontinuous basement membrane, the LECs are attached to extracellular matrix via short anchoring filaments and are connected by discontinuous button-like cell junctions [1]. LECs that are part of collecting lymphatics and lymphatic veins are also surrounded by other cell types, such as lymphatic vascular smooth muscle cells and fibroblasts, have continuous basement membranes, and form lymphatic vessel valves [2]. Disruption

Stewart G. Martin and Peter W. Hewett (eds.), *Angiogenesis Protocols*, Methods in Molecular Biology, vol. 1430,
DOI 10.1007/978-1-4939-3628-1_5, © Springer Science+Business Media New York 2016

of the lymphatic system occurs in a variety of diseases, such as primary and secondary lymphedema, lymph node metastatic cancers, and developmental conditions such as primary lymphedema, arising from mutations in VEGFR-3, FOXC2, and SOX-18 genes [3], and lymphatic malformations (LMs) [4].

Healthy human LECs express a variety of cell surface markers that collectively enable LEC to be distinguished from vascular endothelial cells. These include the pan-endothelial cell marker CD31, a lymphatic vessel endothelial receptor-1 (LYVE-1), a transcription factor PROX-1 and a vascular endothelial growth factor receptor-3 (VEGFR-3). However, LECs do not express CD34, a vascular endothelial cell marker [5]. It is this property of being CD34-negative (CD34^{Neg}) and CD31-positive (CD31^{Pos}) that is used to isolate these cells from the tissue. We recently showed that human lymphatic malformation endothelial cells (LM LECs) can also be isolated using this selection strategy [6].

One limitation to understanding human LEC biology and pathology in vitro is the source of normal LECs used as controls in experiments. Foreskin LECs are not representative of all LECs found in the human body. Therefore, they cannot represent all LEC functions and responses to external stimuli and pathological challenges that might occur in different LECs in different tissues. Researchers are also limited in obtaining other sources of human LECs due to lack of access to a continuous supply of different tissue types in sufficient quantities to enable sufficient cell yield. In contrast, foreskin tissue is easier to obtain due to elective circumcisions. The size of obtained foreskin tissue dictates how many LECs can be isolated. Mostly, following in vitro expansion, sufficient quantities are available by passage 3 to commence experiments.

This chapter will describe the magnetic bead selection method used in our laboratory to isolate foreskin LECs and LM LECs while reducing the presence of contaminating vascular smooth muscle cells, mesenchymal cells, and fibroblasts. The method is modified from method published by Hirakawa et al. [7].

2   Materials

All reagents are stored and prepared as per manufacturer's instructions. Prepare enzyme solutions on the day of cell isolation. Warm the endothelial cell media, cell culture buffers, and enzyme solutions at 37 °C for 30 min prior to use. All solutions stored at 4 °C as per manufacturer's instructions should be warmed to room temperature only for the time needed to use them. All materials should be disposed of after use as specified by one's own institutional safety guidelines.

2.1 Culture Medium and Solutions

1. Endothelial cell media: EGM-2 MV Bullet Kit (Lonza, catalogue number cc-3202) supplemented with 50 ng/mL VEGF-C (R&D, 2179-VC-025). The kit contains human EGF, hydrocortisone, gentamicin (GA-1000), fetal bovine serum, VEGF, human FGF-b, R^3-IGF-1, and ascorbic acid. The media is prepared by warming EGM-2 media and gently thawing the kit components. Once thawed, the individual components are added to 500 mL EGM-2 media bottle in the laminar flow cell culture hood. Add VEGF-C last. Aliquot endothelial cell media into 50 mL sterile tubes, and store at 4 °C until use. This medium is used for tissue collection, cell isolation, and in vitro cell propagation.

2. Calcium and magnesium-free phosphate-buffered saline (PBS): Prepare PBS by dissolving 8.752 g NaCl, 1.416 g $Na_2HPO_4 \cdot 2H_2O$, and 0.395 g KH_2PO_4 in 1000 mL of water. Adjust pH to 7.4. Filter sterilize PBS and store at 4 °C. Prior to use, add 100 U/mL penicillin and streptomycin.

3. Human fibronectin (Sigma-Aldrich, catalogue number F2006): Dissolve 1 mg of fibronectin in 10 mL of sterile H_2O. Store reconstituted solution in 100 µL aliquots at −20 °C. On the day of use, add 10 mL of PBS to 100 µL of fibronectin aliquot (to give a final working concentration of 10 µg/mL). For 25 cm^2 flasks, use 1 mL of fibronectin solution to coat the flask; for 75 cm^2, use 4 mL to coat the flask; and for 150 cm^2, use 7 mL to coat the flask. Note that 10 µg/mL fibronectin solution can be reused as the sterile reconstituted solution is stable at 4 °C for 1 month.

4. StemPro® Accutase® Cell Dissociation Reagent (Life Technologies, Gibco®, catalogue number A11105-01): This reagent is used to detach cells during passaging as it preserves CD34 and CD31 expression and cell viability better than trypsin-based detachment agents.

5. 0.4 % trypan blue solution (Life Technologies, Gibco®, catalogue number 15250-061). To count cells, 10 µL of cell suspension is mixed with 90 µL of trypan blue. 10 µL of this suspension is used in hemocytometer counting.

6. Dimethyl sulfoxide (DMSO, Sigma Aldrich, catalogue number D4540).

2.2 Enzymatic Digestion

1. The following enzymes are required: Dispase II (Roche Applied Biosciences, catalogue number 4942078001), Collagenase Type II (Worthington Lab, catalogue number 4176) and DNAse I (Roche Applied Biosciences, catalogue number 11284932001). Based on tissue weight, prepare an appropriate volume of enzyme media containing 0.04 % dispase II, 0.25 % collagenase II, and 0.01 % DNase I in sterile

PBS. Ten mL of enzymatic media is required per 1 g of tissue. First, weigh required weight of dispase and collagenase into a sterile 50 mL tube, then add the required volume of PBS, and incubate for 30 min with shaking at 37 °C to dissolve. Once dissolved, filter sterilize (0.22 μm filter) dispase/collagenase solution, and in the laminar flow hood, aseptically add the required amount of DNase I. The enzymatic media is always prepared on the day of use and kept at 37 °C until use.

2.3 Magnetic Bead Cell Isolation

In addition to columns (see equipment), the following Miltenyi Biotec reagents are required: CD31 Multisort kit (catalogue number 130-091-935), CD34 Multisort kit (catalogue number 130-056-701), anti-fibroblast beads (catalogue number: 130-050-601), MACS BSA stock solution (catalogue number 130-091-376), and MACS rinse buffer (catalogue number 130-091-222). Aseptically combine MACS BSA stock solution and MACS rinse buffer prior to use. Avoid causing bubbles, and degas the solution to remove existing air bubbles and store at 4 °C.

2.4 In Vitro Cell Characterization and Microscopy

1. Cell fixative: 10 % neutral-buffered formalin.

2. Antibodies and immunohistochemistry reagents: Protein block solution (DAKO Carpinteria, USA). Primary antibodies: mouse anti-human podoplanin antibody (clone D2-40; DAKO), rabbit anti-PROX-1 (ab38692, Abcam) goat anti-human VEGFR-3 (AF349, R&D Systems), and mouse anti-human CD31 (clone JC70A, DAKO). Secondary antibodies: biotinylated goat anti-mouse antibody (DAKO) to detect podoplanin and CD31, swine anti-rabbit IgG (DAKO) to detect PROX-1, and rabbit anti-goat IgG (DAKO) to detect VEGFR-3. Detection of bound antibody complex: Streptavidin–HRP (DAKO) and diaminobenzidine (DAB, Sigma-Aldrich, catalogue number D0426). Mounting media: DPX (Sigma-Aldrich, catalogue number 06522).

3. Immunohistochemistry buffer: 10× Tris-buffered saline (TBS)/0.5 % Tween buffer: Dissolve 24 g Tris–HCl, 5.6 g Tris base, 88 g NaCl, and 5 mL Tween-20 in 950 mL distilled water. Check that the pH is 7.6 at room temperature. If pH is basic, adjust with concentrated HCl, and if acidic, adjust with NaOH. Then add distilled water to a final volume of 1 L.

4. Cell counterstain. Harris' modified hematoxylin solution (Sigma-Aldrich, catalogue number HHS128). Scott's tap water: dissolve 20 g of $MgSO_4 \cdot 7H_2O$ and 3.5 g $NaHCO_3$ in 1 L of distilled water.

2.5 Equipment

All procedures are done aseptically in a Class II laminar flow safety hood equipped with UV light for decontamination. Dissection equipment should be thoroughly washed, cleaned in 70 % ethanol, and autoclaved at 121 °C for 20–30 min (depending on the system used).

Additional equipment:

1. 25 cm² and 75 cm² tissue culture flasks and 90 mm petri dishes.

2. Sterile 5 mL and 10 mL tissue culture pipettes.

3. 15 and 50 mL sterile centrifuge tubes.

4. 100 μm and 70 μm cell strainers (BD Biosciences).

5. 0.22 μm filtration membranes for solutions.

6. Vacuum pump to aid filtration process.

7. 3 mL sterile syringe plunger with black rubber tip.

8. Scalpel handle, scalpel blades, forceps, and small scissors with straight ends.

9. Hemocytometer.

10. 8-Well cell culture slides.

11. Water bath or incubator with a shaker and temperature control.

12. MACS MultiStand (Miltenyi Biotec).

13. MidiMACS Separator (Miltenyi Biotec).

14. MACS LS Columns (Miltenyi Biotec).

15. Centrifuge suitable to hold 15 mL and 50 mL tubes.

16. Cell culture incubator with temperature and gas composition control.

17. Phase contrast microscope and bright field microscope.

18. Plastic container for cell staining.

3 Methods

3.1 Tissue Collection

Prior to collecting the tissue specimen, pre-weigh sterile container with medium into which the specimen will be collected. Foreskin and LM tissues are collected at the time of surgery into a sterile container containing known amount of endothelial cell media and transported to the laboratory on ice. The amount of media used depends on the tissue size. For foreskin samples, 5 mL of media is used, for LMs 10 mL of media is used. The tissue is transported to the laboratory on ice and processed immediately. In the laboratory, weigh the container with the specimen, and then subtract the weight of the empty container and the amount of media in the container to obtain tissue weight. Add 1 mL of enzyme media per 100 mg of tissue.

3.2 Tissue Processing and Cell Isolation

1. Decontaminate the laminar hood with UV light and clean the working space with 70 % ethanol.

2. Transfer tissue into a 90 mm petri dish, and using sterile forceps and scalpel (or scissors), mince the tissue into 1–2 mm

pieces. This is a critical step as better enzyme digestion is achieved with finely minced tissue (*see* **Notes 1** and **2**).

3. Transfer minced tissue aseptically into a sterile 50 mL tube, and add 1 mL of enzyme media (0.04 % dispase II, 0.25 % collagenase II, and 0.01 % DNase I) per 100 mg of tissue. Place the tube at 37 °C with shaking for 30–90 min. Neonatal foreskin samples require approximately 30 min, fibrotic LMs need 60–90 min. To judge if the incubation is sufficient, simply examine the solution for the amount of undigested tissue present.

4. Following incubation, pass the cells through a 100 μm cell strainer placed in a sterile 50 mL tube, and then use a sterile 3 mL syringe black with rubber plunger to grind down the remaining tissue until only traces of extracellular matrix remain.

5. Wash the cell strainer with 10 mL of endothelial cell media (1:1 ratio to enzyme media used) to recover cells attached to the sieve and to inactivate the enzymes in cell suspension.

6. Spin the cells at $300 \times g$ for 5 min, remove the supernatant, then wash the cells three times in sterile PBS/PenStrep, each time discarding the supernatant and gently disrupting the cell pellet prior to a new wash.

7. Add endothelial cell media to resuspended cells (5 mL for 25 cm² flask, 10 mL per 75 cm² flask, and 20 mL for 150 cm² flask), and pass the suspension through a 70 μm strainer to remove cell debris. Seed up to 5×10^5 cells in a 25 cm² flask, 2×10^6 cells per 75 cm² flask, and 5×10^6 cells in 150 cm² flasks.

8. Plate the cells in flasks and incubate at 37 °C, 5 % CO_2, and 21 % O_2 in a humidified incubator. Check cells after 24 h to determine cell attachment.

9. After 24 h, wash away unbound cells (3×5 min PBS/PenStrep) and add new endothelial cell media. Return to the incubator. Change the media every second day (*see* **Note 3**). When cells are approximately 80 % confluent, cell-specific isolations can commence.

3.3 Magnetic Bead Cell Selection

3.3.1 Fibroblast Depletion

We follow the Miltenyi Biotec manufacturer's instruction with this procedure. However, in addition, we repeat purification on a fibroblast and CD34 column twice to ensure maximum removal of fibroblasts and CD34[Pos] cells from the cell suspension (*see* **Note 4**).

1. Remove the media from the flask and wash the cells three times with sterile PBS/PenStrep. To prepare the cell suspensions, add pre-warmed Accutase® to each flask. For 25 cm² flask use 1 mL, for 75 cm² use 5 mL, and for 150 cm² use 7 mL to coat the flask. Generously cover the cells with Accutase® to ensure complete cell detachment from the fibronectin-coated flask.

The cells are incubated for 5 min (maximum 7 min) at 37 °C in Accutase®. Examine under the polarized microscope, and if the cells are not completely detached, carefully tap the flask three to four times to detach them.

2. Add endothelial cell media to the flask (3:1 ratio of media to Accutase®) to inactivate Accutase®. The cell suspension is drawn up and down a sterile cell culture pipette three to four times before transfer to a new sterile 15 mL centrifuge tube.

3. Centrifuge the cells at $300 \times g$ for 5 min, remove the supernatant, and resuspend the cells in 2 mL of endothelial cell media. Count the cells using 0.1 % trypan blue and a hemocytometer (at 1:10 ratio of cells to dye).

4. To deplete fibroblasts from the cell suspension, centrifuge the suspension at $300 \times g$ for 5 min. Aspirate the supernatant completely. Resuspend the cell pellet in 80 μL of MACS buffer per 10^7 total cells, and avoid introducing air bubbles as this slows down the cell isolation.

5. Add 20 μL of anti-fibroblast beads per 10^7 total cells. Mix well with a pipette and incubate for 30 min at room temperature. During this time, prepare the LS column by inserting it into the magnetic field separator. Place a new 15 mL sterile tube at the tip of the column to collect the rinse buffer. Rinse the column with 3 mL of MACS buffer to prepare the column for use. Place a new 15 mL sterile tube at the tip of the column to collect the flow through. The columns will not dry out while you are waiting for the incubation to be completed.

6. At the end of incubation, add 2 mL MACS per 10^7 cells to wash away unbound beads and centrifuge at $300 \times g$ for 10 min. Aspirate supernatant completely then resuspend cells in 500 μL of MACS buffer.

7. Apply the cell suspension to the column. Do not introduce air bubbles as this will either block the column and the sample will be lost, or it will slow the elution and prolong the isolation process. The flow through is collected as it contains fibroblast-depleted cells. Once the flow has ceased, wash the column three times with MACS buffer. Continue to collect the flow through.

8. On the final wash, place the flow through into the centrifuge, spin at $300 \times g$ for 5 min, and then repeat the process for cell suspension and column purification. This step will minimize the fibroblast contamination. Again collect the flow through, centrifuge and remove the supernatant, and resuspend in cell media to count the cells. Depending on the tissue sample, typical yield is 2×10^5–5×10^5 cells at this stage. To prepare the cells for CD34 depletion, centrifuge and remove the supernatant.

3.3.2   CD34 Depletion

1. Resuspend the cell pellet in 80 μL of buffer then add 20 μL of Anti-CD34 MicroBeads. Mix well with a 100 μL pipette, then incubate at 2–8 °C for 30 min. During this time prepare the column for CD34 separation as described above.

2. At the completion of incubation, add 2 mL of MACS buffer to the cells to facilitate removal of the unbound beads then centrifuge at $300 \times g$ for 5 min.

3. Remove the supernatant and add 500 μL of MACS buffer to the cell pellet. Gently resuspend cells with 1 mL pipette and load the cells onto the column to deplete CD34^{Pos} cells. Collect the flow through.

4. Once the cell suspension has gone through the column, wash the column three times using 3 mL of MACS buffer, each time continuing to collect the flow through. The flow through contains CD34^{Neg} cells.

5. Once the CD34^{Neg} cell fraction is collected, centrifuge the cells at $300 \times g$, remove the supernatant, resuspend the cells in 1 mL of buffer, and then repeat the column purification to remove any remaining CD34^{Pos} cells that might have "squeezed" through. This step further reduces CD34^{Pos} cells, which also include CD34^{Pos} nonvascular cells.

6. After the second column purification, centrifuge the CD34^{Neg} negative fraction cells at $300 \times g$ for 5 min, remove the supernatant, and proceed with CD31 purification. At this stage, the cell yield is about 10,000–50,000 cells (depending on the initial specimen size).

3.3.3   CD31 Positive Selection on Fibroblast and CD34-Depleted Cells

1. Following CD34 depletion, resuspend the cell pellet in 80 μL of MACS buffer using 100 μL pipette being careful not to introduce air bubbles.

2. Add 20 μL of Anti-CD31 MicroBeads. Mix well with pipette then incubate at 2–8 °C for 15 min. During this time prepare the MACS column as described above.

3. After the incubation is completed, add 1 mL of MACS buffer to the cells to wash unbound beads then centrifuge at $300 \times g$ for 5 min.

4. Remove the supernatant and resuspend the cells in 500 μL of MACS buffer prior to loading them onto the column to collect CD31^{Pos}cells. This time, the flow through is not needed since when examined microscopically there are too few cells in the flow through to warrant an additional column isolation. Once you have loaded the cells on to the column, wash the column three times using 3 mL of MACS buffer each time.

5. Once the CD31^{Neg} fraction has been collected, remove the column from the separator and place it on a new 15 mL sterile

tube to collect cells. Pipette 5 mL of buffer onto the column, and using the column plunger provided, immediately flush out the magnetically labeled cells ($CD31^{Pos}$ fraction) by firmly and quickly pushing the plunger into the column.

6. Centrifuge the cells at $300 \times g$, remove the supernatant, then gently resuspend the cells in 5 mL of endothelial cell media, and plate the cells in a fibronectin-coated 24-well plate in total volume of 500 μL per well for 24 h. Because any cell purification at any given time will contain a very low percentage of unwanted cells, by seeding the LECs and LM LECs in 24-well plate, just by random chance, some wells will contain pure LECs or LM LECs, and some will contain a mixture of LECs and unwanted cells. Using a 24-well plate maximizes the likelihood of obtaining pure target cells.

7. After 24 h, wash away unbound cells and continue culturing cells until 80 % confluent. At this point it is important to be ruthless in deciding to cull contaminated cultures (*see* **Note 5**). Any well containing fibroblasts and/or vascular smooth muscle cells should be immediately discarded as these cells can quickly overwhelm the LEC cultures. In my experience, LECs do not require these contaminating cells to survive.

8. Once the cells reach 80 % confluence in the 24-well fibronectin-coated dish, detach them from the wells with Accutase® (300 μL per well) and culture into 25 cm² flasks with media change every 48 h.

9. When the cells are about 80 % confluent in 25 cm² flask, they are passaged and immunophenotyped by culturing 100 cells per well in an 8-well slide for staining with LEC markers. The cells can then usually be further subcultured or cryopreserved.

3.4 Phenotyping Cells

Due to low LEC number present at the initial isolation, the cell characterization takes place when the cells are first passaged to second or third passage after isolation. In vitro, confluent monolayers of LECs and LM LECs have a cobblestone morphology (Fig. 1). To confirm LEC identity, seed the cells in 8-well slides and grow to 70–80 % confluency. Remove the cell media and directly add 10 % neutral buffered formalin for 5 min. Once the neutral buffered formalin is removed wash cells in PBS three times prior to staining. As a negative control for antibody staining, cells are incubated in IgG isotype controls. For each step, 200 μL of reagent and 500 μL per well of TBS/Tween wash buffer are used.

1. Remove cell culture medium and wash the cells in TBS buffer three times. Remove buffer and fix cells in 500 μL per well 10 % buffered formal saline for 5 min. Other fixatives such as 4 % paraformaldehyde, acetone, and methanol can also be used to do the immunophenotyping. After fixation, wash cells in TBS/Tween buffer three times for 5 min each time.

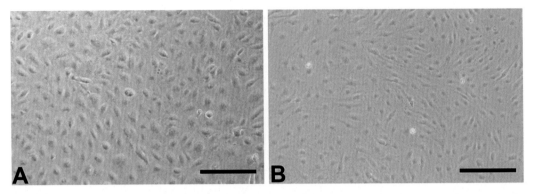

Fig. 1 Foreskin LEC (**a**) and LM LEC (**b**) monolayers have cobblestone morphology. Note the difference in size and shape of two cell types. Scale bar = 100 μm

2. Remove TBS/Tween buffer and block cells in protein block solution for 20 min at room temperature.

3. Remove excess protein solution and add primary antibody (diluted in TBS/Tween buffer) into each well. Antibodies are used at the following concentrations for 12 h at 4 °C: mouse anti-human podoplanin antibody (clone D2-40; working concentration 13.9 μg/mL), rabbit anti-PROX-1 (5 μg/mL), goat antihuman VEGFR-3 (1.5 μg/mL), or mouse antihuman CD31 (clone JC70A, 21 μg/mL).

4. Following incubation, remove the unbound primary antibody and wash cells in three changes of TBS/Tween buffer.

5. Prepare complimentary biotinylated secondary goat anti-mouse antibody (2.6 μg/mL) to detect podoplanin and CD31, swine anti-rabbit IgG (2.1 μg/mL) to detect PROX-1, and rabbit anti-goat IgG (5.3 μg/mL) to detect VEGFR-3.

6. Following 45 min incubation, the unbound secondary antibody is removed and cells washed three times in TBS/Tween buffer.

7. To detect the bound secondary antibodies, the TBS/Tween buffer is removed and streptavidin–HRP (1.8 μg/mL, diluted in TBS/Tween buffer) added to each well for 30 min.

8. To visualize bound streptavidin–HRP, the excess volume is removed and cells washed in TBS/Tween buffer and then incubated in diaminobenzidine (DAB) for 2–5 min. Unbound DAB is washed away in water. At this point the gasket can be removed and the glass slides transferred to a slide rack for counterstaining in hematoxylin.

9. Counterstain cells in hematoxylin (2–3 min). Excess hematoxylin is washed off in three changes of tap water. The hematoxylin is blued in Scott's tap water (1 min) until the nuclei are blue

and slides washed in three changes of tap water. Dehydrate slides in graded ethanol (5 min in 50 %, 75 %, and two changes of 100 %) and two changes of xylene (5 min per solution). The slides are then cover slipped in DPX and examined 24 h later with an optical microscope under ×20 or ×40 magnification. The expected results are shown in Fig. 2.

3.5 Subculturing Human LECs

Once LEC and LM LECs reach 80 % confluency in 25 cm² flask, they can be passaged into 25 cm² or bigger flasks. The cells are split at a ratio of 1:3 (*see* **Note 6**).

1. Remove cells from the incubator and aseptically wash them three times in warm (37 °C) PBS/PenStrep.

2. Warm Accutase® solution is then added to cover the cells. Return the flask to the cell culture incubator for 5–7 min with a first check for cell detachment at 5 min. Examine the flask under a phase contrast microscope to determine if the cells have detached. If the cells have not detached, the flask can be tapped gently to dislodge the cells. If the cells are still attached to the flask, return the flask to the cell culture incubator for a further 2 min.

3. Following cell detachment add 5 mL of endothelial cell media to stop the Accutase® reaction. The cell suspension should be passed two to three times through sterile 5–10 mL cell culture pipette and then transferred to a sterile 15 mL centrifuge tube.

4. Centrifuge the tube at $300 \times g$ for 5 min, remove the supernatant, and resuspend the cells in 15 mL of endothelial cell media. Draw the cells up and down in the pipette before transferring 5 mL of cell suspension to new 25 cm² fibronectin-coated flasks for further culture. Change the media every 2 days. LECs take about 7–10 days to reach confluency, whereas LM LECs take between 3 and 5 days.

3.6 Cryopreservation of Cells (See Note 7)

To cryopreserve LECs and LM LECs, detach the cells as described in Subheading 5. Once the cells are resuspended in endothelial cell media, count the number of cells.

1. Centrifuge the cells at $300 \times g$ for 5 min, remove the supernatant, and resuspend the cell pellet in 90 % FCS/10 % DMSO (v/v) freezing media. We cryopreserve 100,000 cells per mL of freezing media.

2. The cell/freezing media suspension is aliquoted into cryovials and stored at 4 °C for 1 h, followed by −20 °C for 1 h then −80 °C overnight before transfer to liquid nitrogen for long-term storage.

3. To thaw cells, warm the cryovial at 37 °C for 1–2 min, just enough for the frozen content to turn to a slurry. Then transfer the tube contents into 5 mL of warm endothelial cell media.

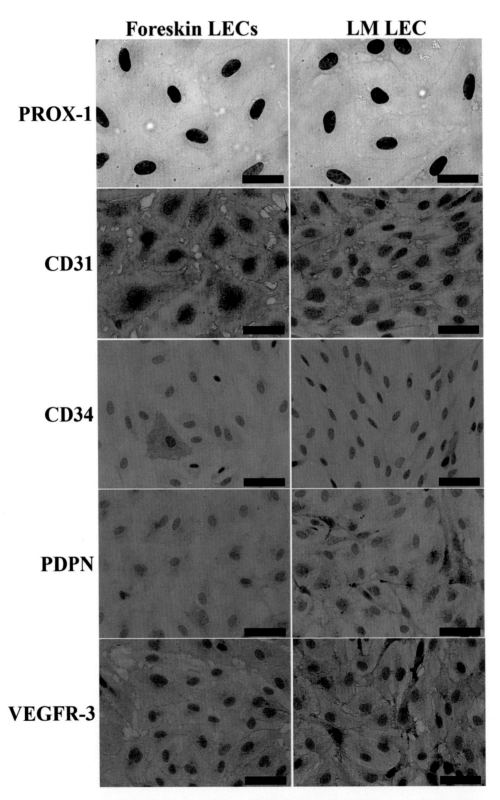

Fig. 2 In vitro, foreskin LEC and LM LEC maintain their CD31^{Pos}podoplanin, ^{Pos}VEGFR-3, and ^{Pos}CD34^{Neg} phenotype (passage 4 shown here). CD34^{Pos} cells are rarely observed in foreskin LEC and LM LEC cell culture. Note variable expression of podoplanin (PDPN) and PROX-1 in both foreskin LECs and LM LECs. Scale bar = 50 μm

4. Centrifuge the suspended cells at $300 \times g$ for 5 min, remove the supernatant containing DMSO, then resuspend the cells in fresh 5 mL of endothelial cell media.

5. Seed the cells into 25 cm² fibronectin-coated flasks for 24 h. On the following day, change cell media to remove nonadherent cells. Note that there will be 20–30 % of dead cells floating after 24 h of culture.

4 Notes

1. The LEC yield will depend on foreskin age. Neonatal skin aged <3 months gives a greater cell yield than foreskin aged >6 months. LM LEC yield will depend on the amount of fibrosis present in the affected tissue.

2. Always prepare fresh enzyme media as this works better when the tissue is more fibrotic.

3. Change the media every second day as longer times between media changes favor survival of fibroblasts and/or vascular smooth muscle cells, even if they are present in small quantities (2–5 %).

4. Podoplanin is not a good marker for isolating LM LECs as LM LECs from macrocystic LM LECs do not uniformly express podoplanin (as detected by clone D2–40) [6].

5. Discard all LEC or LM LEC cultures if they are contaminated with smooth muscle cells or fibroblasts as these cells are detrimental to LEC and LM LEC cultures. Typically, about 75 % of LEC and LM LEC isolations are >95 % purity. The remaining isolations are about 85–90 % pure. Purity of 85–90 % is not sufficient to stop the contaminating cells from overwhelming the LEC and LM LEC cultures by passage 4.

6. Do not use primary foreskin LECs or LM LECs beyond passage 6. Foreskin LECs become increasingly senescent between passages 6–8 and cannot form tubes in vitro. In contrast, some LM LECs proliferate even faster than at earlier passages and can grow on top of each other, a characteristic absent at earlier passages and suggestive of endothelioma transformation. Therefore, it is prudent to freeze cells at the earliest passage possible.

7. Avoid pooling the samples together, in case the HLA incompatibility creates issues for cell survival as suggested for HUVECs [8]. Also, following cryopreservation, new cultures should be passaged once before use in experiments, otherwise the LECs and LM LECs will migrate poorly and will not form tubes on Matrigel™.

5   Summary

Isolating LECs and LM LECs is hindered by fibroblast and vascular smooth muscle cell contamination. This contamination can be reduced if the tissue is processed quickly and cells purified twice to deplete fibroblasts and $CD34^{Pos}$ cells. Although the literature recommends multiple purifications in the event of contamination with fibroblasts and smooth muscle cells, in our hands, it is rare to subsequently obtain a pure culture. From an economic and practical point of view, it is best to discard the wells containing the contamination.

Acknowledgment

This work is supported by Jigsaw Foundation Research Grant and "Women in Science" Fellowship Award.

References

1. Baluk P, Fuxe J, Hashizume H, Romano T, Lashnits E, Butz S et al (2007) Functionally specialized junctions between endothelial cells of lymphatic vessels. J Exp Med 204(10): 2349–2362

2. Alitalo K (2011) The lymphatic vasculature in disease. Nat Med 17(11):1371–1380

3. Ferrell RE, Kimak MA, Lawrence EC, Finegold DN (2008) Candidate gene analysis in primary lymphedema. Lymph Res Biol 6(2):69–76

4. Greene AK, Perlyn CA, Alomari AI (2011) Management of lymphatic malformations. Clin Plastic Surg 38(1):75–82

5. Lokmic Z, Mitchell GM (2011) Visualisation and stereological assessment of blood and lymphatic vessels. Histol Histopathol 26(6):781–796

6. Lokmic Z, Mitchell GM, Koh Wee Chong N, Bastiaanse J, Gerrand YW, Zeng Y et al (2014) Isolation of human lymphatic malformation endothelial cells, their in vitro characterization and in vivo survival in a mouse xenograft model. Angiogenesis 17(1):1–15

7. Hirakawa S, Hong YK, Harvey N, Schacht V, Matsuda K, Libermann T et al (2003) Identification of vascular lineage-specific genes by transcriptional profiling of isolated blood vascular and lymphatic endothelial cells. Am J Pathol 162(2):575–586

8. Baudin B, Bruneel A, Bosselut N, Vaubourdolle M (2007) A protocol for isolation and culture of human umbilical vein endothelial cells. Nat Protoc 2(3):481–485

Chapter 6

Isolation, Culture, and Characterization of Vascular Smooth Muscle Cells

Jessal J. Patel, Salil Srivastava, and Richard C.M. Siow

Abstract

Smooth muscle cells (SMC) are the predominant cell type involved in the pathogenesis of atherosclerosis, vascular calcification and restenosis after angioplasty; however, they are also important in the de novo formation of blood vessels through differentiation of mesenchymal cells under the influence of mediators secreted by endothelial cells. In angiogenesis, vascular SMC are formed by proliferation of existing SMC or maturation and differntiation of pericytes. Experimental findings have demonstrated a potential role of putative smooth muscle progenitor cells in the circulation or within adult tissues and the perivascular adventitia in the development of atherosclerotic plaques, restenosis and angiogenesis. Modulation of vascular smooth muscle phenotype, SMC migration and hypertrophy are now recognized as key events in the development of vascular diseases. This has led to an increase in experimental research on SMC function in response to growth factors, extracellular matrix components, modified lipoproteins, biomechanical forces and other pro-atherogenic and pro-angiogenic mediators to address the cellular mechanisms involved. This chapter highlights well established methodologies used for vascular SMC and pericyte isolation and culture as well as their characterisation. A better understanding of vascular SMC and pericyte biology and their phenotypic modulation is required to identify therapeutic strategies to target angiogenesis and treat cardiovascular diseases.

Key words Smooth Muscle Cell, Pericyte, Angiogenesis, Atherosclerosis, Phenotype, Cell Culture

1 Introduction

Smooth muscle cells (SMC) are the predominant cell type involved in the pathogenesis of atherosclerosis and restenosis after angioplasty [1]; however, they are also important in the formation and development of de novo blood vessels (vasculogenesis) through differentiation of mesenchymal cells under the influence of mediators secreted by the endothelial cells comprising newly formed vessels [2]. In angiogenesis, vascular SMC are formed by proliferation of existing SMC or maturation of pericytes [2, 3]. Experimental findings suggest a potential role of putative smooth muscle progenitor cells in the circulation or within adult tissues and the perivascular adventitia in the development of atherosclerotic plaques

Stewart G. Martin and Peter W. Hewett (eds.), *Angiogenesis Protocols*, Methods in Molecular Biology, vol. 1430, DOI 10.1007/978-1-4939-3628-1_6, © Springer Science+Business Media New York 2016

and biology of angiogenesis [4]. Modulation of vascular smooth muscle phenotype, SMC migration and hypertrophy are now recognized as key events in the development of arterial lesions in vascular diseases [1]. This has led to an increase in experimental research on SMC function in response to growth factors, extracellular matrix, modified lipoproteins, and other pro-atherogenic and pro-angiogenic mediators under controlled in vitro conditions to address the cellular mechanisms involved. Most of the methodologies used for vascular SMC isolation and culture have been developed to accomplish such studies [5].

In vivo, vascular SMC retain the characteristic of phenotypic modulation, ranging between the "contractile" and "synthetic" states. This plasticity allows smooth muscle cells to adapt to local environmental cues within the vessel wall, for example growth factors (platelet-derived growth factor, transforming growth factor-β1), contractile agonists (angiotensin II, endothelins), reactive oxygen species, inflammatory mediators, and biomechanical shear and stretch forces [5–7]. The healthy adult vasculature predominantly consists of the contractile-state SMC, whose main function is in the maintenance of vascular tone. They exhibit a characteristic "muscle-like" appearance, with up to 75 % of their cytoplasm containing contractile filaments. However, in culture, these cells are able to revert to a synthetic and proliferative phenotype, which is normally found in embryonic and young developing blood vessels [8]. These proliferative cells synthesize extracellular matrix components such as elastin and collagen, and consequently contain large amounts of rough endoplasmic reticulum and Golgi apparatus but few myofilaments in their cytoplasm [6–8]. The modulation of vascular SMC from a contractile to a synthetic phenotype is an important event in atherogenesis and restenosis, resulting in myointimal thickening and arterial occlusion. This may arise from damage to the endothelium and exposure of SMC to circulating blood components, such as oxidized lipoproteins, and pro-inflammatory cytokines and stimuli such as reactive oxygen species and growth factors released from endothelial cells, neutrophils, macrophages and platelets [1]. In addition, conditions such as hyperglycemia and hyperlipidemia have been shown to modulate SMC contractile phenotype to a synthetic phenotype, which plays a key role in the development of vascular diseases [9, 10].

The methodology chosen to isolate and culture vascular SMC can determine the initial phenotype of cells obtained in culture [5, 8]. The two main techniques commonly employed in SMC isolation from arterial and venous tissues are enzymatic dissociation, which readily yields a small number of SMC initially retaining a contractile phenotype, and explantation, which yields larger numbers of SMC after 2–3 weeks, in the synthetic and proliferative phenotypes. The lower yield of SMC following enzymatic dissociation is more suited for studies on single dispersed cells while tissue explantation of vascular tissues provides a better potential for obtaining longer-term cultures of confluent SMC monolayers. In addition to the source of SMC, alterations in the methods employed to culture SMC in vitro (e.g., serum, glucose, growth factor concentrations, biomechanical forces and oxygen tensions) can also influence their phenotype. These modifications, which may facilitate the culture of a contractile phenotype, include changes in culture medium source and composition, as well as surface coating of culture plasticware with matrix proteins [11–13]. This chapter describes two alternative techniques for vascular isolation and culture of SMC from arterial tissues, subculture, maintenance, and characterization of SMC phenotype. In addition, a brief description of the isolation and identification of pericytes from human placental microvessels is also provided since these "muscle-like" perivascular cells are recognized to play a key role during angiogenesis for scaffolding, maturation, remodeling, and contraction of microvessels [3].

2 Materials

2.1 Smooth Muscle Cell Cultures

1. The most commonly used growth medium in SMC culture is Dulbecco's modified Eagle's medium (DMEM) containing either 1000 or 4500 mg/L glucose; however, Medium 199 is also suitable (*see* **Note 1**). The following additions to the basal medium are necessary prior to use: (final concentrations) 2 mM L-glutamine, 40 mM bicarbonate, 100 U/ml^{-1} penicillin, 100 μg/ml streptomycin, and 10 % (v/v) fetal calf serum (FCS). Sterile stocks of these components are usually prepared and stored as frozen aliquots as described below. The complete medium can be stored at 4 °C for up to 1 month and is prewarmed to 37 °C prior to use in routine cell culture. Culture medium without the FCS component is used during the isolation procedures.

2. Hanks' balanced salt solution (HBSS) is used as a tissue specimen collection medium. The following additions, from sterile stock solutions, are necessary prior to use (final concentrations): 100 μg ml^{-1} Gentamycin, 0.025 M HEPES and 20 mM bicarbonate. The HBSS can be stored in aliquots at 4 °C for up to 2 weeks.

3. L-Glutamine (200 mM) stock solution: Dissolve 5.84 g L-glutamine in 200 ml tissue culture grade deionized water and sterilize by passing through a 0.22 μm filter. Aliquots of 5 ml are stored at –20 °C and 4 ml used in 400 ml of medium.

4. Bicarbonate (4.4 %, 0.52 M) solution: Dissolve 44 g NaHCO$_3$ in 1000 ml tissue culture grade deionized water, and sterilize by autoclaving for 10 min at 115 °C. Aliquots of 15 ml are

stored at 4 °C for up to 6 months and 2 aliquots used in 400 ml of medium.

5. Penicillin and streptomycin stock solution (80× concentrate): Dissolve 480 mg penicillin (G sodium salt) and 1.5 g streptomycin sulfate in 200 ml tissue culture grade deionized water and sterilize by passing through a 0.5 μm pre-filter and a 0.22 μm filter. Aliquots of 5 ml are stored at –20 °C and 1 aliquot used in 400 ml of medium.

6. Gentamycin solution (80× concentrate): Dissolve 750 mg gentamycin sulfate in 100 ml tissue culture grade deionized water and sterilize by passing through a 0.22 μm filter. Aliquots of 5 ml are stored at –20 °C and 1 aliquot used in 400 ml of HBSS.

7. HEPES solution (1 M): Dissolve 47.6 g of HEPES in 200 ml tissue culture grade deionized water and sterilize by passing through a 0.22 μm filter. Aliquots of 5 ml can be stored at –20 °C and 2 aliquots used in 400 ml of HBSS.

8. Trypsin solution (2.5 %): Trypsin from porcine pancreas is dissolved (2.5 g 100 ml^{-1}) in PBS-A and sterilized by passing through a 0.22 μm filter. Aliquots of 10 ml are stored at –20 °C.

9. EDTA solution (1 %): EDTA disodium salt is dissolved (500 mg/50 ml^{-1}) in tissue culture grade deionized water and sterilized through a 0.22 μm filter. Aliquots of 5 ml are stored at 4 °C.

10. Trypsin (0.1 %)–EDTA (0.02 %, 0.5 mM) solution is prepared by adding 10 ml trypsin (2.5 %) and 5 ml EDTA (1 %) to 250 ml sterile PBS-A. This solution is prewarmed to 37 °C before use to detach cells from culture flasks and stored at 4 °C for up to 2 months.

2.2 Enzymatic Dissociation of SMC

1. Collagenase, Type II. Dissolve collagenase in serum free medium (3 mg ml^{-1}) on ice. Particulate material is removed by filtering the solution through a 0.5 μm pre-filter and then sterilize by passing through a 0.22 μm filter. This enzyme solution can be stored long term as 5–10 ml aliquots at –20 °C until use.

2. Elastase, type IV from porcine pancreas. Immediately before use, elastase is dissolved in serum free medium (1 mg/ml) and the pH of solution adjusted to 6.8 with 1 M HCl. This solution is then sterilized by passing through a 0.22 μm filter and kept on ice.

2.3 Immunofluorescence Microscopy

1. Antibody to specific SMC antigen, e.g., monoclonal mouse α-smooth muscle actin.

2. Normal rabbit serum.

3. Fluorescein isothiocyanate conjugated rabbit anti-mouse secondary antibody.

4. Methanol (100 %).

2.4 Equipment

All procedures should be carried out in a Class II laminar flow safety cabinet using aseptic technique. Dissection equipment should be thoroughly washed and kept sterilized by immersion in 70 % ethanol or by autoclaving at 121 °C for 20 min.

1. 25 cm^2 and 75 cm^2 tissue culture flasks and 90 mm petri dishes.

2. Sterile Pasteur, 5 and 10 ml pipettes.

3. Sterile 30 ml universal containers and 10 ml centrifuge tubes.

4. Scalpel handles and blades, small scissors, watchmaker's forceps, and hypodermic needles.

5. Cork board for dissection covered with aluminum foil, both sterilized by thorough spraying with 70 % ethanol.

6. Sterile conical flasks of various sizes.

7. Lab-Tek chamber slides (Nunc).

3 Methods

3.1 Collection of Tissue Samples

In this laboratory, cells are routinely isolated from human umbilical arteries, a readily available source of human vascular SMC. As soon as possible after delivery, the whole umbilical cord, obtained with prior ethical approval and consent of mothers, is placed in the HBSS collection medium and stored at 4 °C. Cords collected and stored in this way can be used for SMC isolation up to 48 h after delivery. SMC can also be isolated from arteries following harvesting of endothelial cells from the corresponding umbilical vein [14]. Immediately prior to proceeding with SMC isolation, 5 cm lengths of the umbilical artery should be carefully dissected out from the cord, ensuring minimal surrounding connective tissue remains, and stored in new collection medium. Other common sources of arterial tissue are from human, mouse, rat, or porcine aortae, which should be carefully dissected out from the body, cleaned of extraneous tissues and stored in the collection medium at 4 °C as soon as possible (*see* **Note 2**). Isolation of SMC/pericytes from microvessels can be performed using human placental or bovine retinal tissue [15].

3.2 Isolation of Smooth Muscle Cells by Enzymatic Dispersion

1. As much surrounding connective tissue as possible should be dissected away from around the artery and the tissue washed with new HBSS collection medium. The artery is then placed on the sterile dissection board and covered with HBSS to keep it moist.

2. The artery is fixed to the dissection board at one end using a hypodermic needle and then cut open longitudinally, using small scissors, with the luminal surface upward.

3. The endothelium is removed along the whole length of the artery by scraping the cell layer off with a sterile scalpel blade and the tissue re-moistened with HBSS.

4. The thickness and nature of the arterial wall will vary depending on the source of tissue; however, in general the arterial intima and media are peeled into 1–2 mm width transverse strips using watchmaker's forceps and a scalpel. Muscle strips are transferred in to a 90 mm petri dish containing HBSS. This procedure is repeated for the whole surface of the vessel.

5. Most of the HBSS medium is then aspirated off and the muscle strips cut into 1–2 mm cubes using scissors or a scalpel blade. The cubes are then washed in new HBSS and transferred into a sterile conical flask of known weight and the mass of tissue measured to determine volume of enzyme solution needed for digestion.

6. Collagenase solution in serum free culture medium is next added to the tissue to give a ratio of tissue (g) to enzyme solution (ml) of 1:5 (w/v). The flask is then covered with sterile aluminum foil and shaken in a water bath at 37 °C for 30 min.

7. Elastase solution is then prepared and directly added to the solution containing the tissue and collagenase. The flask is returned to the shaking water bath at 37 °C and every 30 min during the following 2–5 h the suspension mixed by pipetting in the sterile safety cabinet.

8. At each 30 min interval, a 10 μl sample of suspension is transferred to a hemocytometer to check for the appearance of single cells. This is repeated until the tissue is digested and there are no large cell aggregates visible. Digestion should not proceed for longer than 5 h to avoid loss of cell viability (*see* **Note 3**).

9. The cell suspension is finally divided into 10 ml centrifuge tubes and centrifuged at 50–100×*g* for 5 min. The supernatants are carefully aspirated and 2–5 ml of prewarmed complete culture medium added to resuspend the cells using a sterile pipette. Cells are then seeded into 25 cm² culture flasks at a density of about 8×10^5 cells ml^{-1} and placed into a 37 °C, 5 % CO$_2$ incubator with the flask cap loose.

10. Viable SMC should adhere to the flask wall within 24 h and all the medium is then replaced with fresh prewarmed complete culture medium. Half of the culture medium is replaced every 2–3 days until a confluent SMC monolayer is obtained.

3.3 Isolation of Smooth Muscle Cells by Explant Culture

Explant cultures are suitable if limited vascular tissue is available, for example, from human aortas or carotid arteries. The sample is first treated exactly as described in **steps 1–3** of Subheading 3.2 above and then the following procedure is adopted:

1. The artery is cut into 2 mm cubes with a scalpel blade and placed on to the surface of a 25 cm^2 culture flask using a sterile Pasteur pipette, ensuring that the luminal surface is in contact with the flask wall. The artery should be kept continually moist with the HBSS and a small drop of serum containing medium should be placed on each cube when placed in the flask.

2. The cubes are distributed evenly on the surface with a minimum of 12–16 cubes per 25 cm^2 flask. The flask is then placed upright and 5 ml serum containing medium added directly to the bottom of the flask before transferring into a 37 °C, 5 % CO_2 incubator with its cap loose. To facilitate adherence of the explanted tissue to the culture plastic substrate, the flask is kept upright for 2–4 h in the incubator before the flask is carefully placed horizontally such that the medium completely covers the attached muscle cubes.

3. The explants should be left undisturbed for 4 days, inspecting daily for infections (*see* **Note 4**). Every 4 days any unattached explant cubes should be removed and half the medium replaced with fresh prewarmed serum containing medium. Cells will initially migrate out from the explants within 1–2 weeks.

4. After 3–4 weeks there should be sufficient density of SMC around the explants for removal of the tissue. Using a sterile Pasteur pipette, the cubes are gently dislodged from the plastic flask surface and aspirated off with the culture medium. The culture medium is replaced and cells are then left for a further 2–4 days to proliferate and form a confluent SMC monolayer (*see* **Note 5**).

3.4 Isolation of Smooth Muscle Cells/Pericytes from Microvessels

Due to the relative availability of human placentas, isolation from this tissue will be the described below. Use of placental tissue yields larger quantities of pericytes due to the high amount of villi present. Alternatively bovine retinas are a suitable tissue source for pericyte isolation.

1. Dissect a central section of the placenta and wash thoroughly in serum-free medium. Ensure that the section chosen is distant from any large blood vessels and the outer membrane.

2. Manually dissect and cut the tissue into small 5 mm^2 pieces and incubate in serum-free media containing 3 mg/ml collagenase for 3 h at 37 °C in a shaking water bath.

3. Separate the microvessels by passing the suspension through a 70 μM mesh filter (Falcon) and wash through two times with serum-free media.

4. Remove the microvessels from the mesh filter and place into a 25 cm^2 culture flask containing medium supplemented with 20 % serum.

5. After 24 h the medium is removed and attached cells washed once with warmed sterile PBS to remove floating debris. Fresh growth medium is added and following 5–6 days, pericytes and endothelial cells proliferate out from the microvessel fragments.

6. As the culture medium does not contain endothelial cell growth supplements, any endothelial cells present initially no longer survive after two rounds of trypsinization leaving a pure culture of pericytes which can be characterized by their morphology and antigen expression detected by immunofluorescence.

3.5 Subculture of Smooth Muscle Cells

1. Once a confluent monolayer has been attained in a 25 cm² flask by either isolation method, the SMC can be subcultured (passaged) into further 25 cm² flasks or a 75 cm² flask. The culture medium is removed and cells are washed twice with prewarmed sterile PBS-A to remove traces of serum.

2. Prewarmed trypsin–EDTA solution (0.5 ml for 25 cm² flask or 1 ml for a 75 cm² flask) is added to cover the cells and the flask incubated at 37 °C for 2–4 min. The flask is then examined under the microscope to ensure cells have fully detached. This can also be facilitated by vigorous tapping of the side of the flask three to six times to break up cell aggregates (*see* **Note 6**).

3. Serum containing medium (5 ml) is added to stop the action of the trypsin which can reduce SMC viability through prolonged exposure. The cell suspension is then drawn up and down a sterile Pasteur pipette four to six times to further break up any cell clumps.

4. The cells are then transferred into new culture flasks at a split ratio of 1:3 and sufficient serum containing medium added to the new flasks (5 ml in a 25 cm² and 10 ml in a 75 cm² flask). The flasks are returned to the 37 °C, 5 % CO_2 incubator and the culture medium changed as described in Subheading 3.2, **step 10**.

Smooth muscle cells can be passaged between 10 and 20 times, depending on species, before their proliferation rate significantly decreases. Phenotypic changes of enzyme dispersed SMC to the "proliferative" state occurs following passaging, the extent of which depends on the vessel type and species from which the SMC are derived, the culture medium and their seeding density (further discussed below) [5, 6, 8, 16].

3.6 Characterization and Maintenance of Smooth Muscle Cell Phenotype

In culture, smooth muscle cell phenotype can be determined by changes in morphology, as well as phenotype marker expression. Morphologically, synthetic SMCs exhibit a characteristic "hill and valley" appearance, whereas contractile SMCs cells appear more elongated and spindle-like [5, 8, 17]. Contractile SMCs have been shown to express a number of specific phenotype markers, which

Table 1
Markers for characterization of vascular smooth muscle cells (*see* refs. [8, 16, 19–21])

Specific SMC protein markers	Function
Smooth muscle myosin heavy chain	SMC contractile protein
Calponin	Contractile regulator
SM22α	Cytoskeletal protein
Desmin	Intermediate filament
H-Caldesmon	Actin binding protein
Metavinculin	Actin binding protein
Smoothelin	Cytoskeletal protein
α-Smooth muscle actin	Contractile protein

are summarized in Table 1. Smooth muscle myosin heavy chain (SMMHC) is the most specific marker; however, expression is not always maintained in long-term SMC cultures. Markers, which are indicative of the synthetic phenotype are uncommon; thus, a decline in expression of contractile markers, or increased production of proteins such as type I collagen or osteopontin are generally accepted as a switch to a more synthetic phenotype [17]. Although these markers are considered "specific" for genes regulating SMC contraction, no one marker has been identified to be a sole indicator of a contractile SMC phenotype; therefore, it is considered best practice that a range of markers are assessed [17]. The absence of endothelial cells in cultures can be confirmed by negative staining for von Willebrand factor or the lack of uptake of acetylated low density lipoproteins, both endothelial cell specific markers [14, 18].

To maintain a contractile phenotype in vitro, numerous modifications in SMC culture conditions can be employed. These include the use of selective culture medium, serum reduction or deprivation, addition of heparin, surface coating of cultureware with collagen (either in the monomeric or fibrillar state), and the use of micropatterned-grooved surfaces [11–13, 22, 23]. Our assessment of expression of SMC phenotype markers demonstrated that monomeric collagen coating of cultureware promoted expression of contractile phenotype genes, whereas cells grown in a proprietary SMC culture medium (containing growth factors such as insulin, epidermal growth factor and basic fibroblast factor) exhibited a more proliferative and synthetic phenotype (Fig. 1). Cultures supplemented with heparin expressed contractile phenotype markers to a greater extent than those cultured on monomeric collagen (data not shown); though given heparin could inhibit cellular contraction [24], monomeric collagen may

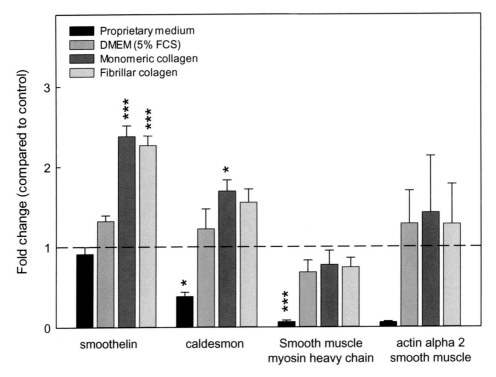

Fig. 1 Effect of cell culture conditions on SMC phenotype marker gene expression. Commercially sourced human aortic SMC isolated using enzymatic dissociation were cultured for 2 days in basal medium (DMEM supplemented with 10 % FCS), and compared to cells cultured in either proprietary medium, DMEM containing lower serum (5 % FCS), or cells cultured in basal DMEM on cultureware coated with either fibrillar or monomeric collagen. Expression of mRNA for smoothelin and caldesmon, determined by qPCR, was significantly upregulated in cells cultured on monomeric collagen, compared to basal medium (indicated by the *dashed line*). Expression of myosin heavy chain and smooth muscle alpha-actin 2 was also enhanced in monomeric collagen coated cultures. Cells cultured on fibrillar collagen also exhibited a similar pattern of mRNA expression to monomeric collagen, although to a lesser extent. SMC grown in proprietary medium exhibited an mRNA marker profile consistent with a synthetic phenotype, compared to the basal DMEM medium. (*$p < 0.05$, ***$p < 0.001$, $n = 5$)

be the more suitable choice. From these findings, the proprietary SMC medium is suggested for expanding cell populations, which are then subsequently transferred on to culture plastic-ware coated with monomeric collagen to induce a contractile SMC phenotype suitable for in vitro studies of cellular contraction.

The following section describes the method by which culture plastic-ware can be coated with monomeric collagen to maintain a contractile phenotype in vitro (*see* **Note 7**):

1. In sterile conditions, dilute stock type I rat tail collagen (Corning®—354236) to a concentration of 50 μg/ml in 0.02 N (0.02 M) acetic acid. Make enough of this diluted solution to cover the culture wells sufficiently (~40 μl/well in a 96-well tray; ~1 ml/well in a 6-well tray).

2. Add diluted collagen solution to cultureware and incubate for 1 h at room temperature.

3. After 1 h, remove collagen solution and rinse twice in sterile PBS.

4. The plates can be used immediately, or dried and stored at 4 °C until required.

   In addition to assessing gene expression, immunohistochemistry is routinely utilized in identifying SMC phenotype. The following method briefly describes a staining method using a fluorescein iso-thiocyanate (FITC)-labeled primary antibody against smooth muscle α-actin.

1. Smooth muscle cells are subcultured into Lab-Tek slide wells and characterized after 48 h.

2. The culture medium is removed from the wells and cells gently washed three times with serum free culture medium before being fixed with ice-cold methanol (100 %) for 45 s, and then further washed three times with ice-cold PBS-A.

3. Cells are then incubated with a mouse monoclonal anti-smooth muscle α-actin antibody at 1:50 dilution with PBS-A for 60 min at room temperature. As a negative control, some cells are incubated with PBS-A only at this stage.

4. The primary antibody or PBS-A is then removed and cells washed three times with PBS-A and incubated for 5 min with normal rabbit serum at 1:20 dilution.

5. After a single wash with PBS, cells are then further incubated for 30 min at room temperature with FITC-conjugated rabbit anti-mouse IgG (Santa Cruz) diluted 1:50 in PBS-A.

6. Finally cells are washed three times with PBS-A and viewed under a microscope equipped for epifluorescence with appropriate filters for FITC.

   Visualization of positive staining with this technique should reveal cells with a three-dimensional network of long, straight, and uninterrupted α-actin filaments running in parallel to the longer axis of the cells and an underlying row of parallel filaments along the smaller cell axis, with no cytoplasmic staining between filaments.

3.7 Characterization of Pericytes

Although pericytes are related to vascular smooth muscle cells, pericytes can to a degree be distinguished by marker expression [25]. There is as yet, no specific molecular marker for pericytes, although, a number of markers that are commonly present in pericytes, albeit not exclusively, can be used for detection [26]. Commonly used markers include: desmin and α-smooth muscle actin, both contractile filaments; regulator of G protein signaling 5 (RGS-5), a GTPase-activating protein; neuron-glial 2 (NG2), a chondroitin sulfate proteoglycan; and platelet-derived growth factor receptor beta (PDGFRβ), a tyrosine-kinase receptor [3, 25–27]. In a similar manner to SMC phenotype determination, assessment of multiple markers would be required to give a positive presence of pericytes. It must be stressed that the markers mentioned above in cases cannot identify all pericytes and that expression can vary between species, which may lead to greater difficulty in true identification of pericytes. One example of this is α-smooth muscle actin, which is not expressed to an extent in skin or CNS pericytes under normal circumstances but is upregulated during retinopathy and in subcutaneously transplanted tumors [25].

3.8 Cryopreservation of Smooth Muscle Cells

Vascular smooth muscle cells can be cryopreserved with a recovery > 50 %. Explant cultures of SMC do not appear to be adversely affected by freezing; however, enzyme dispersed SMC may have a reduced proliferation rate and passaging efficiency on thawing. The following protocol is suggested; however, other techniques of cryopreservation are also available.

1. Confluent SMC cultures should be detached from one 75 cm² flask as described in Subheading 3.5.

2. Following centrifugation of the cell suspension for 5 min at 1000 rpm, the supernatant is aspirated and the cell pellet resuspended well in serum containing culture medium with an additional 10 % (v/v) dimethyl sulfoxide (DMSO) and transferred to a suitable cryovial.

3. To facilitate gradual freezing, the cryovial is then stored at 4 °C for 1 h, transferred to –20 °C for 1 h min and –70 °C for 1 h before being immersed into liquid nitrogen for long term storage. Alternatively, cryovials can be placed in a dedicated "freezing chamber" containing isopropanol that lowers their temperature by 1 °C per minute when placed directly in a –70 °C freezer.

4. To defrost cells, the cryovial should be rapidly warmed to room temperature by placing at 37 °C and the cell suspension transferred to a 25 cm² culture flask. 20 ml of prewarmed serum containing culture medium is then added to the flask and the medium changed after 24 h.

5. Alternatively, on defrosting, the cells can be centrifuged at 50–100 $\times g$ for 5 min in serum containing medium. The supernatant is aspirated to remove the DMSO, the cell pellet resuspended well in prewarmed serum containing medium and then transferred into a 25 cm² flask. Cells should be passaged once prior to use in experiments.

3.9 Summary

The three techniques for SMC isolation described yield cells with very different proliferative properties in culture. If larger quantities of SMC in culture are required, the explant isolation technique is recommended, although initially slower to yield cells. Enzymatic dispersion may provide more cells in the contractile phenotype, but the initial yield may be low and subcultures less readily proliferate through as many passages. Future studies are likely to address the isolation and characterization of stem cells which can differentiate into smooth muscle progenitor cells. It remains to be elucidated whether these progenitor cells participate in processes leading to angiogenesis and the pathogenesis of vascular diseases. A better understanding of vascular SMC and pericyte biology can be achieved when cells are cultured under physiological conditions that encompass biomechanical forces (e.g. fluid shear stress and stretch) and oxygen levels (e.g. 1–10 kPa) found in vivo of relevance to arterial and venous circulations and the microcirculation in metabolically active tissues and tumours (*see* **Note 8**). Elucidation of the molecular mechanisms underlying their phenotypic modulation is required to identify therapeutic strategies to target angiogenesis and treat cardiovascular diseases.

4 Notes

1. Other "smooth muscle cell optimized" proprietary culture media are commercially available such as Smooth Muscle Basal and Growth Media (Lonza, or PromoCell), which are based on MCDB131 medium. These media may help to promote more rapid outgrowth of cells from tissue explants, but it should be noted that they may contain components which could alter SMC function, such as insulin, basic fibroblast growth factor (bFGF) and hydrocortisone.

2. For maximal SMC yield and viability, cells should be isolated from as soon as possible after harvesting the vessels. This also reduces the risk of infections since tissues are often handled and excised under nonsterile conditions.

3. If the SMC yield and viability is low following enzyme dispersion, soybean trypsin inhibitor, at a final concentration of 0.1 mg/ml^{-1}, can be added to the enzyme solution to inhibit the action of nonspecific proteases which may contaminate commercial elastase.

4. Should infections frequently occur following explantation, additional antibiotics and fungicides can be supplemented to the culture medium for the initial 24 h following isolation and then the medium replaced. Gentamycin (25 μg ml^{-1}) and Amphotericin B (2 μg/ml^{-1}) are commonly used and 5 ml aliquots of these can be stored at −20 °C as 2.5 mg/ml^{-1} and 0.2 mg/ml^{-1} stocks respectively.

5. When the explanted tissue is removed, SMC can also be detached by trypsinization and redistributed evenly in the same flask as described in Subheading 3.4. This is advisable if cells have grown in a very dense pattern around the explants and will facilitate obtaining the confluent monolayer of cells.

6. The trypsin–EDTA solution should be prewarmed to 37 °C only immediately prior to use and not left in a heated water

bath for extended periods to prevent loss of activity. Cells should not need incubation with trypsin–EDTA at 37 °C for longer than 5–7 min to detach from the flask and this may indicate that a fresh solution should be prepared.

7. Coating of culture vessels with monomeric collagen may provide a physiologically relevant matrix for supporting SMC growth, as well as, maintenance of a cell phenotype of relevance to the experimental model.

8. Long term adaptation and culture of vascular smooth muscle cells and pericytes under physiological oxygen tensions of relevance to tissue conditions in vivo can be achieved using a workstation with temperature, humidity, oxygen and carbon dioxide control (e.g. SCI-tive, Baker Ruskinn, Bridgend, Wales, UK).

Acknowledgements

The vascular smooth muscle cell research in this laboratory is supported by the British Heart Foundation, Heart Research UK, Innovate UK and Unilever. We are grateful to all the midwives in the Maternity Unit and Birth Centre at St. Thomas' Hospital, Guy's and St Thomas' NHS Foundation Trust for supply of human umbilical cords obtained with prior ethical approval and consent of mothers.

References

1. Lusis AJ (2000) Atherosclerosis. Nature 407(6801):233–241

2. Carmeliet P (2003) Angiogenesis in health and disease. Nat Med 9(6):653–660

3. Bergers G, Song S (2005) The role of pericytes in blood-vessel formation and maintenance. Neuro-Oncology 7(4):452–464

4. Liu C, Nath KA, Katusic ZS et al (2004) Smooth muscle progenitor cells in vascular disease. Trends Cardiovasc Med 14(7):288–293

5. Campbell JH, Campbell GR (1993) Culture techniques and their applications to studies of vascular smooth muscle. Clin Sci 85(5):501–513

6. Thyberg J, Hedin U, Sjolund M et al (1990) Regulation of differentiated properties and proliferation of arterial smooth muscle cells. Arteriosclerosis 10(6):966–990

7. Owens GK, Kumar MS, Wamhoff BR (2004) Molecular regulation of vascular smooth muscle cell differentiation in development and disease. Physiol Rev 84(3):767–801

8. Campbell JH, Campbell GR (1987) Phenotypic modulation of smooth muscle cells in culture. In: Campbell JH, Campbell GR (eds) Vascular smooth muscle cells in culture. CRC Press, Boca Raton, Florida, USA, pp 39–55

9. Madi HA, Riches K, Warburton P et al (2009) Inherent differences in morphology, proliferation, and migration in saphenous vein smooth muscle cells cultured from nondiabetic and Type 2 diabetic patients. American journal of physiology. Cell Physiol 297(5):C1307–C1317

10. Shen H, Eguchi K, Kono N et al (2013) Saturated fatty acid palmitate aggravates neointima formation by promoting smooth muscle phenotypic modulation. Arterioscler Thromb Vasc Biol 33(11):2596–2607

11. Han M, Wen JK, Zheng B et al (2006) Serum deprivation results in redifferentiation of human umbilical vascular smooth muscle cells. American journal of physiology. Cell Physiol 291(1):C50–C58

12. Orlandi A, Ropraz P, Gabbiani G (1994) Proliferative activity and alpha-smooth muscle actin expression in cultured rat aortic smooth muscle cells are differently modulated by transforming growth factor-beta 1 and heparin. Exp Cell Res 214(2):528–536

13. Orr AW, Hastings NE, Blackman BR et al (2010) Complex Regulation and Function of the Inflammatory Smooth Muscle Cell Phenotype in Atherosclerosis. J Vasc Res 47(2):168–180

14. Morgan DM (1996) Isolation and culture of human umbilical vein endothelial cells. Methods Mol Med 2:101–109

15. Schlingemann RO, Rietveld FJ, de Waal RM et al (1990) Expression of the high molecular weight melanoma-associated antigen by pericytes during angiogenesis in tumors and in healing wounds. Am J Pathol 136(6):1393–1405

16. Thyberg J, Nilsson J, Palmberg L et al (1985) Adult human arterial smooth muscle cells in primary culture. Modulation from contractile to synthetic phenotype. Cell Tissue Res 239(1):69–74

17. Rensen SS, Doevendans PA, van Eys GJ (2007) Regulation and characteristics of vascular smooth muscle cell phenotypic diversity. Neth Hear J 15(3):100–108

18. Voyta JC, Via DP, Butterfield CE et al (1984) Identification and isolation of endothelial cells based on their increased uptake of acetylated-low density lipoprotein. J Cell Biol 99(6):2034–2040

19. Skalli O, Ropraz P, Trzeciak A et al (1986) A monoclonal antibody against alpha-smooth muscle actin: a new probe for smooth muscle differentiation. J Cell Biol 103(6 Pt 2):2787–2796

20. Proudfoot D, Shanahan CM (2001) Vascular Smooth Muscle. In: Koller MR, Palsson BO, Masters JRW (eds) Human Cell Culture. Kulwer academic press, Great Britain

21. Hao H, Gabbiani G, Bochaton-Piallat ML (2003) Arterial smooth muscle cell heterogeneity: implications for atherosclerosis and restenosis development. Arterioscler Thromb Vasc Biol 23(9):1510–1520

22. Campbell JH, Campbell GR (2012) Smooth muscle phenotypic modulation--a personal experience. Arterioscler Thromb Vasc Biol 32(8):1784–1789

23. Chang S, Song S, Lee J et al (2014) Phenotypic modulation of primary vascular smooth muscle cells by short-term culture on micropatterned substrate. PLoS One 9(2), e88089

24. Lobov GI, Pan'kova MN (2010) Heparin inhibits contraction of smooth muscle cells in lymphatic vessels. Bull Exp Biol Med 149(1):4–6

25. Armulik A, Abramsson A, Betsholtz C (2005) Endothelial/pericyte interactions. Circ Res 97(6):512–523

26. Paquet-Fifield S, Schluter H, Li A et al (2009) A role for pericytes as microenvironmental regulators of human skin tissue regeneration. J Clin Invest 119(9):2795–2806

27. Armulik A, Genove G, Betsholtz C (2011) Pericytes: developmental, physiological, and pathological perspectives, problems, and promises. Dev Cell 21(2):193–215

Chapter 7

Isolation and Transfection of Primary Culture Bovine Retinal Pericytes

Vincent A. Primo and Joseph F. Arboleda-Velasquez

Abstract

This protocol describes an enzymatic approach for isolating homogeneous cultures of pericytes from retinas of bovine source. In summary, retinas are dissected, washed, digested, filtered, cultured in specific media to select for pericytes, and finally expanded for a low passage culture of about 14 million bovine retinal pericytes (BRP) within 4–6 weeks. This protocol also describes a liposomal-based technique for transfection of BRPs.

Key words Pericyte, Isolation, Bovine, Retina, Primary culture

1 Introduction

Pericytes, the cells that constitute the outer layer of the vasculature, play a crucial role in vascular development and stability [1, 2]. Retina capillaries have the highest ratio of pericyte to endothelial cells (EC), approximately 1:1 in humans [3]. Loss of pericytes, especially in capillaries, has been linked to vascular instability and leakage, a phenotype common in many small vessel diseases such as diabetic retinopathy and CADASIL (Cerebral Autosomal Dominant Arteriopathy with Subcortical Infarcts and Leukoencephalopathy) [4–6]. Furthermore, pericyte loss or dysfunction is thought to disrupt normal cell communication with neighboring endothelial cells, furthering vascular complications [4]. Pericyte cultures are thus a valuable model for understanding the mechanism and possible therapeutic targets for small vessel-degenerative diseases [7].

Because of its high ratio of pericytes, ease of dissection, and large size, the bovine retina is a preferred tissue source for pericyte isolation. In the literature, there are a multitude of techniques for isolating pericytes from a wide range of tissues such as skeletal muscle, brain, ear, and retina from species such as mouse, rat, bovine, hamster, and human [7–15]. Some methods involve fluorescence

Stewart G. Martin and Peter W. Hewett (eds.), *Angiogenesis Protocols*, Methods in Molecular Biology, vol. 1430,
DOI 10.1007/978-1-4939-3628-1_7, © Springer Science+Business Media New York 2016

activated cell sorting (FACS), magnetic beads, or antibody-based approaches to obtain pure cultures. While these approaches are effective in increasing the purity of pericyte cultures, they tend to reduce cell yield or involve high passaging. The protocol presented here allows for isolation of high quantity, homogeneous pericyte cultures as early as the first passage (P1). Pericytes are known to change their phenotype and gene expression pattern upon culture; thus, it is important to perform experiments with early passage pericytes (P1–P3) [8, 15–20]. Other protocols have been published claiming to generate pericytes with stable marker expression and phenotype over time; however, we have not evaluated these protocols [15, 19].

In this protocol, bovine retinas are carefully dissected and washed to reduce residual retinal pigmented epithelium (RPE) cells. Next, the retina is digested with collagenase, resulting in single cell suspension. The cell suspension is pelleted, washed multiple times, and then plated into large T175 flasks. This protocol requires the testing several batches of bovine calf serum (BCS) to identify specific lots which best support pericyte selection and proliferation. Most companies will provide small aliquots of serum for this purpose. An appropriate BCS for selection will allow for growth of numerous pericyte colonies with only few colonies of ECs present. When trypsinized, EC will lift off plastic before the pericyte suggesting differential strength of attachment [3]. We take advantage of this phenomenon by using a diluted Dulbecco's phosphate-buffered saline solution (PBS)–ethylenediaminetetraacetic acid (EDTA) solution to remove EC contaminants and leave behind only the pericytes, resulting in a pure, or near pure, P0 culture. However, P0 pericytes stay in discrete colonies and will not spread out to cover the entire surface area of the flask (Fig. 1a). To overcome this, we then split the BRPs, generally 1:1.5, which allows for equal distribution of pericytes in the flask at P1 (Fig. 1b). A more potent BCS is then used to expand the pericytes resulting

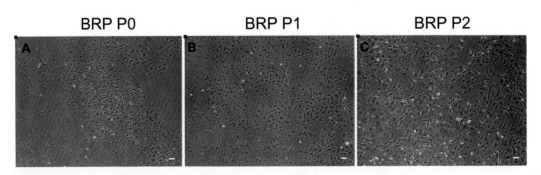

Fig. 1 (**a**) Bovine retinal pericytes (BRP) will mostly stay in segregated colonies before the first passage. (**b**) Passaging BRP will allow a more even distribution. (**c**) Growing them in growth culture medium containing a more potent bovine calf serum (10 %–20 % BCS), will result in rapid expansion. Scale bar: 100 μm

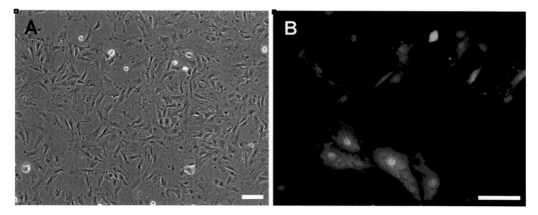

Fig. 2 (**a**) Bovine retinal pericytes (BRP) plated at an optimal transfection density of 44,000 cells/cm². (**b**) Expression of GFP following liposomal transfection of a GFP expression plasmid into BRP. Scale bar: 100 μm

in higher confluence throughout the entire surface area of the flask, although it is well established that 100 % confluence is never reached (Fig. 1c) [7].

Plating BRPs at a density of 44,000–50,000 cells/cm² (Fig. 2a) will result in a fairly confluent culture of BRPs suitable for studying signaling mechanisms that require direct cell-to-cell contact. Furthermore, it is an optimal density for transfection of BRPs. In this protocol, we also describe a successful strategy for liposome-based transfection of BRP monocultures. The DNA plasmid is incorporated into liposomes and delivered into the cytoplasm of the cell by fusion with the plasma membrane. Figure 2b shows BRPs following liposomal transfection of a GFP plasmid.

2   Materials

2.1   Equipment

1. No. 15 surgical blades (two per calf eye).

2. Surgical/utility/sterile drape: One required during removal of the orbital fat and then one per eye dissected.

3. Opthalmic PVA spears: One-to-two per eye.

4. Scissors and forceps: Two sets, autoclave prior to use.

5. Culture dishes (150 mm × 25 mm): One for dissection to prevent rolling, then one per eye for collagenase digestion.

6. 50 mL centrifuge tubes (two per eye).

7. Beakers: Two 500 mL or greater, cover with foil and autoclave.

8. 100 μm cell strainer (one per eye).

9. 70 μm cell strainer (one per eye).

2.2 Reagents   Unless otherwise stated, all user prepared reagents listed below should be sterilized by filtration (0.22 μm pore size).

1. ~4 L tissue culture grade Dulbecco's phosphate-buffered saline (PBS) without $CaCl_2$ or $MgCl_2$ (autoclaved or sterile-filtered). This is sufficient for 16 eyes and is required at 4 °C for Subheading 3.1 and at 37 °C for Subheading 3.2.

2. Enzyme solution (15 mL per eye): PBS (without $CaCl_2$ or $MgCl_2$) containing 1021.02 U/mL collagenase type-2 and 0.25 % fetal bovine serum (FBS). Prepare fresh every time (*see* **Note 1**).

3. Wash medium (50 mL per eye): Dulbecco's modified Eagle's medium (DMEM) 1 g/L glucose, 100 U/mL penicillin/100 mg/mL of streptomycin, 0.025 mg/mL nystatin, and 5 % FBS. Store at 4 °C.

4. Betadine 10× solution: Add 40 g of poly(vinylpyrrolidone)–iodine complex to 1 L of PBS (without $CaCl_2$ or $MgCl_2$) store at room temperature. Prepare 1 L of 1× Betadine solution fresh, before each isolation.

5. Selection culture medium: DMEM with 1 g/L glucose containing 100 U/mL penicillin/100 mg/mL streptomycin, 2 mM L-glutamine, 10 % bovine calf serum (BCS) and store at 4 °C (*see* **Note 2**).

6. Growth Culture Medium: DMEM 1 g/L glucose, 100 U/mL penicillin/100 mg/mL of streptomycin, 2 mM L-glutamine containing either 10 %, 20 % or 1 % bovine BCS as indicated in the text. Store at 4 °C.

7. PBS–EDTA wash: 500 mL PBS + 2.5 mL of 0.5 M EDTA pH 8.0.

8. 0.2 % gelatin: 0.2 % (w/v) tissue culture grade gelatin added to PBS and autoclaved to dissolve and sterilize. Store at RT.

9. Gelatinized T175 flasks: Add 15 mL of 0.2 % gelatin to each T175 flask and incubate at 37 °C in a 5 % CO_2 incubator for at least 1 h. Aspirate the gelatin solution prior to use.

10. 70 % ethanol.

2.3 Transfection Reagents

1. Lipofectamine 2000 kit (Life Technologies).

2. Opti-MEM I, reduced serum medium.

3. DNA Plasmids (prepared endotoxin-free).

4. Growth culture medium containing 20 % and 1 % BCS (*See* Subheading 2.2, **item 6, Note 3**).

3 Methods

3.1 Pericyte Isolation

1. Obtain the desired number of calf eyes from donor cows (*see* **Note 4**).

2. Set up a sterile work area in a laminar flow hood. Manipulate the eyes on the lid of a 150 mm × 25 mm culture dish to prevent them from rolling and contaminating other areas of the work area. Use the bottom half of the 150 mm × 25 mm culture dish to collect tissue waste. Place two sets of scissors and forceps, two large beakers, PVA spears, surgical blades, PBS (cold), enzyme solution, wash medium, and selection culture medium in the tissue culture hood.

3. Sterilize the eyes by immersing them in a tub of 1× Betadine solution for 15 min—longer incubations may be toxic to the retina.

4. Carefully wipe the outside of the tub with 70 % ethanol and place in the tissue culture hood. For all further steps work inside the tissue culture hood.

5. Half fill the two large beakers with cold PBS and wash the eyes twice, carefully transferring them one by one, from the betadine solution into the first beaker and then into the second beaker of PBS.

6. Working on a sterile drape, remove the surrounding orbital fat and muscle from all the eyes. Keep the eyes in cold PBS when not being dissected (*see* **Note 5**). Change the sterile drape.

7. Hold the eye and make an incision with a #15 surgical blade approximately 5 mm posterior of the iris.

8. Using a clean pair scissors, cut straight around the whole eyeball, maintaining that same 5 mm distance from the iris (*see* Fig. 3a). Remove the anterior segment and discard the vitreous (*see* **Note 6**).

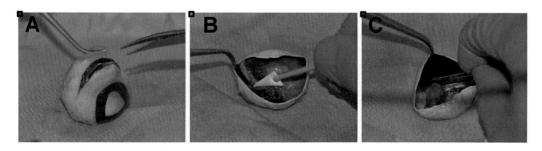

Fig. 3 (**a**) An incision is made approximately 5 mm posterior of the corneal limbus and cut around. (**b**) The retina is gently scraped away with a PVA spear, taking care not to touch the underlying retinal pigmented epithelium (RPE). (**c**) The retina is fully detached from the eye upon cutting the retina from the optic nerve

9. Using a surgical PVA spear, gently peel the retina away from the black retinal pigment epithelium (RPE), taking care not to touch the RPE with the spear (Fig. 3b) (*see* **Note 7**). The retina will remain attached to the eyecup via the optic nerve. Use clean forceps and a sterile #15 surgical blade to cut the retina away from the optic nerve (Fig. 3c). Remove the retina ensuring that the RPE is left intact. Discard the remaining tissue and change the sterile drape ready for the next eye (*see* **Note 8**).

10. Using clean forceps transfer the retina to a 50 mL tube containing ~25 mL of cold PBS.

11. Gently invert the tube five to ten times to wash the retina. Let the retina settle at the bottom of tube and aspirate as much of the PBS as possible (*see* **Note 9**).

12. Add 25 mL of fresh PBS to the tube and repeat the wash step above (*see* **step 11**).

13. Dump the retina and any residual PBS into the cap of the 50 mL tube. Using clean forceps, place the retina into a new 150 mm × 25 mm culture dish.

14. Add 1 mL of enzyme solution using a p1000 pipette and pipette up and down forcing the tissue through the 1 mL pipette tip ~15 times until it flows easily (*see* **Note 10**).

15. Add a further 13 mL of enzyme solution with a 10 mL serological pipette and pipette up and down three to four times.

16. Place the culture dish in a 37 °C (5 % CO_2) incubator for 1 h. During this incubation step, return to **step 7** and proceed with dissection of the next eye (*see* **Note 11**).

17. After 1 h incubation, pipette up and down using a 10 mL pipette 10–14 times. Observe under the phase contrast microscope to ensure that the retina is now in single cell suspension (*see* **Note 12**).

18. Pass the retinal digest through a 100 μm cell strainer into a fresh 50 mL centrifuge tube.

19. Add 25 mL of wash medium and gently invert to mix.

20. Centrifuge at 2000 RPM for 4 min.

21. Aspirate the supernatant and wash twice more with 25 mL of wash medium.

22. Resuspend the cells in 25 mL of selection culture medium and pass through a 70 μm cell strainer into a T175 flask. Incubate overnight at 37 °C (5 % CO_2).

23. The following day, aspirate medium, wash twice with 20 mL of pre-warmed PBS and add 25 mL of selection culture medium (*see* **Note 13**).

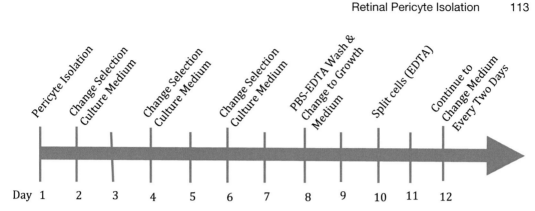

Fig. 4 General timeline for isolation of bovine retinal pericytes

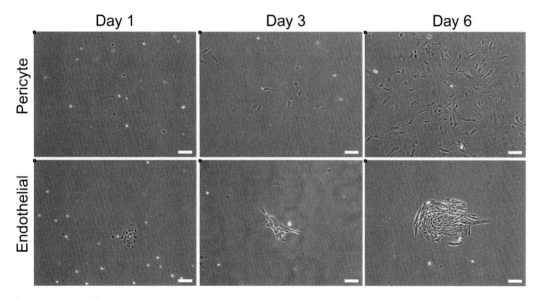

Fig. 5 Within the first week of isolation, bovine pericytes will appear as flat, dispersed, and irregular-shaped cells with many processes. Bovine endothelial cells will appear more elongated and congregate together. Scale bar: 100 μm

3.2 Pericyte Purification and Maintenance

For an overview of the process of pericyte purification and maintenance *see* Fig. 4.

1. Continue to wash and change selection culture medium every 2 days. After 8 days in culture identify and mark the contaminating cells (*see* **Note 14**). Images showing the typical morphology of endothelial versus pericyte colonies in the early stages of culture are shown in Fig. 5. On *day 8*, aspirate the selection culture medium and wash twice with 10 mL of PBS.

2. Add 15 mL of PBS–EDTA wash and immediately monitor cells under a microscope. After about 2.5 min, vigorously slap the side of the flask three times; you should notice some cells will

detach from the flask. Observe and wait another minute and slap the flask three more times. At this point you may not be able to tell the difference between the pericytes and contaminants as they will be rounding up, but use your markings to confirm the contaminants have detached. If many cells are floating, proceed to the next step immediately to prevent loss of pericytes. This entire step should not exceed 5 min (*see* **Note 15**).

3. Immediately wash the flask two to three times with PBS and add 25 mL of growth culture medium with 20 % BCS.

4. Split cells ~1:1.5 on *day 10*. Add 5 mL of trypsin–EDTA solution and incubate at 37 °C for 3–5 min maximum. Tap the side of the flask to help lift the cells. Once cells have lifted off the flask, immediately add 5 mL of growth culture medium with 20 % BCS.

5. Centrifuge cells at 1.3 RPM for 3.5 min, aspirate the supernatant, and resuspend cells in growth culture medium with 20 % BCS.

6. The cells should be passaged at a ratio 1:1.5, or less (e.g., split 8 flasks to 12 flasks or 9 flasks into 13). For optimal pericyte adhesion, use gelatinized T175 flasks washed once with PBS before adding the cells. Ensure a final volume of 25 mL of medium in T175 flasks.

7. On day 12 change to growth culture medium with 10 % BCS, or continue to expand the cells more rapidly in growth culture medium with 20 % BCS. In the latter medium, BRPs will require passaging again within a week.

3.3 Pericyte Transfection

1. Remove the BRPs from a T175 flask using 5 mL of trypsin–EDTA solution this should only take 3–5 min maximum (*see* Subheading 3.2, **step 4**). Once cells have detached add growth culture medium with 20 % BCS.

2. Spin cells at 1.3 RPM for 3.5 min, aspirate supernatant, and resuspend cells in growth culture medium with 20 % BCS. Plate the cells at a density of 44,000–50,000 cells/cm^2 (Fig. 2a) on gelatinized flasks (*see* Subheading 2.2, **item 9** and Subheading 3.2, **step 6**).

3. Twenty minutes after plating the cells, gently rock the flask to ensure even cell distribution (*see* **Note 16**) and incubate overnight in a 37 °C, 5 % CO_2 incubator.

4. The next day wash the cells twice with PBS and transfect using Lipofectamine 2000 according to the manufacturer's protocol. Maintain the cells in growth culture medium with 20 % BCS overnight (*see* **Note 17**).

5. Approximately 16 h later, reduce the growth culture medium BCS content to 1 % (*see* **Note 18**).

6. The cells should be transfected with the DNA plasmid and at an optimal density for cell-to-cell communication (*see* **Note 19**).

4   Notes

1. Each batch of collagenase will have a different level of activity expressed in U/mg. To determine how much collagenase to use, calculate the weight (mg) of collagenase required to contain 1021.02 U and dissolve in 1 mL of PBS. Although one eye requires 15 mL of enzyme solution, we recommend that you do not make less than 50 mL. The collagenase powder is very light and weighing small amounts will increase the chances of introducing a weighing error.

2. BCS of this quality was found to reduce EC growth and promote pericyte growth.

3. This is the same as the growth culture medium, but with 1 % BCS as opposed to 10 % or 20 % BCS.

4. In our experience, eyes taken from male bovine calves at 2–14 days old yield a much higher quantity of pericytes than adult bovine eyes. Also, for first timers, we suggest starting with 4–5 eyes, while 8 or 9 eyes are more feasible with practice. In our experience, using 8 eyes provide approximately 14 million BRPs within 4–6 weeks.

5. This trimming step will save time, which becomes more important the more eyes you are preparing. In addition, leaving a piece of optic nerve or fat near the posterior of the eye is useful for holding and manipulating the eye for the next two steps.

6. The vitreous can usually be tipped or pulled out, but sometimes this is not the case. If it does no come out easily, use forceps and scissors to pull and cut fragments of the vitreous out. Also, use the PVA spears to try to scrape/pull out vitreous. Sometimes the retina will detach during this step. This is fine, just try to keep the retina within the eyecup and proceed to next step.

7. If you touch the RPE black will appear on the PVA spear. Use a new PVA spear if this happens to reduce RPE contamination.

8. To avoid cross contamination, ideally, you should replace the drape, but if you are careful it is possible to use two or three drapes for all dissections.

9. Be extremely careful to avoid aspirating the retina.

10. Depending on the tip used, you may need to widen the bore by cutting 1–2 mm off the tip. Perform subsequent steps with an uncut pipette tip.

11. You now have 1 h to dissect as many eyes as you can before you need to proceed to the next step. With practice, we are able to dissect a total of 8–9 eyes in this time. We suggest that in the first instance you start with 4–5 eyes.

12. If not, incubate for an extra 10–15 min, no longer. We have noticed longer incubations (>90 min) result in greater cell contamination.

13. In our experience incubations of more than 16 h before changing the medium can result in more cell contaminants

14. Use a fine tip marker to mark the bottom of the flask where contaminant colonies reside. This will make it easier to confirm they have lifted off the tissue culture plastic (*see* Subheading 3.2, **step 2**).

15. The longer you wash, the more likely you will remove EC; however, very long incubations may also remove pericytes. On *day 8* you should notice your flask is predominantly covered with pericytes and most if not all ECs will have been removed.

16. When plating, we have noticed uneven distribution of BRPs throughout the flask. This step has helped us to achieve a more uniform distribution.

17. The lipofectamine transfection procedure is somewhat toxic and a few BRPs will die. We use 20 % growth culture medium to reduce this effect.

18. The reduction in serum concentration is necessary to reduce cell proliferation and maintain cells at their present density.

19. We have not studied the activity of transfected BRPs beyond 3 days post-transfection.

Acknowledgements

We thank Dr. Patricia D'Amore for her valuable discussion, insights, and guidance with the development of this method. This work was supported by NIH grants EY021624 and EY021624-03 (J.A.-V.), and American Heart Association Scientist Development Grant 12SDG8960025 (J.A.-V.).

References

1. Fruttiger M (2007) Development of the retinal vasculature. Angiogenesis 10:77–88

2. Kuwabara T, Cogan DG (1963) Retinal vascular patterns. VI. Mural cells of the retinal capillaries. Arch Ophthalmol 69:492–502

3. Shepro D, Morel NM (1993) Pericyte physiology. FASEB J 7:1031–1038

4. Goldberg RB (2009) Cytokine and cytokine-like inflammation markers, endothelial dysfunction, and imbalanced coagulation in development of diabetes and its complications. J Clin Endocrinol Metab 94:3171–3182

5. Arboleda-Velasquez JF, Manent J et al (2011) Hypomorphic Notch 3 alleles link Notch signaling to ischemic cerebral small-vessel disease. Proc Natl Acad Sci U S A 108:E128–E135

6. Speiser P, Gittelsohn AM, Patz A et al (1968) Studies on diabetic retinopathy. 3. Influence of diabetes on intramural pericytes. Arch Ophthalmol 80:332–337

7. Bryan BA, D'Amore PA (2008) Pericyte isolation and use in endothelial/pericyte coculture models. Methods Enzymol 443:315–331

8. Mogensen C, Bergner B, Wallner S et al (2011) Isolation and functional characterization of pericytes derived from hamster skeletal muscle. Acta Physiol (Oxf) 201:413–426

9. Antonetti DA, Wolpert EB (2003) Isolation and characterization of retinal endothelial cells. Methods Mol Med 89:365–374

10. Boroujerdi A, Tigges U, Welser-Alves JV et al (2014) Milner R. Isolation and culture of primary pericytes from mouse brain. Methods Mol Biol 1135:383–392

11. Dore-Duffy P (2003) Isolation and characterization of cerebral microvascular pericytes. Methods Mol Med 89:375–382

12. Dorovini-Zis K, Prameya R, Huynh H et al (2003) Isolation and characterization of human brain endothelial cells. Methods Mol Med 89:325–336

13. Maier CL, Shepherd BR, Yi T et al (2010) Pober JS. Explant outgrowth, propagation and characterization of human pericytes. Microcirculation 17:367–380

14. Neng L, Zhang W, Hassan A et al (2013) Isolation and culture of endothelial cells, pericytes and perivascular resident macrophage-like melanocytes from the young mouse ear. Nat Protoc 8:709–720

15. Tigges U, Welser-Alves JV, Boroujerdi A et al (2012) A novel and simple method for culturing pericytes from mouse brain. Microvasc Res 84:74–80

16. Newcomb PM, Herman IM (1993) Pericyte growth and contractile phenotype: modulation by endothelial-synthesized matrix and comparison with aortic smooth muscle. J Cell Physiol 155:385–393

17. Armulik A, Abramsson A, Betsholtz C (2005) Endothelial/pericyte interactions. Circ Res 97:512–523

18. D'Amore PA (1992) Capillary growth: a two-cell system. Semin Cancer Biol 3:49–56

19. Sundberg C, Kowanetz M, Brown LF et al (2002) Stable expression of angiopoietin-1 and other markers by cultured pericytes: phenotypic similarities to a subpopulation of cells in maturing vessels during later stages of angiogenesis in vivo. Lab Invest 82:387–401

20. Hughes S, Chan-Ling T (2004) Characterization of smooth muscle cell and pericyte differentiation in the rat retina in vivo. Invest Ophthalmol Vis Sci 45:2795–2806

Part III

In Vitro Techniques

Chapter 8

In Vitro Assays for Endothelial Cell Functions Required for Angiogenesis: Proliferation, Motility, Tubular Differentiation, and Matrix Proteolysis

Suzanne A. Eccles, William Court, and Lisa Patterson

Abstract

This chapter deconstructs the process of angiogenesis into its component parts in order to provide simple assays to measure discrete endothelial cell functions. The techniques described will be suitable for studying stimulators and/or inhibitors of angiogenesis and determining which aspect of the process is modulated. The assays are designed to be robust and straightforward, using human umbilical vein endothelial cells, but with an option to use other sources such as microvascular endothelial cells from various tissues or lymphatic endothelial cells. It must be appreciated that such reductionist approaches cannot cover the complexity of the angiogenic process as a whole, incorporating as it does a myriad of positive and negative signals, three-dimensional interactions with host tissues and many accessory cells including fibroblasts, macrophages, pericytes and platelets. The extent to which in vitro assays predict physiological or pathological processes in vivo (e.g., wound healing, tumor angiogenesis) or surrogate techniques such as the use of Matrigel™ plugs, sponge implants, corneal assays etc remains to be determined.

Key words Endothelial cells, Tubular differentiation, Migration, Chemotaxis, Haptotaxis, Invasion, Matrix metalloproteinases, Image analysis

1 Introduction

Since Folkman's seminal work in the early 1970s [1] and the later identification of a myriad of molecular mediators, there has been a burgeoning interest in all aspects of neoangiogenesis and lymphangiogenesis [2–15]. In particular, it is recognized that successful inhibition of tumor angiogenesis (on which sustained malignant growth and spread depends) could have significant therapeutic benefits [4, 7, 16–41]. Such strategies were expected to have wide application in many solid tumors, and to be less susceptible to the development of resistance [29]. However, it has become clear that bypass resistance mechanisms to monotherapy, particularly in advanced malignancy, limits their efficacy [42–46] and increasingly antiangiogenic (or antivascular) agents are used in

Stewart G. Martin and Peter W. Hewett (eds.), *Angiogenesis Protocols*, Methods in Molecular Biology, vol. 1430,
DOI 10.1007/978-1-4939-3628-1_8, © Springer Science+Business Media New York 2016

combination with standard cytotoxic or targeted therapies. Other conditions that may also be treatable by antiangiogenic agents include diabetic retinopathy, arthritis, and psoriasis [41, 47–53]. On the other hand, the ability to induce angiogenesis in a controlled manner may assist healing of skin ulcers or wounds due to injury or surgery [54, 55], or ameliorate certain cardiovascular pathologies [56, 57].

In order to develop effective angiogenesis modulators, it is advantageous to gain a deeper understanding of the complex gene expression changes and molecular mechanisms underlying the process. We also need informative in vitro (and in vivo) assays for hypothesis testing and screening of potential therapeutic agents [7, 58–72]. This is particularly important as it is now recognized that, as well as "classical" neoangiogenesis mediated largely by capillary sprouting in response to gradients of angiogenic factors derived from tumor cells or host inflammatory cells, there are other significant contributory mechanisms including vascular co-option, recruitment of endothelial precursors and "vasculogenic mimicry" [73–81].

Endothelial cells (EC) in the adult are normally quiescent unless activated by angiogenic factors such as VEGFs, FGF2, HGF, angiopoietin 2, or cytokines including IL-8, many of which are induced by hypoxia in the solid tumor microenvironment [2, 3, 61, 69, 82–88]. EC activation results in enhanced proliferation, migration, and "invasion" of surrounding tissues. This is mediated by upregulation of specific integrins (such as $\alpha v \beta 3$ and $\alpha v \beta 5$) [18, 89–92] and secretion of matrix metalloproteinases (MMPs) [93–98] and urokinase plasminogen activator (uPA) [99–101]. Many of these molecular mediators have been used as targets for therapy and also for imaging the angiogenic process and its response to therapeutic intervention [102, 103].

Angiogenesis can potentially be controlled at many levels: inhibition of the production [104] or sequestration of circulating angiogenic factors (e.g., Avastin and Aflibercept [105, 106]; antibody-mediated blocking of binding to EC receptors such as VEGFR-2 and -3 or Tie-2 [21, 107–109]; inhibition of integrin binding [89, 110] and targeting the major angiogenic factor receptors or their downstream signaling pathways with small molecule tyrosine kinase inhibitors [111–119]. Recently new possibilities have emerged from nonclassical signaling pathways including Delta-notch [120–124], voltage gated sodium channels and Na/Ca++ exchange via NXC1 [125, 126].

Depending on the point of intervention in the angiogenic process, different assays may be required [7, 59, 62, 63, 70, 127–133], but (following biochemical assays to establish target inhibition) the simplest and most informative of them aim to assess the ability of EC to perform one or more of the following key functions:

- Proliferation in response to an angiogenic stimulus (e.g., VEGF, bFGF, Ang-2)

- Chemotaxis (directional motility) in response to an attractant gradient
- Haptotaxis (integrin-mediated non-directional motility) on matrix proteins
- Tubular differentiation on matrix proteins
- Proteolytic enzyme production (e.g., gelatinase MMPs by zymography or ELISA)

Assays to measure each of these functions will now be described in turn.

2   Materials

2.1   Cell Culture and Maintenance

1. Normal Human Umbilical Vein Endothelial Cells (pooled) (ZHC-2101) (Cellworks, Buckingham (http://www.cellworks.co.uk/) (*see* **Notes 1** and **2**).
2. Large Vessel Endothelial Cell Growth Medium Package which contains: Large Vessel Endothelial Cell Basal Medium and Large Vessel Endothelial Cell Growth Supplement (ZHM-2953) CellWorks CalTag MedSystems.
3. Phosphate buffered saline (PBS) containing 1 mm ethylenediamine tetra acetic acid (EDTA).
4. Trypsin–EDTA solution—0.025 %/1 mM in PBS. Alternatively TrypLE™ Express trypsin for cell detachment (Invitrogen Paisley UK) (*see* **Note 3**).
5. Trypan blue 0.4 % (w/v) in PBS.
6. Tissue culture flasks.
7. Disposable pipettes.
8. Hemocytometer.
9. Humidified incubator set at 37 °C with an atmosphere of 5 % CO_2 in air.

2.2   Cell Proliferation Assay

1. 96-well tissue culture plates.
2. Multichannel pipette.
3. Alkaline phosphatase substrate containing buffered p-nitrophenylphosphate (pNPP) (Sigma-Aldrich UK). Protect from light and keep at 4 °C.
4. Solution of 1 M sodium hydroxide (NaOH).
5. Solution of trypsin–EDTA or TrypLE™ Express trypsin.
6. Plate reader or spectrophotometer capable of reading absorption at 405 nm or luminescence.
7. CellTiter-Glo® G7571 Promega.

2.3 Chemotaxis Assays (HUVEC)

2.3.1 Human Umbilical Cells (HUVEC) Pooled Donors (CellWorks Buckingham UK)

1. Large Vessel Endothelial Cell Growth Medium.

2. Solution of trypsin–EDTA or TrypLE Express (Invitrogen).

3. Olympus Ix70 inverted fluorescent microscope fitted with both a 20× and 40× objective and cooled CCD digital camera.

4. U-MWB filter set cube to detect the fluorophore (Olympus UK Ltd).

5. Image acquisition software—Image Pro Plus (Media Cybernetics UK).

6. BD FluoroBlok™ 24-well plates (BD Biosciences) 3 μM pore (cat no 351156). *See* Fig. 3 or Falcon plates and separate inserts.

7. Fibronectin matrix protein coated 24-well 3 μM FluoroBlok™ plates (BD Biosciences. Store at 4 °C.

8. Recombinant human epidermal growth factor for use as chemoattractant (EGF). Make a stock of 1 mg/ml EGF by dissolving lyophilized EGF in 0.2 μm-filtered 10 mM acetic acid containing 0.1 % bovine serum albumin (BSA). Store small aliquots at –20 °C.

9. CellTracker™ Green dye CMFDA (Molecular Probes Invitrogen Paisley UK).

2.4 Invasion Assay

1. BDBiocoat™ FluoroBlok™ Matrigel coated 24-well plates (BD Biosciences).

2.5 Haptotaxis Assay

1. Endothelial cells (e.g., human umbilical vein endothelial cells (HUVEC) pooled donors (CellWorks Buckingham UK).

2. Large Vessel Endothelial Cell Growth Medium.

3. Solution of trypsin–EDTA or TrypLE Express (Invitrogen).

4. Sterile 0.1–10 μl pipette tips for Gilson P20.

5. Inverted microscope fitted with 20× objective and digital camera.

6. Mitomycin C Sigma Aldrich prepared immediately before use.

7. 24-Well tissue culture plates.

8. Sterile Gilson pipette tip for scraping a wound in the cell monolayer.

2.6 Tubular Differentiation

1. Matrigel™ (BD Biosciences) stored at –20 °C. Place the gel on ice in a 4 °C refrigerator to thaw overnight before use.

2. 24-Well tissue culture plates.

3. Gilson pipette tips for P1000.

4. CellTracker™ Green dye CMFDA (Molecular Probes, Invitrogen).

2.7 Quantitation of Migration and Invasion	1. Olympus IX70 inverted fluorescence microscope fitted with an environmental chamber (Solent Scientific Limited UK). 2. Retiga EXi cooled CCD camera (Q Imaging). 20× and 40× objectives. 3. Image Pro Plus 7.0 (Media Cybernetics Europe).
2.8 Quantitation of Tubularization	1. Olympus IX70 inverted fluorescence microscope fitted with an environmental chamber (Solent Scientific Limited UK). 2. Retiga EXi cooled CCD camera (Q Imaging). 20× and 40× objectives. 3. Image Pro Plus 7.0 (Media Cybernetics Europe). 4. Celigo® cytometer (Nexcelom Bioscience).
2.9 Gelatin Zymography	1. 10 % Novex® Tris-Glycine SDS Gels containing either casein or collagen (Invitrogen). 2. Tris-Glycine SDS running buffer (10×). Prepare 1× Tris-Glycine SDS running buffer by adding 100 ml of 10× Novex® Tris-Glycine SDS running buffer to 900 ml of deionized water. 3. Novex® Tris-Glycine SDS sample buffer (2×) stored at −20 °C. 4. Protein molecular weight markers stored at −20 °C. 5. MMP-2/MMP-9 standards (CHEMICON Europe, Ltd). 6. Novex® Zymogram renaturing buffer (10×). Prepare 1× renaturing buffer by adding 100 ml of 10× Novex® 10× renaturing buffer to 900 ml of deionized water. 7. Novex® Zymogram developing buffer (10×). Prepare 1× developing buffer by adding 100 ml of 10×Novex® 10× Developing Buffer to 900 ml of deionized water. 8. Colloidal Blue Staining Kit (Invitrogen). 9. Methanol. 10. Distilled water. 11. XCell II Mini cell electrophoresis tank (Novex® Invitrogen). 12. Power pack. 13. Growth factors for stimulation of cells: e.g., vascular endothelial growth factor (VEGF), epidermal growth factor (EGF), basic fibroblast growth factor (bFGF), hepatocyte growth factor (HGF). 14. Phorbol 12-myristate 13-acetate (PMA).
2.10 Relative Quantitation of Matrix Metalloproteinases in Zymogram	1. GelPro analyzer 3.1 (Media Cybernetics Europe). 2. Digital camera.

3 Methods

3.1 Cell Culture and Maintenance

The culture of normal primary and early passage human endothelial cells requires specialized culture media. Careful culture techniques are necessary to ensure cell survival and maintenance of expression of key proteins. Normal primary endothelial cells have a finite lifespan. In practice, this means that, depending on the cell type, they can be cultured in vitro for a maximum of 10–15 passages (*see* **Note 4**) before becoming senescent at which point the cells become etiolated and stop proliferating.

1. Thaw media supplements rapidly at 37 °C and add to basal media using a sterile Gilson tip. Pre-warm the mixture to 37 °C.

2. Add the requisite amount of medium to the tissue culture flask, e.g., 5 ml per flask for T25, 15 ml per flask for T75.

3. Thaw a vial of cryopreserved endothelial cells rapidly at 37 °C and wipe the vial with ethanol to ensure sterility. Calculate the number of cells required to seed the minimum density of cells in the appropriate tissue culture flask. This will depend upon the cell line. Microvascular endothelial cells require a minimum of 5×10^3 cells/cm^2; however, large vessel endothelial cells can be seeded at half this density (*see* **Note 4**).

4. Add cells to the flask and place in an incubator at 37 °C set at 5 % CO_2 in air. The procedure should take no more than 5–8 min to ensure maximum cell viability.

5. Change medium after 24 h to remove cryoprotectant.

6. Feed the cells with fresh medium every other day.

7. Harvest or subculture cells when they reach 60–80 % confluence.

3.2 Subculture

1. Aspirate medium from the flasks.

2. Rinse the cell layer gently with 5 ml PBS–EDTA.

3. Add 1 ml trypsin–EDTA or TrypLE™ Express to the cell layer ensuring that the entire surface of the cell sheet monolayer is covered. Pour off the excess fluid, leaving a thin layer covering the cells and return the flask to the incubator.

4. After about 2 min, examine the cells microscopically to monitor detachment. If the cells appear to be lifting off the surface tap the flask gently to complete the process. If not, return and monitor at 30–60 s intervals until this occurs (*see* **Note 4**).

5. Add 5 ml of fresh medium and resuspend the cells carefully by aspirating the cells using a disposable pipette and transferring to a suitable sized vial.

6. To determine the number of viable cells per ml, remove 20 μl cell suspension from the vial, dilute with 20 μl trypan blue and count the viable (dye excluding) cells using a hemocytometer.

7. Reseed flasks at the minimum density as above.

3.3 Cell Proliferation Assays

3.3.1 Screening Compounds for Growth Inhibitory Activity

Most current assays for measuring cell proliferation are based upon the reduction of the yellow tetrazolium salt MTT (3-(4, 5-dimethylthiazolyl-2)-2, 5-diphenyltetrazolium bromide) by the action of dehydrogenase enzymes from metabolically active cells. The resulting intracellular purple formazan can be solubilized and quantified by reading the absorbance of the product on a spectrophotometer. An alternative cheaper metabolic assay is based upon a ready-to-use buffered alkaline phosphatase substrate containing p-nitrophenylphosphate (pNPP). Prior to use, the substrate appears as a colorless to pale yellow solution. After incubation with cellular derived alkaline phosphatase the compound turns a bright yellow and this color intensity can be measured on a spectrophotometer. A further simple alternative method to assess cell survival is the Sulforhodamine-B (SRB) assay. SRB is a water-soluble dye that binds to the basic amino acids of the cellular proteins. The colorimetric measurement of the bound dye provides an estimate of the total protein mass that is related to the cell number. Whilst this method is the suggested preferred technique for high through-put screening, it is reportedly less sensitive for the measurement of endothelial cells which, compared with tumor cells, have relatively low protein levels [134]. Rapid, surrogate measurements of proliferation can be performed using estimations of total surface area using contrast algorithms on high content screening machines such as the Celigo® cytometer (Fig. 1). This has the advantage that cell counts can be made in real time at frequent intervals, rather than at a single end-point. Another popular method is CellTiter-Glo® which measures the amount of ATP present in metabolically active cells by the addition of a luminescent substrate. Comparisons of both of these methods show excellent agreement (Fig. 2)

1. Remove growth medium from a T75 flask of endothelial cells.

2. Rinse with 5 ml of PBS–EDTA solution.

3. Add 3 ml TrypLE Express and pour off excess. When cells have detached, add 10 ml endothelial cell basal medium and resuspend cells gently until a single cell suspension is produced.

4. Count viable cells using a hemocytometer and adjust cell concentration to 2×10^4 cells/ml.

5. Using a multichannel pipette, dispense a 100 μl volume of cells into each well of a 96-well plate with the exception of the outside wells. To the outside wells add 100 μl of distilled water.

6. Make doubling dilutions of the agent of interest in a replicate plate before transferring to the cell plate as follows (see **Note 5**).

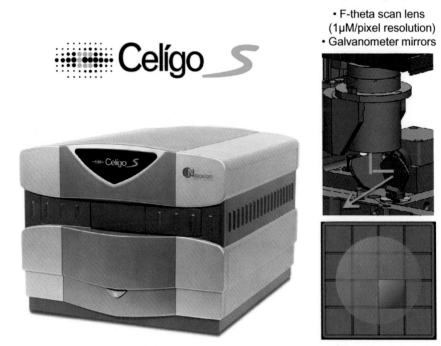

Fig. 1 Illustration of the Celigo multichannel cytometer with LED-based bright-field and fluorescence options showing proprietary galvanometric image acquisition set up

7. Add 100 μl tissue culture medium to the wells of a 96-well plate except the outside wells and the first column.

8. Add 200 μl of the agent at its maximum concentration to each of three wells in the first empty column of wells. Repeat this for any further compounds to be tested by adding the agent to the next group of three wells.

9. Fit a multi-dispensing pipette with tips and set to 100 μl. Collect 100 μl of compound from the first row, add to the second row, mix and take 100 μl on to the next row. Repeat the procedure to the penultimate row. Leave the last row as a negative control.

10. Remove the medium from the plate containing the cells and transfer the drug dilutions across and dispense into the wells using the multipipette.

11. Incubate the plates at 37 °C and 5 % CO_2 for 4 days, or until cells in control wells appear confluent.

12. Remove the medium by gently tapping the inverted plate on an absorbent paper towel.

13. Add 100 μl of paranitrophenylphosphate (3 mg/ml in 0.1 M sodium acetate containing 0.1 % Triton X-100 and incubate for 2 h at 37 °C. Stop the reaction by adding 50 μl 1 M NaOH.

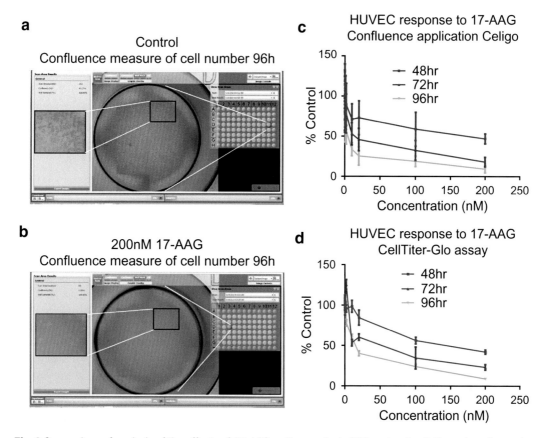

Fig. 2 Comparison of analysis of the effects of 17-AAG on the survival of ECs using the Celigo cytometer analysis (confluence application) software (Control **a** versus treated **b**). Graphical representation of the Celigo data (**c**) compared with a CellTiter-Glo® endpoint assay using luminescence (**d**)

14. Measure absorbance at 405 nm in a spectrophotometer.

15. If using CellTiter-Glo® analysis method, after **step 11** remove 100 μl medium from the wells and then add 100 μl substrate and read the luminescence using a suitable plate reader.

3.4   Chemotaxis

3.4.1   Cell Labeling with Cell Tracker Fluorescent Dye

1. Add CMFDA to cell culture flask medium containing growth medium to give 5 μM final concentration of dye (*see* **Note 6**).

2. Leave cells at 37 °C for a minimum of 45 min. Remove medium and wash once with PBS. Apply trypsin to cells as previously described (*see* Subheading 3.2, **step 1**, **step 6** to remove cells from the flask) and use in chemotaxis assay.

3.5   Screening Compounds for Effects on Migration in a FluoroBlok™ Transwell Assay

1. Remove growth medium from a labeled T75 flask of endothelial cells.

2. Rinse with 5 ml of PBS–EDTA solution.

3. Add 3 ml TrypLE Express and pour off excess.

Part of FluoroBlok™ plate   Single insert

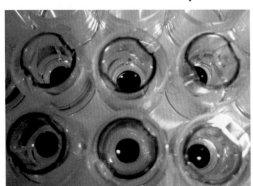 

Fig. 3 Photograph of a FluoroBlok™ plate (BD Biosciences) showing one of the removable filter inserts. The blue (polyethylene terephthalate coated filter) is specifically designed to absorb visible light within the 490–700 nm range which prevents transmission of fluorescence from the cells on the top of the filter

4. When cells have detached add 10 ml endothelial cell basal medium and resuspend gently until a single cell suspension is produced.

5. Count viable cells using a hemocytometer and adjust cell concentration to 4×10^5 cells/ml.

6. Add 150 μl of the diluted cell suspension to the top chamber of the Transwell insert (Fig. 3) placed in a companion plate. If the experiment involves an investigation of single growth factor mediated chemotaxis then the use of a fibronectin-coated transwell is recommended (*see* **Note 7**). Otherwise, if FCS is used as a chemoattractant, use an uncoated Fluoroblok™ insert.

7. Dilute compound to be screened to double the concentration desired. Add 150 μl to the cells in the top chamber of the insert to give a final volume of 300 μl (*see* **Note 8**).

8. Add 150 μl serum-free endothelial cell basal medium (containing drug vehicle at the highest concentration that is on the compound plate) to one or more inserts, which will then serve as untreated controls.

9. To the bottom chamber of each insert add 800 μl of serum-free medium (*see* Subheading 3.5, **step 8**) containing both the chemoattractant and compound at the same concentration as the upper chamber. Include in the experiment at least one insert that contains cells in the top chamber but does not contain any chemoattractant. Unless a specific attractant is required, HUVEC will migrate well to 5 % FCS.

10. Add 150 μl of the suspension of labeled cells to one well of a 24-well plate. Add 150 μl of growth medium. This is a control to confirm that the cells are viable for the duration of the experiment and can be counted separately at the end of the migration period.

11. Incubate the plates at 37 °C in a humidified atmosphere of 5 % CO_2 in air for 6 h.

12. View the cells that migrate in real time and photograph under UV illumination by selecting the U-MWB filter set cube and the 10× objective on the microscope. Monitor the cells migrating onto the lower surface of the membrane every 2 h (*see* **Note 9**).

13. After 6 h, either count the cells immediately or fix by removing the medium in the lower chamber and replacing it with 4 % paraformaldehyde in PBS.

14. Acquire images of the migrated cells using the CCD camera at 10× magnification in at least three random representative areas under each of the membranes.

15. Count migrated cells as described in Subheadings 3.9 or 3.10.

3.6 Invasion Assay

Invasion assays are performed as above, but the observation time needs to be increased as the cells also have to pass through the Matrigel™ barrier above the filter (e.g., allow ~18 h for HUVEC invasion assays).

Coating of the upper surface of the FluoroBlok™ inserts with a matrix barrier may be performed in the laboratory for the sake of economy but can be problematic. Matrigel™ solutions solidify rapidly at room temperature and have a tendency to form a meniscus on the membranes. This can lead to uneven invasion of cells through the gel. To avoid these problems it may be preferable to purchase the pre-coated inserts from Becton Dickinson marketed as the BD BioCoat™ Tumor Invasion System.

Rehydrate the inserts as follows:

1. Remove the package from −20 °C storage and allow it to come to room temperature.

2. Add 0.5 ml warm (37 °C) PBS to the interior of the insert wells.

3. Allow the plate to rehydrate for 2 h at 37 °C in non-CO_2 environment.

4. After rehydration, carefully remove the medium from the insert wells without disturbing the layer of BD Matrigel™ on the membrane.

5. Pre-label the cells with CMFDA cell tracker as described above (Subheading 3.4, **steps 1–2**) and seed them into the upper chamber (*see* **Note 10**).

6. Proceed as described for chemotaxis assays (Subheading 3.3) but with a longer observation period.

7. Observe the underside of the membranes under ultraviolet illumination (*see* **Note 11**).

8. Acquire images of the migrated cells using the CCD camera at 40× magnification in three random representative areas under each of the membranes.

9. Count migrated cells as described in Subheading 3.9.

3.7 Haptotaxis Assays

1. Harvest HUVEC from flasks as described in Subheading 3.1 and plate in 24-well plates at 5×10^4 cells/well. Incubate at 37 °C in a humidified incubator at 5 % CO_2 in air. Allow cells to become 90 % confluent.

2. Prepare cell culture medium with appropriate inhibitors and controls and keep at 37 °C in a heated water bath.

3. Using the sterile tip on a Gilson pipette, carefully score the cell monolayers vertically down the center of each well. Wash the surface once with PBS to remove detached cells.

4. Add fresh growth medium containing inhibitors or an equivalent volume of vehicle as a negative control. To ensure that wound closure is due solely to migration, inhibit cell proliferation by adding mitomycin C at 2.5 µg/ml.

5. Monitor 'wound' closure every hour by acquiring duplicate images of the wound in each well under phase contrast from time 0 (to measure the initial wound width) for up to 24 h. This procedure is simplified by the use of a mechanized stage and image acquisition software (Image Pro Plus Media Cybernetics UK).

6. Express results as percentage change in wound width at each time point. In Image Pro Plus the images acquired can also be compressed and then displayed as a movie file. Running the movie of untreated cultures against the treated wells will provide qualitative illustrations of the effects of the inhibitors (*see* **Note 12**).

3.8 Quantitation of EC Migration and Invasion

1. Open image file using ImagePro software.

2. Select "Measure, Count/Size" from the pull-down menu on the control bar at the top of the screen.

3. In the "Count/Size" window select "Automatic Bright Objects", select the Options button and ensure that the Options selected read Outline style—Outline, Label Style None and Four connect.

4. Perform a count. This initial cell count will include all bright objects in the image including the pores in the membrane. Filter these out as follows:

5. Select "Measurements" in the "Count/Size" window, highlight the Area default setting and then select the Edit range option.

6. Drag the vertical bar slowly to the right until the pores are excluded from the final count. Click the filter objects option.

7. The final cell count, along with any statistics, may be transferred by DDE (Dynamic Data Exchange) to Excel as follows.

8. Click "view" on the top toolbar and select "statistics". Select "file" and then "DDE to Excel".

3.9 Quantitation of EC Migration Using a Celigo Cytometer

1. Use the Direct Cell counting application : settings: 40 % well mask, Algorithm = Brightfield, Intensity threshold = 2, Precision = High, Cell diameter Pixel = 80, Separate touching objects yes, Cell Area 700–10,000, Cell Intensity 0–255, minimum aspect ratio = 0.

3.10 Tubular Differentiation

1. Place Matrigel™ on ice in an insulated container at 4 °C and leave overnight to thaw completely.

2. Place endothelial growth medium at 4 °C.

3. Place any tips or plates that will be coming into contact with Matrigel at 4 °C until cooled.

4. All subsequent procedures should be performed in a tissue culture hood.

5. Mix Matrigel™ to homogeneity by gentle pipetting and add to an equal volume of serum-free medium.

6. Add 200 μl of diluted gel to each well of a 24-well plate.

7. Place plate at 37 °C for at least 30 min to gel.

8. Label endothelial cells with 5 μM CMFDA cell tracker green dye for 1 h.

9. Remove growth medium from labeled T75 flask of HUVEC and rinse cells with 5 ml of PBS–EDTA solution.

10. Add 1 ml TrypLE Express, remove excess and incubate briefly at 37 °C to release cells.

11. When cells have detached, add 10 ml endothelial cell basal medium and resuspend gently until a single cell suspension is produced.

12. Count cells using a hemocytometer and adjust cell concentration with fresh medium to 6×10^4 cells per ml.

13. Add 500 μl of cell suspension to each Matrigel-coated well. Include an uncoated well as a negative control.

14. Incubate plate for ~6 h at 37 °C in a humidified incubator at 5 % CO_2 in air (*see* **Note 13**).

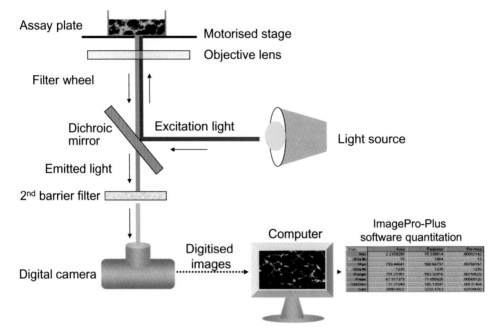

Fig. 4 Diagram of the image analysis set up used to quantify cell migration, invasion, haptotaxis, and tubular differentiation, with an example of tubular differentiation

15. Acquire images of tubule formation using a cooled CCD camera on an inverted fluorescence microscope using a U-MWB filter set cube and 10× objective (Fig. 4).

16. Use Image Pro Plus or a Celigo cytometer to analyze tubule formation as described in Subheadings 3.11 or 3.12.

3.11 Quantitation of Tubularization Using Image Pro Software (See Fig. 5)

1. If the images are acquired in color it will be necessary to select the color channel that contains the highest contrast of image (i.e., red, green, or blue).

2. Select Process, Colour Channel, and Extract from the menu bar. Click the G box in Generate Channel and then OK.

3. Select Measure, Count/size, and Measure (select Measurements).

4. There is an option within this menu to select the parameters that are used to discriminate the images from each other. We find that the best option to use is area (polygon). This choice may be cell type dependent and optimal discrimination should be established by selecting a number of parameters and seeking the greatest ratio between test and control images.

3.12 Quantitation of Tubularization Using a Celigo Cytometer (See Fig. 6)

1. Use the Confluence application settings: 60 % well mask, Algorithm = Brightfield, Intensity threshold = 0, Saturated Intensity = 0, Precision = Normal, Diameter = 8, Background Correction = Yes, Minimum Thickness = 3, minimum cluster size = 500.

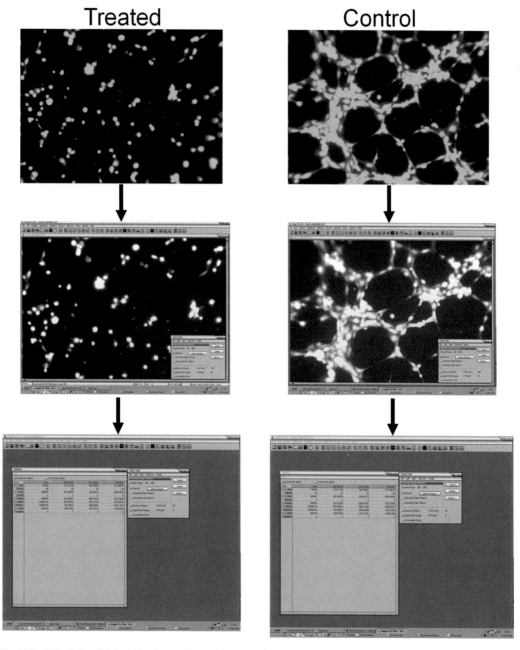

Fig. 5 Endothelial cell tubularization and quantitation using Image-Pro Plus software (tubule area, number of branch points, branch length etc.)

3.13 Zymography   This technique involves the electrophoresis of secreted protease enzymes through discontinuous polyacrylamide gels containing enzyme substrate (either type III gelatin or B-casein). After electrophoresis, removal of sodium dodecyl sulfate (SDS) from the gel by washing in 2.5 % Triton X-100 solution allows enzymes to

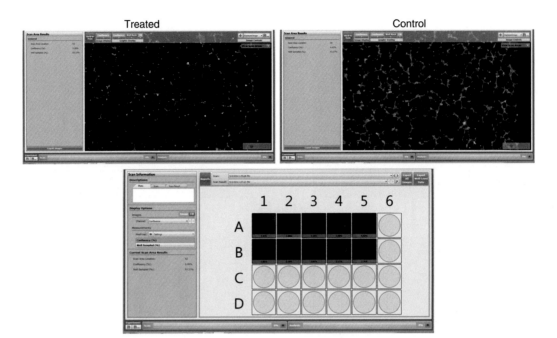

Fig. 6 Image analysis of endothelial cell tubularization and quantitation using Celigo cytometer software (confluence application)

renature and degrade the protein substrate. Staining of the gel with a protein dye such as Colloidal Blue or Coomassie Blue allows the proteolytic activity to be detected as clear bands of lysis against a blue background. Alternative methods that employ an enzyme linked immunosorbent assay (ELISA) can also be used to detect both MMP2 and MMP9 (Merck Millipore, Life Technologies etc.), e.g., effects of different substrates on matrix metalloproteinase expression and activation in endothelial cells (*see* Fig. 7)

1. Remove growth medium from a T75 flask of endothelial cells.

2. Rinse with 5 ml of PBS–EDTA solution.

3. Add 3 ml TrypLE™ Express and pour off excess. When cells have detached, add 10 ml endothelial cell basal medium and resuspend cells gently until a single cell suspension is produced.

4. Count viable cells using a hemocytometer and adjust cell concentration to 2×10^4 cells/ml.

5. Using a multichannel pipette, dispense a 100 μl volume of cells into each well of a 96-well plate with the exception of the outside wells. Repeat this procedure with plates that have been coated with different extracellular matrix proteins (*see* **Note 14**). Add 100 μl of distilled water to the outside wells to maintain humidity.

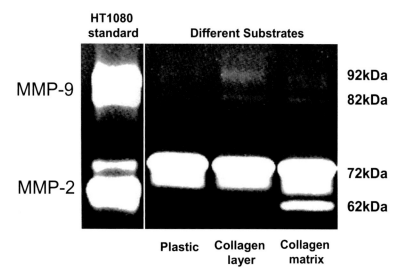

Fig. 7 Zymogram showing active and latent forms of MMP2 (72 kDa and 62 kDa) and MMP-9 (92 kDa and 82 kDa) respectively. Supernatants of PMA stimulated HT1080 cells can be used as a standard since this contains both latent and active forms of MMP-2 and MMP-9 and serves in effect as a molecular weight marker. Note that endothelial cells grown on plastic often express little MMP-9 and that activated forms (of both enzymes) are more readily detected on physiological substrates such as collagen matrix

6. Remove conditioned medium from plate after 24 or 48 h.

7. Place samples on ice and mix 1:1 with 2× Tris glycine SDS sample buffer and incubate at room temperature for 10 min.

8. Place a 10 % Novex® Tris-Glycine SDS gel containing gelatin in an XCell II Minicell electrophoresis tank as described in the manufacturer's instructions. Fill tank with Tris glycine SDS running buffer to recommended level.

9. Add samples to the wells in the gel. Include MMP-2/MMP-9 standards. Alternatively use PMA stimulated HT1080 conditioned medium (*see* **Note 15**).

10. Connect electrophoresis equipment to power pack and run at 125 V for 90 min.

11. Dilute 1× zymogram renaturing buffer and 10× developing buffer 1:9 with deionized water.

12. Remove gel from tank and incubate with renaturing buffer for 30 min with gentle agitation.

13. Remove buffer and replace with developing buffer and repeat incubation with agitation.

14. Remove buffer, replace with fresh developing buffer and incubate for at least 4 h or preferably overnight.

15. Stain with either Colloidal Blue Stain (Invitrogen) or Coomassie Blue (40 %(v/v) methanol, 10 % (v/v) acetic acid containing 0.5 % w/v Coomassie Blue for 1 h with gentle agitation) (*see* **Note 16**).

16. De-stain using either distilled water for the Colloidal Blue Stain (7 h) or destain solution (30 % (w/v) methanol, 10 % (v/v) acetic acid replaced every 4 h for the Coomassie Blue stained gels.

17. Place the gel to on a lightbox and acquire an image of the clear areas of lysis using a digital camera.

18. Analyze the areas of proteolysis in Gel Pro as follows.

19. Open the image file select "1D gels "and "show toolbar". Click the "Lane" option.

20. A window will pop up with "Extract Intensity". Accept this option and the gel image will appear as black and white.

21. Click "Find Lanes" and adjust the bars that appear over the lanes to reduce the area of interest to just the areas of lysis and click "OK".

22. On the 1D Gel window select Mass/IOD option and then select IOD.

23. The numbers that appear will give an absolute integrated intensity measurement of the bands. By comparing these figures it is possible to obtain relative quantitation of the areas of lysis in different lanes.

24. For testing potential inhibitors of MMP production, follow **steps 1–5**, then after 24 h, incubate the cells in serum-free medium for a further 24 h.

25. Make dilutions of the agent of interest in a replicate plate in serum free medium before transferring to the cell plate.

26. Remove the medium from the plate containing the cells and transfer the drug dilutions across and dispense into the wells using the multipipette.

27. Incubate the plates at 37 °C at 5 % CO_2 in air for 2 h.

28. Stimulate the cells with 20 ng VEGF and then follow **steps 6–28**.

4 Notes

1. Endothelial cells (EC) can be obtained from many different species and human EC are readily commercially available. EC can be obtained from large and small veins and arteries, from capillaries, or from specialized vascular areas such as the umbilical vein of newborns, blood vessels in specific organs (e.g.,

ovary, lung, brain [135–137]) or from solid tumors. Most commonly they are derived from human umbilical veins from pooled donors and for the purpose of most assays this is the most reliable and economical source. However, it should be recognized that neoangiogenesis generally originates from microvascular (rather than macrovascular) EC, and it is advisable to perform key studies in such cells, for example commercially available human dermal microvascular cells (HDMEC) (ZHC-2226 Cellworks Caltag Med Systems). In addition, human lymphatic endothelial cells are now also available and can be used in much the same way as vascular EC, although requiring specific growth medium (SC-2500 Cellworks CalTag Med Systems). Both of these endothelial cell lines and appropriate media for their propagation are also available from Cambrex Bio Science Nottingham Ltd. (http://www.cambrex.com).

2. It is of the utmost importance to maintain cell lines free of any contaminating species of mycoplasma, and this should be regularly monitored. Many methods are available to do this it may be by direct culture or by indirect assay. Indirect methods include DNA staining, biochemical detection, nucleic acid hybridization, immunoassays, and polymerase chain reaction (PCR) DNA amplification. Indirect methods have the advantage of detecting those mycoplasma species that are not easily cultivated. Assay methods can typically detect several, if not all, of the most common cell culture contaminants (*Mycoplasma hyorhinis, M. arginini, M. orale, M. fermentans, M. salivarium, and Acholesplasma laidlawii*) as well as some of the more unusual species.

3. Trypsin–EDTA treatment is the standard method for cell detachment. A prolonged exposure of cells to trypsin, however, can result in a loss of viability. We find that TrypLE™ Express treatment is significantly less damaging to cells and results in fewer cell clumps.

4. Subculture the cells when they reach 60–80 % confluence (if the recommended seeding densities are used, this usually takes 7–10 days, depending on the doubling time). Generally speaking the cells will remain viable for at least 10 passages before approaching senescence. It is important to ensure that the cells are still proliferating well if they are to be used in functional assays and if there is evidence of etiolation the cells should be discarded. When subculturing endothelial cells it is preferable to avoid pelleting them by centrifugation as they are very easily damaged in this process.

5. Before setting up the assay to evaluate inhibitors of proliferation it is always advisable to ensure that the assay will give linear measurements for either color change technique by the

construction of a standard curve. This is done by seeding the cells at a range of densities that reflect those that may occur during the course of the assay. A 96-well plate should be seeded in triplicate at the appropriate range of cell seeding densities and allowed to adhere. The substrate should be the added and the assay run as previously described. A graph drawn of cell seeding density against mean absorbance should appear linear.

6. One of the advantages of using fluorescent dyes such as CMFDA is that it is possible to stain different cell lines (or cells treated with different agents) with alternative dyes and monitor these mixed cell populations in cocultures. For example, differentially labeled EC and tumor cells can be co-migrated to examine 2-way modulating effects; cells pretreated with drugs or siRNA can be mixed with controls and separately counted etc. There are a variety of similar cell dyes available from Molecular Probes possessing differing absorption and emission maxima that correspond to red, orange, and blue. Details of these compounds are available from the website at http://probes.invitrogen.com/. The selection of the appropriate filter sets will enable the visualization of these mixed cell populations under the microscope ultraviolet light source. Quantum dots are also increasingly being considered for these types of applications as they can provide high intensity luminescent signals [138, 139] and luciferase-mediated luminescence is also a further alternative [140, 141]. The latter requires transient transfection of a luciferase gene (e.g., from the firefly, *Photinus pyralis*). Similarly green fluorescent protein from *Aequoria victoria* or *Renilla* spp. could be used to generate fluorescence, although all transfection procedures are most commonly used in immortal cells rather than normal cells with a finite life-span.

7. In the majority of cases cells will migrate to defined growth factors such as VEGF or FGF more readily when a substrate that promotes adhesion is present; indeed in some cases it is not possible to obtain any migration to a soluble ligand if a natural substrate is lacking. The appropriate substrate and optimal dilution for coating will need to be determined empirically for each cell line and ligand pair. A range of substrates is commercially available (e.g., collagen 1, collagen IV, laminin, fibronectin) and coating procedures are supplied with each product data sheet. FCS works well as a chemoattractant because it not only contains a plethora of potent chemoattractant factors, but also contains high levels of proteins such as fibronectin, which coat the membrane naturally. In most cases the rate and extent of migration to a defined ligand will be less than when using FCS. On the other hand, if inhibitors under test are targeting a specific signaling pathway, for instance, che-

motaxis to FCS may only be partially inhibited as other mechanisms will contribute.

8. In some instances the cells may require pre-incubation with the inhibitory compound prior to stimulation with a chemoattractant. If this is necessary it may be convenient to add the inhibitor to a flask of cells for a predetermined length of time before seeding. Fresh compound should be added once the cells are in the chamber to ensure continued activity. However, if the compound acts rapidly both the cell suspension and compound can be added simultaneously to the top chamber, and medium and compound to the bottom chamber. The addition of the chemoattractant to the lower chamber is then delayed until initiation of chemomigration is desired. If this approach is used the length of time necessary for inhibition of the appropriate target (and its duration) should be established in pilot studies.

9. Endothelial cells will generally migrate to 5 % FCS in adequate numbers within 6 h and to perform a cell count at intervals may not be necessary if an endpoint assay is sufficient.

10. The cells may also be labeled with the lipophilic dye DiI (*see* http://probes.invitrogen.com/) according to the manufacturer's instructions.

11. It is important to observe the lower surface of the membrane under UV illumination before capturing images of the invading cells. If there is an uneven coating of Matrigel™ which will sometimes arise due to problems in manufacture or manual coating, the cells may tend to preferentially migrate around the edge of the membrane instead of uniformly across its surface. Alternatively, if there is a meniscus effect, such that the Matrigel is thicker at the circumference, then migration will preferentially occur in the center where the layer is thinnest. The use of Matrigel in chemoinvasion assays has been discussed in several publications [58, 142]. Recently, assays have been developed that can accommodate shear stress in endothelial functional assays [143].

12. Dynamic (as opposed to endpoint only assays) will also indicate if inhibition is sustained throughout the observation period, or if any recovery is taking place. This may be useful to determine the stability of test compounds.

13. Although EC can normally be cultured on Matrigel™ for up to 24 h, tubularization usually reaches an optimum level (before partially degrading) at less than 10 h. There are some reports that 24–48 h after plating on Matrigel, viability of EC is decreased [144].

14. A thin layer of collagen I is prepared by diluting rat tail collagen type 1–50 μg/ml with 0.02 M acetic acid and 50 μl to

each well. Incubate at room temperature for 1 h. Remove solution and wash each well with PBS. A collagen matrix may be prepared by diluting the collagen to 3 mg/ml. 100 μl of a mixture containing serum-free growth medium, 0.1 M sodium bicarbonate buffer, 3 mg/ml rat tail collagen type 1, and distilled water (2:1:4:2) is added to each well and allowed to gel at 37 °C for 30 min.

15. If a semiquantitative analysis of the zymography samples is to be performed the protein content of the samples loaded in each lane will need to be standardized. For generating control supernatant containing high levels of gelatinases, serum-starve an 80 % confluent 75 cm³ flask of HT1080 human sarcoma cells overnight. Add fresh serum-free growth medium containing 20 ng/ml PMA, leave for a further 24 h and then collect the conditioned medium. A sample of this medium run on the zymogram will show both active and latent forms of MMP2 and MMP9. These bands will appear at 62 kDa and 72 kDa (MMP2) and at 82 kDa and 92 kDa (MMP9).

16. Either method of staining the gels is suitable although a five-fold higher sensitivity may be obtained using the Colloidal Blue technique. If the gels are to be stored, incubate them in 100 ml of 2 % glycerol for 30 min then dry overnight using a gel drier system (Gibco-Invitrogen).

References

1. Folkman J (1971) Tumor angiogenesis: therapeutic implications. N Engl J Med 285(21):1182–1186

2. Benelli R, Lorusso G, Albini A et al (2006) Cytokines and chemokines as regulators of angiogenesis in health and disease. Curr Pharm Des 12(24):3101–3115

3. Bosisio D, Salvi V, Gagliostro V et al (2014) Angiogenic and antiangiogenic chemokines. Chem Immunol Allergy 99:89–104

4. Bzowska M, Mezyk-Kopec R, Prochnicki T et al (2013) Antibody-based antiangiogenic and antilymphangiogenic therapies to prevent tumor growth and progression. Acta Biochim Pol 60(3):263–275

5. Cao Z, Shang B, Zhang G et al (2013) Tumor cell-mediated neovascularization and lymphangiogenesis contrive tumor progression and cancer metastasis. Biochim Biophys Acta 1836(2):273–286

6. Carmeliet P (2005) Angiogenesis in life, disease and medicine. Nature 438(7070):932–936

7. Eccles SA (2004) Parallels in invasion and angiogenesis provide pivotal points for therapeutic intervention. Int J Dev Biol 48(5-6):583–598

8. Farnsworth RH, Lackmann M, Achen MG et al. (2013) Vascular remodeling in cancer. Oncogene 33(27):3496–3505

9. Favre J, Terborg N, Horrevoets AJ (2013) The diverse identity of angiogenic monocytes. Eur J Clin Investig 43(1):100–107

10. Gomes FG, Nedel F, Alves AM et al (2013) Tumor angiogenesis and lymphangiogenesis: tumor/endothelial crosstalk and cellular/microenvironmental signaling mechanisms. Life Sci 92(2):101–107

11. Jeltsch M, Leppanen VM, Saharinen P et al. (2013) Receptor tyrosine kinase-mediated angiogenesis. Cold Spring Harb Perspect Biol 5(9) pii: a009183

12. Kushner EJ, Bautch VL (2013) Building blood vessels in development and disease. Curr Opin Hematol 20(3):231–236

13. Marcelo KL, Goldie LC, Hirschi KK (2013) Regulation of endothelial cell differentiation and specification. Circ Res 112(9):1272–1287

14. Ribatti D (2014) History of research on angiogenesis. Chem Immunol Allergy 99:1–14

15. Semenza GL (2013) Cancer-stromal cell interactions mediated by hypoxia-inducible

factors promote angiogenesis, lymphangiogenesis, and metastasis. Oncogene 32(35):4057–4063

16. Bates DO, Harper SJ (2005) Therapeutic potential of inhibitory VEGF splice variants. Future Oncol 1(4):467–473

17. Bisacchi D, Benelli R, Vanzetto C et al (2003) Anti-angiogenesis and angioprevention: mechanisms, problems and perspectives. Cancer Detect Prev 27(3):229–238

18. Cai W, Chen X (2006) Anti-angiogenic cancer therapy based on integrin alphavbeta3 antagonism. Anti Cancer Agents Med Chem 6(5):407–428

19. De Bonis P, Marziali G, Vigo V et al (2013) Antiangiogenic therapy for high-grade gliomas: current concepts and limitations. Expert Rev Neurother 13(11):1263–1270

20. El-Kenawi AE, El-Remessy AB (2013) Angiogenesis inhibitors in cancer therapy: mechanistic perspective on classification and treatment rationales. Br J Pharmacol 170(4):712–729

21. Eroglu Z, Stein CA, Pal SK (2013) Targeting angiopoietin-2 signaling in cancer therapy. Expert Opin Investig Drugs 22(7):813–825

22. Fakih M (2013) The evolving role of VEGF-targeted therapies in the treatment of metastatic colorectal cancer. Expert Rev Anticancer Ther 13(4):427–438

23. Feliz LR, Tsimberidou AM (2013) Anti-vascular endothelial growth factor therapy in the era of personalized medicine. Cancer Chemother Pharmacol 72(1):1–12

24. Gerald D, Chintharlapalli S, Augustin HG et al (2013) Angiopoietin-2: an attractive target for improved antiangiogenic tumor therapy. Cancer Res 73(6):1649–1657

25. Girard N (2013) Antiangiogenic agents and chemotherapy in advanced non-small cell lung cancer: a clinical perspective. Expert Rev Anticancer Ther 13(10):1193–1206

26. Hadler-Olsen E, Winberg JO, Uhlin-Hansen L (2013) Matrix metalloproteinases in cancer: their value as diagnostic and prognostic markers and therapeutic targets. Tumour Biol 34(4):2041–2051

27. Huang Y, Goel S, Duda DG et al (2013) Vascular normalization as an emerging strategy to enhance cancer immunotherapy. Cancer Res 73(10):2943–2948

28. Katoh M (2013) Therapeutics targeting angiogenesis: genetics and epigenetics, extracellular miRNAs and signaling networks (Review). Int J Mol Med 32(4):763–767

29. Kerbel RS (2006) Antiangiogenic therapy: a universal chemosensitization strategy for cancer? Science 312(5777):1171–1175

30. Lambrechts D, Lenz HJ, de Haas S et al (2013) Markers of response for the antiangiogenic agent bevacizumab. J Clin Oncol 31(9):1219–1230

31. Liang G, Chen G, Wei X et al (2013) Small molecule inhibition of fibroblast growth factor receptors in cancer. Cytokine Growth Factor Rev 24(5):467–475

32. Mansfield AS, Markovic SN (2013) Inhibition of angiogenesis for the treatment of metastatic melanoma. Curr Oncol Rep 15(5):492–499

33. Moserle L, Jimenez-Valerio G, Casanovas O (2014) Antiangiogenic therapies: going beyond their limits. Cancer Discov 4(1):31–41

34. Mountzios G, Pentheroudakis G, Carmeliet P (2014) Bevacizumab and micrometastases: revisiting the preclinical and clinical rollercoaster. Pharmacol Ther 141(2):117–124

35. Oh WK, McDermott D, Porta C et al (2014) Angiogenesis inhibitor therapies for advanced renal cell carcinoma: toxicity and treatment patterns in clinical practice from a global medical chart review. Int J Oncol 44(1):5–16

36. Patel A, Sun W (2014) Ziv-aflibercept in metastatic colorectal cancer. Biologics 8:13–25

37. Robison NJ, Campigotto F, Chi SN et al (2014) A phase II trial of a multi-agent oral antiangiogenic (metronomic) regimen in children with recurrent or progressive cancer. Pediatr Blood Cancer 61(4):636–642

38. Shahneh FZ, Baradaran B, Zamani F et al (2013) Tumor angiogenesis and anti-angiogenic therapies. Human Antibodies 22(1-2):15–19

39. Sie M, den Dunnen WF, Hoving EW et al (2014) Anti-angiogenic therapy in pediatric brain tumors: an effective strategy? Crit Rev Oncol Hematol 89(3):418–432

40. Tomao F, Papa A, Rossi L et al (2014) Beyond bevacizumab: investigating new angiogenesis inhibitors in ovarian cancer. Expert Opin Investig Drugs 23(1):37–53

41. Yoo SY, Kwon SM (2013) Angiogenesis and its therapeutic opportunities. Mediat Inflamm 2013:127170

42. Clarke JM, Hurwitz HI (2013) Understanding and targeting resistance to anti-angiogenic therapies. J gastrointest Oncol 4(3):253–263

43. Giuliano S, Pages G (2013) Mechanisms of resistance to anti-angiogenic therapies. Biochimie 95(6):1110–1119

44. Li D, Xie K, Ding G et al (2014) Tumor resistance to anti-VEGF therapy through up-regulation of VEGF-C expression. Cancer Lett 346(1):45–52

45. Limaverde-Sousa G, Sternberg C, Ferreira CG (2014) Antiangiogenesis beyond VEGF inhibition: a journey from antiangiogenic single-target to broad-spectrum agents. Cancer Treat Rev 40(4):548–557

46. Soda Y, Myskiw C, Rommel A et al (2013) Mechanisms of neovascularization and resistance to anti-angiogenic therapies in glioblastoma multiforme. J Mol Med 91(4): 439–448

47. Dupont E, Savard PE, Jourdain C et al (1998) Antiangiogenic properties of a novel shark cartilage extract: potential role in the treatment of psoriasis. J Cutan Med Surg 2(3):146–152

48. Griffin RJ, Molema G, Dings RP (2006) Angiogenesis treatment, new concepts on the horizon. Angiogenesis 9(2):67–72

49. Griggs J, Skepper JN, Smith GA et al (2002) Inhibition of proliferative retinopathy by the anti-vascular agent combretastatin-A4. Am J Pathol 160(3):1097–1103

50. Lavie G, Mandel M, Hazan S et al (2005) Anti-angiogenic activities of hypericin in vivo: potential for ophthalmologic applications. Angiogenesis 8(1):35–42

51. Sauder DN, Dekoven J, Champagne P et al (2002) Neovastat (AE-941), an inhibitor of angiogenesis: Randomized phase I/II clinical trial results in patients with plaque psoriasis. J Am Acad Dermatol 47(4):535–541

52. Stellmach V, Crawford SE, Zhou W et al (2001) Prevention of ischemia-induced retinopathy by the natural ocular antiangiogenic agent pigment epithelium-derived factor. Proc Natl Acad Sci U S A 98(5):2593–2597

53. Tong Y, Zhang X, Zhao W et al (2004) Antiangiogenic effects of Shiraiachrome A, a compound isolated from a Chinese folk medicine used to treat rheumatoid arthritis. Eur J Pharmacol 494(2-3):101–109

54. Callaghan MJ, Chang EI, Seiser N et al (2008) Pulsed electromagnetic fields accelerate normal and diabetic wound healing by increasing endogenous FGF-2 release. Plast Reconstr Surg 121(1):130–141

55. Saaristo A, Tammela T, Farkkila A et al (2006) Vascular endothelial growth factor-C accelerates diabetic wound healing. Am J Pathol 169(3):1080–1087

56. Detillieux KA, Cattini PA, Kardami E (2004) Beyond angiogenesis: the cardioprotective potential of fibroblast growth factor-2. Can J Physiol Pharmacol 82(12):1044–1052

57. Levy AP, Levy NS, Loscalzo J et al (1995) Regulation of vascular endothelial growth factor in cardiac myocytes. Circ Res 76(5):758–766

58. Albini A, Benelli R, Noonan DM et al (2004) The "chemoinvasion assay": a tool to study tumor and endothelial cell invasion of basement membranes. Int J Dev Biol 48(5-6):563–571

59. Blosser W, Vakana E, Wyss LV et al (2014) A method to assess target gene involvement in angiogenesis in vitro and in vivo using lentiviral vectors expressing shRNA. PLoS One 9(4), e96036

60. Buchanan CF, Voigt EE, Szot CS et al (2014) Three-dimensional microfluidic collagen hydrogels for investigating flow-mediated tumor-endothelial signaling and vascular organization. Tissue Eng Part C Methods 20(1):64–75

61. Carmeliet P (2005) VEGF as a key mediator of angiogenesis in cancer. Oncology 69(Suppl 3):4–10

62. Guo S, Lok J, Liu Y et al (2014) Assays to examine endothelial cell migration, tube formation, and gene expression profiles. Methods Mol Biol 1135:393–402

63. Lee H, Kim S, Chung M et al (2014) A bioengineered array of 3D microvessels for vascular permeability assay. Microvasc Res 91:90–98

64. Ling TY, Liu YL, Huang YK et al (2014) Differentiation of lung stem/progenitor cells into alveolar pneumocytes and induction of angiogenesis within a 3D gelatin-microbubble scaffold. Biomaterials 35(22):5660–5669

65. Mousseau Y, Mollard S, Qiu H et al (2014) In vitro 3D angiogenesis assay in egg white matrix: comparison to Matrigel, compatibility to various species, and suitability for drug testing. Lab Investig 94(3):340–349

66. Park KP, Du P, Subbiah R et al (2014) Vascular morphogenesis of human umbilical vein endothelial cells on cell-derived macromolecular matrix microenvironment. Tissue Eng A 20(17-18):2365–2377

67. Prisco AR, Bukowy JD, Hoffmann BR et al (2014) Automated quantification reveals hyperglycemia inhibits endothelial angiogenic function. PLoS One 9(4), e94599

68. Qin Q, Qian J, Ge L et al (2014) Effect and mechanism of thrombospondin-1 on the angiogenesis potential in human endothelial progenitor cells: an in vitro study. PLoS One 9(2), e88213

69. Richardson MR, Robbins EP, Vemula S et al (2014) Angiopoietin-like protein 2 regulates endothelial colony forming cell vasculogenesis. Angiogenesis 17(3):675–683

70. Rojas-Rodriguez R, Gealekman O, Kruse ME et al (2014) Adipose tissue angiogenesis assay. Methods Enzymol 537:75–91

71. Seo BR, Delnero P, Fischbach C (2014) In vitro models of tumor vessels and matrix: engineering approaches to investigate transport limitations and drug delivery in cancer. Adv Drug Deliv Rev 69–70:205–216

72. Voltan R, Zauli G, Rizzo P et al (2014) In vitro endothelial cell proliferation assay reveals distinct levels of proangiogenic cytokines characterizing sera of healthy subjects and of

patients with heart failure. Mediat Inflamm 2014:257081

73. Dome B, Hendrix MJ, Paku S et al (2007) Alternative vascularization mechanisms in cancer: pathology and therapeutic implications. Am J Pathol 170(1):1–15

74. Donnem T, Hu J, Ferguson M et al (2013) Vessel co-option in primary human tumors and metastases: an obstacle to effective anti-angiogenic treatment? Cancer Med 2(4):427–436

75. Elzarrad K, Haroon A, Reed D et al (2009) Early incorporated endothelial cells as origin of metastatic tumor vasculogenesis. Clin Exp Metastasis 26(6):589–598

76. Fan YL, Zheng M, Tang YL et al (2013) A new perspective of vasculogenic mimicry: EMT and cancer stem cells (Review). Oncol Lett 6(5):1174–1180

77. Hillen F, Griffioen AW (2007) Tumour vascularization: sprouting angiogenesis and beyond. Cancer Metastasis Rev 26(3-4):489–502

78. Janic B, Arbab AS (2010) The role and therapeutic potential of endothelial progenitor cells in tumor neovascularization. TheScientificWorld JOURNAL 10:1088–1099

79. Leenders WP, Kusters B, de Waal RM (2002) Vessel co-option: how tumors obtain blood supply in the absence of sprouting angiogenesis. Endothelium 9(2):83–87

80. Qian CN (2013) Hijacking the vasculature in ccRCC--co-option, remodelling and angiogenesis. Nat Rev Urol 10(5):300–304

81. Zhao C, Yang H, Shi H et al (2011) Distinct contributions of angiogenesis and vascular co-option during the initiation of primary microtumors and micrometastases. Carcinogenesis 32(8):1143–1150

82. Alevizakos M, Kaltsas S, Syrigos KN (2013) The VEGF pathway in lung cancer. Cancer Chemother Pharmacol 72(6):1169–1181

83. Brkovic A, Pelletier M, Girard D et al (2007) Angiopoietin chemotactic activities on neutrophils are regulated by PI-3K activation. J Leukoc Biol 81(4):1093–1101

84. Chang LK, Garcia-Cardena G, Farnebo F et al (2004) Dose-dependent response of FGF-2 for lymphangiogenesis. Proc Natl Acad Sci U S A 101(32):11658–11663

85. Han EC, Lee J, Ryu SW et al (2014) Tumor-conditioned Gr-1(+)CD11b(+) myeloid cells induce angiogenesis through the synergistic action of CCL2 and CXCL16 in vitro. Biochem Biophys Res Commun 443(4):1218–1225

86. Nold-Petry CA, Rudloff I, Baumer Y et al (2014) IL-32 promotes angiogenesis. J Immunol 192(2):589–602

87. Strieter RM, Burdick MD, Mestas J et al (2006) Cancer CXC chemokine networks and tumour angiogenesis. Eur J Cancer 42(6):768–778

88. Valiente M, Obenauf AC, Jin X et al (2014) Serpins promote cancer cell survival and vascular co-option in brain metastasis. Cell 156(5):1002–1016

89. Alghisi GC, Ruegg C (2006) Vascular integrins in tumor angiogenesis: mediators and therapeutic targets. Endothelium 13(2):113–135

90. Dong Y, Xie X, Wang Z et al (2014) Increasing matrix stiffness upregulates vascular endothelial growth factor expression in hepatocellular carcinoma cells mediated by integrin beta1. Biochem Biophys Res Commun 444(3):427–432

91. Jahangiri A, Aghi MK, Carbonell WS (2014) beta1 integrin: Critical path to antiangiogenic therapy resistance and beyond. Cancer Res 74(1):3–7

92. Serini G, Valdembri D, Bussolino F (2006) Integrins and angiogenesis: a sticky business. Exp Cell Res 312(5):651–658

93. Dossi R, Frapolli R, Di Giandomenico S et al (2014) Antiangiogenic activity of trabectedin in myxoid liposarcoma: Involvement of host TIMP-1 and TIMP-2 and tumor thrombospondin-1. Int J Cancer 136(3):721–729

94. Lakka SS, Gondi CS, Rao JS (2005) Proteases and glioma angiogenesis. Brain Pathol 15(4):327–341

95. Ramer R, Fischer S, Haustein M et al (2014) CANNABINOIDs INHIBIT angiogenic capacities of Endothelial cells via release of Tissue inhibitor of matrix metalloproteinases-1 from lung cancer cells. Biochem Pharmacol 91(2):202–216

96. Rundhaug JE (2005) Matrix metalloproteinases and angiogenesis. J Cell Mol Med 9(2):267–285

97. van Hinsbergh VW, Engelse MA, Quax PH (2006) Pericellular proteases in angiogenesis and vasculogenesis. Arterioscler Thromb Vasc Biol 26(4):716–728

98. Zhang G, Miyake M, Lawton A et al (2014) Matrix metalloproteinase-10 promotes tumor progression through regulation of angiogenic and apoptotic pathways in cervical tumors. BMC Cancer 14(1):310

99. Rabbani SA, Mazar AP (2001) The role of the plasminogen activation system in angiogenesis and metastasis. Surg Oncol Clin N Am 10(2):393–415, x

100. Stefansson S, McMahon GA, Petitclerc E et al (2003) Plasminogen activator inhibitor-1 in tumor growth, angiogenesis and vascular remodeling. Curr Pharm Des 9(19):1545–1564

101. Stojkovic S, Kaun C, Heinz M et al (2014) Interleukin-33 induces urokinase in human

endothelial cells-possible impact on angiogenesis. J Thromb Haemost 12(6):948–957

102. Hu J, Yan M, Pu C et al (2014) Chemically synthesized matrix metalloproteinase and angiogenesis-inhibiting peptides as anticancer agents. Anti Cancer Agents Med Chem 14(3):483–494

103. Sheldrake HM, Patterson LH (2014) Strategies to inhibit tumor associated integrin receptors: rationale for dual and multi-antagonists. J Med Chem 57(15):6301–6315

104. Luangdilok S, Box C, Harrington K et al (2011) MAPK and PI3K signalling differentially regulate angiogenic and lymphangiogenic cytokine secretion in squamous cell carcinoma of the head and neck. Eur J Cancer 47(4):520–529

105. Aravantinos G, Pectasides D (2014) Bevacizumab in combination with chemotherapy for the treatment of advanced ovarian cancer: a systematic review. J Ovarian Res 7:57

106. Ciombor KK, Berlin J (2014) Aflibercept—a decoy VEGF receptor. Curr Oncol Rep 16(2):368

107. Chen H, Ding X, Gao Y et al (2013) Inhibition of angiogenesis by a novel neutralizing antibody targeting human VEGFR-3. mAbs 5(6):956–961

108. Quagliata L, Klusmeier S, Cremers N et al (2014) Inhibition of VEGFR-3 activation in tumor-draining lymph nodes suppresses the outgrowth of lymph node metastases in the MT-450 syngeneic rat breast cancer model. Clin Exp Metastasis 31(3):351–365

109. Yu Y, Lee P, Ke Y et al (2013) Development of humanized rabbit monoclonal antibodies against vascular endothelial growth factor receptor 2 with potential antitumor effects. Biochem Biophys Res Commun 436(3):543–550

110. Khusal R, Da Costa Dias B, Moodley K et al (2013) In vitro inhibition of angiogenesis by antibodies directed against the 37kDa/67kDa laminin receptor. PLoS One 8(3), e58888

111. Bitting RL, Healy P, Creel PA et al (2013) A Phase Ib Study of Combined VEGFR and mTOR Inhibition With Vatalanib and Everolimus in Patients With Advanced Renal Cell Carcinoma. Clin genitourin Cancer 12(4):241–250

112. Chien MH, Lee LM, Hsiao M et al (2013) Inhibition of metastatic potential in breast carcinoma in vivo and in vitro through targeting VEGFRs and FGFRs. Evid Based Complement Alternat Med 2013:718380

113. Davis SL, Eckhardt SG, Messersmith WA et al (2013) The development of regorafenib and its current and potential future role in cancer therapy. Drugs Today 49(2):105–115

114. Grailer JJ, Steeber DA (2013) Vascular endothelial growth factor receptor inhibitor SU5416 suppresses lymphocyte generation and immune responses in mice by increasing plasma corticosterone. PLoS One 8(9), e75390

115. Hyams DM, Chan A, de Oliveira C et al (2013) Cediranib in combination with fulvestrant in hormone-sensitive metastatic breast cancer: a randomized Phase II study. Investig New Drugs 31(5):1345–1354

116. Kim YS, Li F, O'Neill BE et al (2013) Specific binding of modified ZD6474 (Vandetanib) monomer and its dimer with VEGF receptor-2. Bioconjug Chem 24(11):1937–1944

117. Landry JP, Fei Y, Zhu X et al (2013) Discovering small molecule ligands of vascular endothelial growth factor that block VEGF-KDR binding using label-free microarray-based assays. Assay Drug Dev Technol 11(5):326–332

118. Mayer EL, Scheulen ME, Beckman J et al (2013) A Phase I dose-escalation study of the VEGFR inhibitor tivozanib hydrochloride with weekly paclitaxel in metastatic breast cancer. Breast Cancer Res Treat 140(2):331–339

119. Schmieder R, Hoffmann J, Becker M et al (2013) Regorafenib (BAY 73-4506): Antitumor and antimetastatic activities in preclinical models of colorectal cancer. Int J Cancer 135(6):1487–1496

120. Oon CE, Harris AL (2011) New pathways and mechanisms regulating and responding to Delta-like ligand 4-Notch signalling in tumour angiogenesis. Biochem Soc Trans 39(6):1612–1618

121. Real C, Remedio L, Caiado F et al (2011) Bone marrow-derived endothelial progenitors expressing Delta-like 4 (Dll4) regulate tumor angiogenesis. PLoS One 6(4), e18323

122. Rodriguez P, Higueras MA, Gonzalez-Rajal A et al (2012) The non-canonical NOTCH ligand DLK1 exhibits a novel vascular role as a strong inhibitor of angiogenesis. Cardiovasc Res 93(2):232–241

123. Thurston G, Kitajewski J (2008) VEGF and Delta-Notch: interacting signalling pathways in tumour angiogenesis. Br J Cancer 99(8):1204–1209

124. Watson O, Novodvorsky P, Gray C et al (2013) Blood flow suppresses vascular Notch signalling via dll4 and is required for angiogenesis in response to hypoxic signalling. Cardiovasc Res 100(2):252–261

125. Andrikopoulos P, Baba A, Matsuda T et al (2011) Ca2+ influx through reverse mode Na+/Ca2+ exchange is critical for vascular endothelial growth factor-mediated extracellular signal-regulated kinase (ERK) 1/2 acti-

vation and angiogenic functions of human endothelial cells. J Biol Chem 286(44): 37919–37931

126. Andrikopoulos P, Fraser SP, Patterson L et al (2011) Angiogenic functions of voltage-gated Na+ Channels in human endothelial cells: modulation of vascular endothelial growth factor (VEGF) signaling. J Biol Chem 286(19):16846–16860

127. Blacher S, Erpicum C, Lenoir B et al (2014) Cell invasion in the spheroid sprouting assay: a spatial organisation analysis adaptable to cell behaviour. PLoS One 9(5), e97019

128. Eccles SA, Box C, Court W (2005) Cell migration assays and their application in cancer drug discovery. Biotech Ann Rev 11:391–421

129. Edgar LT, Hoying JB, Utzinger U et al (2014) Mechanical interaction of angiogenic microvessels with the extracellular matrix. J Biomech Eng 136(2):021001

130. Poitevin S, Cussac D, Leroyer AS et al (2014) Sphingosine kinase 1 expressed by endothelial colony-forming cells has a critical role in their revascularization activity. Cardiovasc Res 103(1):121–130

131. Sanderson S, Valenti M, Gowan S et al (2006) Benzoquinone ansamycin heat shock protein 90 inhibitors modulate multiple functions required for tumor angiogenesis. Mol Cancer Ther 5(3):522–532

132. Stapor PC, Azimi MS, Ahsan T et al (2013) An angiogenesis model for investigating multicellular interactions across intact microvascular networks. Am J Physiol Heart Circ Physiol 304(2):H235–H245

133. Vinci M, Gowan S, Boxall F et al (2012) Advances in establishment and analysis of three-dimensional tumor spheroid-based functional assays for target validation and drug evaluation. BMC Biol 10:29

134. Connolly DT, Knight MB, Harakas NK et al (1986) Determination of the number of endothelial cells in culture using an acid phosphatase assay. Anal Biochem 152(1):136–140

135. He QW, Xia YP, Chen SC et al (2013) Astrocyte-derived sonic hedgehog contributes to angiogenesis in brain microvascular endothelial cells via RhoA/ROCK pathway after oxygen-glucose deprivation. Mol Neurobiol 47(3):976–987

136. Parra-Bonilla G, Alvarez DF, Alexeyev M et al (2013) Lactate dehydrogenase a expression is necessary to sustain rapid angiogenesis of pulmonary microvascular endothelium. PLoS One 8(9), e75984

137. Xu Y, Wang D, Zhao LM et al (2013) Endoglin is necessary for angiogenesis in human ovarian carcinoma-derived primary endothelial cells. Cancer Biol Ther 14(10): 937–948

138. Barnett JM, Penn JS, Jayagopal A (2013) Imaging of endothelial progenitor cell subpopulations in angiogenesis using quantum dot nanocrystals. Methods Mol Biol 1026:45–56

139. Kwon H, Lee J, Song R et al (2013) In vitro and in vivo imaging of prostate cancer angiogenesis using anti-vascular endothelial growth factor receptor 2 antibody-conjugated quantum dot. Korean J Radiol 14(1):30–37

140. Gildea JJ, Harding MA, Gulding KM et al (2000) Transmembrane motility assay of transiently transfected cells by fluorescent cell counting and luciferase measurement. Biotechniques 29(1):81–86

141. Spessotto P, Giacomello E, Perri R (2002) Improving fluorescence-based assays for the in vitro analysis of cell adhesion and migration. Mol Biotechnol 20(3):285–304

142. Benelli R, Albini A (1999) In vitro models of angiogenesis: the use of Matrigel. Int J Biol Markers 14(4):243–246

143. Mahler GJ, Frendl CM, Cao Q et al (2014) Effects of shear stress pattern and magnitude on mesenchymal transformation and invasion of aortic valve endothelial cells. Biotechnol Bioeng 111(11):2326–2337

144. Ranta V, Mikkola T, Ylikorkala O et al (1998) Reduced viability of human vascular endothelial cells cultured on Matrigel. J Cell Physiol 176(1):92–98

Chapter 9

Tube-Forming Assays

Ryan M. Brown, Christopher J. Meah, Victoria L. Heath, Iain B. Styles, and Roy Bicknell

Abstract

Angiogenesis involves the generation of new blood vessels from the existing vasculature and is dependent on many growth factors and signaling events. In vivo angiogenesis is dynamic and complex, meaning assays are commonly utilized to explore specific targets for research into this area. Tube-forming assays offer an excellent overview of the molecular processes in angiogenesis. The Matrigel tube forming assay is a simple-to-implement but powerful tool for identifying biomolecules involved in angiogenesis. A detailed experimental protocol on the implementation of the assay is described in conjunction with an in-depth review of methods that can be applied to the analysis of the tube formation. In addition, an ImageJ plug-in is presented which allows automatic quantification of tube images reducing analysis times while removing user bias and subjectivity.

Key words Angiogenesis, Endothelial cell, Matrigel, Tube formation, Sprouting, ImageJ

1 Introduction

Angiogenesis is a complex, multi-step process in which new vessel formation begins from the existing vasculature [1]. When unregulated, this process plays a crucial role in the disease progression of solid tumors and the metastasis of cancers, meaning inhibition of this pathological process could result in a range of new anticancer therapies [2]. Angiogenic models provide an elegant route to better understand this molecular process within a setting that is well controlled and rapid. While in vivo models offer the best representation of the various stages of angiogenesis, in vitro models are often favored because they are easier to implement, control, and understand [3].

The Matrigel tube forming in vitro assay has been used extensively and is favored due to its rapid generation of well-defined data [3]. While an in vitro assay cannot fully replicate the in vivo situation, the Matrigel assay is very useful because it can easily characterize important proteins involved in angiogenesis, making it an

Stewart G. Martin and Peter W. Hewett (eds.), *Angiogenesis Protocols*, Methods in Molecular Biology, vol. 1430,
DOI 10.1007/978-1-4939-3628-1_9, © Springer Science+Business Media New York 2016

excellent diagnostic tool. The assay takes 24 h to perform, as opposed to 1 or 2 weeks for more complex assays, and 2D images are easily captured with conventional brightfield microscopy. This removes the need for fixing and staining, allowing sequential images to be acquired reducing the number of cells needed. The use of brightfield microscopy also means that there are no other compounds or processes that may affect the cellular dynamics of endothelial cells (ECs); which can be an issue when interpreting other angiogenesis assays. It also provides a wealth of data in quantifying the tube formation, including the number of branch points and total area of the network. However, the method by which this data is collected and interpreted is very important and will be discussed in more detail later in this chapter.

The Matrigel tube formation assay focuses solely on the differentiation stage of angiogenesis, whereby endothelial cells migrate from the existing blood vessel to form the vessel sprout [3]. Matrigel is derived from mouse Engelbreth-Holm-Swarm sarcoma and contains a combination of matrix proteins that encourage the ECs to migrate and differentiate into tube-like networks that simulate the in vivo process. Growth factors, such as VEGF, are absent meaning no proliferation occurs. To understand how metastasis achieves pathological angiogenesis, EC gene knockdown is often undertaken to examine its effect on migration [2]. Experiments usually span over 24 h and begin with a monolayer of endothelial cells being seeded onto a layer of Matrigel. Upon exposure to the matrix proteins, ECs spontaneously migrate to form tubular networks over the first few hours of the experiment. These networks then gradually degrade as the cells undergo apoptosis during the next 24 h. The presence of a lumen in these networks is still debated due to the results of conflicting studies which utilized electron and light microscopy (electron microscopy fixation can affect biological structures) [4, 5]. The Matrigel assay was a crucial element in our identification of RhoJ as an important regulator of the differentiation stage of angiogenesis [6].

The central theme of an angiogenesis assay is to provide a model of in vivo dynamics that is easy to implement, understand, and control. However, a reductionist approach to developing these models can lead to assays that are not particularly biomimetic. The Matrigel tube-formation assay, for example, gives networks of ECs that resemble a vascular network but do not include key features of "true" angiogenesis, such as directed EC invasion into the extracellular matrix (ECM), fluid flow, or polarization of abluminal sides of ECs [7, 8]. In order to include more of these key features, novel three-dimensional assays that can better mimic native systems are being developed. One such assay was recently developed by Nguyen et al. [9] that permits the directed invasion of ECs, as well as fluid flow, which can have a profound effect on EC behavior [9, 10]. While more complex, biomimetic models may provide greater insight into in vivo angiogenesis, they are time consuming and so best suited to the detailed analysis of the modes of action of anti-angiogenic agents, rather than the screening of new agents. In this chapter, we provide detailed protocols for performing Matrigel gel tube formation assays and their automated analysis which allow high-throughput screening.

2   Materials

2.1   Matrigel Assay

1. Matrigel: This is commercially available in native or growth factor-reduced forms; the selection of which depends on whether the effects of angiogenic growth factors or inhibitors are being investigated. Use without dilution.

2. Human umbilical vein endothelial cells (HUVEC) are frequently used and can be purchased from commercial sources or extracted from umbilical cords.

3. Dulbecco's phosphate-buffered saline (PBS): Calcium and magnesium-free PBS (10 mM Na_2HPO_4, 1.76 mM KH_2PO_4, 2.7 mM KCl, and 0.14 M NaCl, pH 7.4).

4. 12-well tissue culture plates.

5. Brightfield microscope equipped with a 10× objective.

6. Trypsin-EDTA solution: A standard cell culture grade trypsin-EDTA (Ethylenediaminetetraacetic acid) solution for passaging cells.

7. Endothelial cell growth medium: Medium 199 supplemented with L-glutamine (4 mM/L), heparin (90 µg/mL), 10 % (v/v) fetal calf serum (FCS) supplemented with bovine brain extract.

3   Methods

3.1   Matrigel Assay

1. Culture HUVECs to 80–90 % confluence in endothelial cell growth medium at 37 °C with 5 % CO_2.

2. Thaw the Matrigel on ice at 4 °C overnight taking care not to let it warm to room temperature (see **Note 1**).

3. Using a 12-well plate, wet the wells with PBS before adding 70 µL of Matrigel to each well.

4. Leave the Matrigel to solidify at 37 °C for 30 min.

5. Take a confluent dish of ECs, aspirate off the medium, and wash with PBS (10 mL, 3×).

6. Add 1 mL of trypsin and incubate for 5 min at 37 °C until the cells have lifted off the plate.

7. Dilute cells into 5 mL of medium and centrifuge at 195×*g* for 5 min.

8. Remove the supernatant and suspend the pellet in 10 mL of culture media.

9. Count a 10 µL aliquot of cells to calculate the cell density using a hemocytometer.

10. Seed 1.4×10^5 cells on top of the Matrigel layer in 1 mL of culture medium.

11. Incubate the cells at 37 °C in 5 % CO_2.

12. Images should be captured using a brightfield microscope at regular time intervals during this 24 h period (*see* **Notes 2** and **3**).

3.2 Image Analysis   Although the method described generates images containing tubular networks of high integrity, analyzing these networks is less straightforward. Figure 1 highlights this problem. Formulating qualitative descriptors is very simple when comparing these images; the control clearly exhibits greater tube network integrity than RhoJ-knockdown ECs. In order to quantify the networks, manual labeling is often employed. This can include counting the number of nodes, evaluating the number of branch points (3-, 4-, or 5-way), and counting loops [8]. It is generally good practice to use as many variables as possible as it has been shown that some factors

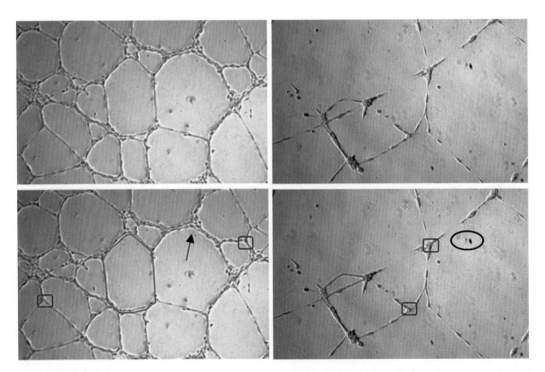

Fig. 1 Tube formation assay of negative control cells 6 h post-seeding (**a**) and RhoJ knockdown cells 24 h post-seeding (**b**). *Blue boxes* indicate examples of nodes (**c, d**) and the *green rings* identify loops formed by the cells. *Arrow* (**c**) points to an ambiguous region that could be a node or undifferentiated cells. *Black circle* (**d**) highlights undifferentiated endothelial cells

will affect these variables differently [8]. Manual methods are highly subjective, for example, in Fig. 1c the arrow points to a region which could be interpreted as a branch point or ignored as an area of undifferentiated cells. Furthermore, manual analysis is time consuming and not scalable. By using software to automatically analyze the networks, the efficiency and throughput of the assay can be increased. A number of platforms exist, such as Angiosys (*see* **Note 4**) and Wimasis (*see* **Note 5**), which can be used to achieve this to varying extents. A systematic comparison of these programs revealed that Wimasis is more refined and accurate than Angiosys, but both incur significant cost and offer limited user control over key features [11]. For detailed description of alternative angiogenesis analysis programs, *see* **Notes 4–6**.

As an example of a typical workflow which can be used to automatically and rapidly quantify the networks, a program which is freely available as an ImageJ plug-in is described in the remainder of this chapter. The main benefits of this program over that of Angiosys and Wimasis is that it is open-source software and ImageJ is a widely used and freely available image-processing platform. It operates in a different manner to Angiogenesis Analyzer (*see* **Note 6**) and achieves similar results. Staining is not necessary as the segmentation is compatible with images of low contrast and a non-uniform background. To reduce the amount of specialized imaging equipment needed, all computation was performed in post-processing. The plug-in is available for download from the Bicknell Group website.

1. Differentiate between background and regions of interest on the image using the software. This segmentation must account for nonuniform background, which is a common artifact, in order to be a robust tool. Our program has an initial smoothing step (using a Gaussian with a user-defined size which should represent feature size) to account for this (Fig. 2a, b), obtained through flat-field correction by obtaining the difference between the original and blurred images.

2. Perform edge detection to identify the morphology of the network. This is an intensity-based thresholding algorithm where the gradient of the intensity in the horizontal and vertical directions is used to identify the location of tubules. In order to connect these pixels in the binary image, the image is dilated, thus completing the segmentation. It is at this stage that the number of loops and the area of the loops are determined. Skeletonization provides a simplified representation of the network by reducing tubules to one pixel thickness, making all nodes connect at a single pixel allowing for easy determination of the location of nodes and termini by analyzing pixel neighborhoods. This morphological analysis can be extended using existing ImageJ plug-ins, an excellent example of which can be found by Landini et al. [12].

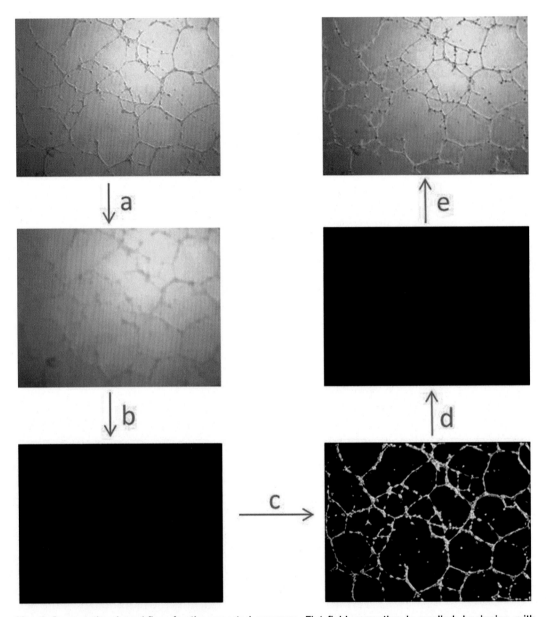

Fig. 2 Computational workflow for the reported program. Flat-field correction is applied, beginning with Gaussian blurring (**a**). Two images blurred by different-sized Gaussians are subtracted giving the feature-detected image (**b**). Edge detection then gives a binary image giving the tubules (**c**). Dilation followed by skeletonization gives tubules that are one-pixel thick, simplifying network quantification (**d**). Parameters, such as nodes, are located and quantified; the results are generated and plotted onto the original image (**e**)

3. The image should be cleared of undifferentiated cells that are not part of networks by size-based thresholding so that they are not counted as nodes or termini.

4. The result is then calculated giving the number of loops, average area of the loops, number of and connectivity of the nodes,

number of termini, and the number of pixels for each connected component of the skeleton.

3.3 Interpretation of Results

The angiogenesis software described allows for rapid analysis of large data sets. However, careful interpretation of this data and knowledge of its limitations are required to generate reliable and meaningful results. We are interested in genes which may have an anti-angiogenic effect once knocked down; these include RhoJ and β-PIX. RhoJ is a GTPase involved in cytoskeletal rearrangement. Reducing the levels of these proteins in cells will affect how they migrate and therefore give networks that are fragmented and poorly formed, relative to a negative control [6]. In each case, the cells apoptose over 24 h, leading to fewer tubules and network fragmentation. This is apparent from the reduced number of nodes observed over 24 h in both the control and knockdown experiments. Using the number of nodes, Kruskal-Wallis tests confirm that RhoJ and β-PIX knockdown ECs do give networks of lower integrity after 24 h; however, this is not true at the early stages of the experiment.

The number of loops formed by tubules offers the best indication of network fragmentation, and also gave the highest significance - *see* Fig. 3. The number of loops between control cells β-PIX and RhoJ knockdown cells shows a dramatic difference at early stages of the assay. This indicates that the cells do not migrate and form networks of high integrity when RhoJ or β-PIX expression is knocked down. Determining the number of loops is a much simpler computational problem, with less selectivity issues than the nodes, meaning that this parameter is likely to be the most effective and reliable to measure.

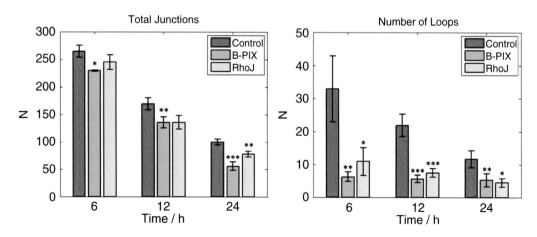

Fig. 3 Quantitative information highlighting network integrity is higher in controls in comparison to RhoJ and β-PIX. The number of loops gave results with the highest significance. Kruskal-Wallis (relative to control) *p*-values key—*p* = 0.1–0.05: **p = 0.05–0.01: ***p < 0.01: ***

4 Notes

1. Planning is essential to ensure images can be captured at times suitable for the researcher. It is usually best to begin experiments in the morning to leave time to take images after 12 h.

2. The selection of time points used to capture images is important as the morphology of the network is as a function of time. Typically, three intervals at 6, 12, and 24 h can easily capture the morphological changes.

3. Images should also be taken at least in triplicate. Images should be taken from the central region of the well, avoiding the well edges which can affect tube formation and image quality.

4. Angiosys is available from TCS Cellworks and is able to quantify properties such as the number of nodes, termini, tubule length, and area but cannot count loops. The segmentation method used by Angiosys is intensity thresholding, this means good contrast is required in the image leading the suppliers to suggest using a stain to enhance contrast; staining is not usually attractive as it may interfere with the network formation. It is therefore preferential to segment the tubules before loading the images into Angiosys; however, a significant amount of preprocessing is required to do this. The software has been demonstrated to be effective in extracting the aforementioned parameters, but it is slow (100 images is reported to take 48–72 h) and expensive; these factors must be considered when deciding on an analytical tool.

5. Wimasis is a pay-per-image online platform where images are uploaded to the Wimasis website. Results are generated and can then be downloaded by the user. An important feature of this tool is that the company offers specialized services to researchers, meaning Wimasis can be contracted to create bespoke analytical tools for a range of image analysis problems. The standard angiogenesis software offers more parameters and requires less user input than Angiosys making this option more streamlined. Wimasis is able to automatically segment the networks and provide accurate numbers of a range of parameters including nodes, loops, and nets. It successfully segments networks even with poor contrast and is able to ignore undifferentiated cells. The main drawback of Wimasis is cost of processing the images which would be prohibitive when handling large datasets on a regular basis. Another issue with Wimasis is that the method by which the tools achieve the quantification is not published. The method utilized is important when verifying results and their significance and therefore represents a major drawback with using this tool.

6. Angiogenesis Analyzer is another program designed to quantify features with the Matrigel tube formation assay, which is written in Fiji [13]. This program is freely available to download and can operate on phase contrast and fluorescence images. Firstly, the image is skeletonized (the workflow for how this is achieved is not reported) and the resulting network is treated as a tree consisting of branches. These branches are quantified by the pixel connectivity for each branch, leading to the generation of the number of nodes with connectivity and number of loops and termini. Artifacts from the skeletonization process are ignored by a size-based thresholding of smaller branches, which operates well, resulting in only major branches being counted. The sensitivity used in the thresholding and many other values are also open to modification by the user to increase effectiveness across a range of images. Angiogenesis Analyzer's ability to quantify the Matrigel assay with freely available plug-ins makes it the strongest tool of the three aforementioned programs.

References

1. Adams RH, Alitalo K (2007) Molecular regulation of angiogenesis and lymphangiogenesis. Nat Rev Mol Cell Biol 8:464–478

2. Heath VL, Bicknell R (2009) Anticancer strategies involving the vasculature. Nat Rev Clin Oncol 6:395–404

3. Auerbach R, Lewis R, Shinners B et al (2003) Angiogenesis assays: a critical overview. Clin Chem 49:32–40

4. Bikfalvi A, Cramer EM, Tenza D et al (1991) Phenotypic modulations of human umbilical vein endothelial cells and human dermal fibroblasts using two angiogenic assays. Biol Cell 72:275–278

5. Connolly JO, Simpson N, Hewlett L et al (2002) Rac regulates endothelial morphogenesis and capillary assembly. Mol Cell Biol 13:2474–2485

6. Kaur S (2011) Characterisation of the expression profile and endothelial function of Rho GTPase RhoJ. PhD Thesis. University of Birmingham

7. Donovan D, Brown NJ, Bishop ET et al (2001) Comparison of three in vitro human "angiogenesis" assays with capillaries formed in vivo. Angiogenesis 4:113–121

8. Staton CA, Reed MWR, Brown NJ (2009) A critical analysis of current in vitro and in vivo angiogenesis assays. Int J Exp Pathol 90:195–221

9. Nguyen DHT, Stapleton SC, Yang MT et al (2013) Biomimetic model to reconstitute angiogenic sprouting morphogenesis in vitro. Proc Natl Acad Sci U S A 110:6712–6717

10. Kang H, Bayless KJ, Kaunas R (2008) Fluid shear stress modulates endothelial cell invasion into three-dimensional collagen matrices. Am J Physiol Heart Circ Physiol 3120:2087–2097

11. Khoo CP, Micklem K, Watt SM (2011) A comparison of methods for quantifying angiogenesis in the Matrigel assay in vitro. Tissue Eng Part C Methods 17:895–906

12. Landini G (2008) Advanced shape analysis with ImageJ. In: Second ImageJ user developer conference proceedings. Luxembourg. p 116–121

13. Carpentier G, Martinelli M, Courty J et al (2012) Angiogenesis analyzer for ImageJ. In: Fourth ImageJ user developer conference proceedings. Mondorf-les-Bains, Luxembourg, p 198–201

Chapter 10

In Vitro Coculture Assays of Angiogenesis

Mark Richards and Harry Mellor

Abstract

During angiogenesis, endothelial cells invade into the stromal matrix: a complex, structured array of extracellular matrix proteins. This three-dimensional deformable substrate also contains a mixture of angiogenic factors as well as embedded stromal cells. Interactions between endothelial cells and the stromal tissue make complex and important contributions to the process of angiogenesis; however, the composition of the stromal matrix is hard to replicate in vitro. The coculture angiogenesis assay is a long-term assay that uses fibroblasts to secrete and condition a stromal matrix that more closely mimics tissue than a simple collagen gel. Like all in vitro assays of angiogenesis, it has both strengths and weaknesses. Here we give protocols for the two of the most useful applications of the assay: screening for regulators of angiogenesis and high-resolution imaging.

Key words Endothelial cells, Angiogenesis, Coculture, Imaging, Fibroblasts, VEGF

1 Introduction

The coculture assay of angiogenesis was first described by Bishop et al. [1], who developed the assay to screen for both inhibitors and activators of angiogenesis. In the original assay, primary human umbilical endothelial cells (HUVEC) are mixed with normal human dermal fibroblasts (NHDF) in equal ratio and plated on fibronectin-coated dishes. Over 14 days, the HUVEC form a network of branching tubules that resemble capillaries in vivo [2]. The primary role of the fibroblasts in the assay is to produce a three-dimensional matrix that mimics the stromal matrix. Analysis of matrix from the assay shows that it is rich in collagen type I and also contains fibronectin, tenascin-C, decorin, and versican [1, 3]. The mature assay is 3–5 cells deep, with endothelial tubes growing between layers of fibroblasts and matrix. This is deep enough for tube formation, but shallow enough to facilitate imaging. Importantly, the fibroblasts remodel the secreted matrix, producing fibrillar collagen. This matrix is a much closer approximation

Stewart G. Martin and Peter W. Hewett (eds.), *Angiogenesis Protocols*, Methods in Molecular Biology, vol. 1430, DOI 10.1007/978-1-4939-3628-1_10, © Springer Science+Business Media New York 2016

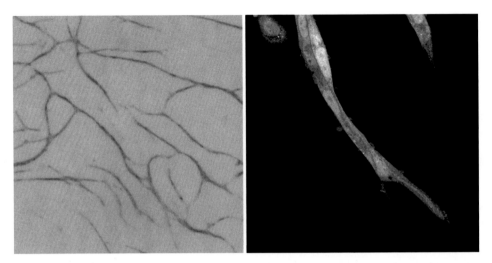

Fig. 1 Imaging of the coculture assay. The *left panel* shows the assay stained with NBT and imaged at low magnification for quantification. The *right panel* shows the coculture with ECs pre-labeled with CellTracker Green and imaged live at high magnification

of the in vivo state than simple gels of collagen or Matrigel. On maturation, the endothelial cells secrete a basement lamina that is rich in laminin and collagen IV [1, 3]. Studies using electron microscopy show that the endothelial cells form a patent lumen bounded by cell-cell junctions, although, in the absence of flow, this is mainly closed [1].

Mavria et al. made an important modification to the original assay by plating the endothelial cells directly onto a confluent layer of fibroblasts. This shortens the time spent by the endothelial cells in the assay significantly, allowing the use of oligonucleotide-based siRNA silencing [4]. This greatly facilitates screening for novel regulators of angiogenesis, with a high-content image-based readout (Fig. 1, [5–7]). The shorter assay does not affect the overall outcome, but rather eliminates the early stages of the assay, where the two populations of cells segregate and proliferate. The methods below give our current protocol for the basic assay, together with modifications for quantitative screening and high-resolution imaging.

2   Materials

2.1   Cell Culture

1. Human fibronectin solution: Add 10 μg/mL fibronectin to Dulbecco's phosphate-buffered saline (PBS) without calcium and magnesium.

2. Fibronectin-coated flasks: Add sufficient human fibronectin solution to cover the entire surface of the flask and incubate at

37 °C for 30 min. Aspirate the fibronectin and allow the flasks to dry in a sterile laminar flow hood.

3. Prepare HUVEC from human umbilical cords using standard methods [8], or purchase from a commercial source. Maintain HUVEC in endothelial cell growth medium 2, with the recommended supplements (EGM-2 medium, Lonza) passaging them as required when confluent onto fibronectin-coated flasks. HUVEC may be used up to passage 5.

4. Normal human dermal fibroblasts (NHDF) can be obtained from various commercial suppliers. NHDF are used up to passage 12.

5. NHDF medium: Dulbecco's modified Eagle medium (DMEM) containing 10 % fetal calf serum (FCS), 100 U/mL penicillin, 100 µg/mL streptomycin, and 292 µg/mL L-glutamine.

6. HEK-293T cells: HEK-293T can be obtained from ATCC and maintained in DMEM containing 10 % FCS, 100 U/mL penicillin, 100 µg/mL streptomycin, and 292 µg/mL L-glutamine.

7. Ham F12/DMEM + 20 % FCS: Ham's F12 nutrient mixture/DMEM containing 20 % FCS.

8. Trypsin/ethylenediaminetetraacetic acid (EDTA): standard 0.05 % trypsin/EDTA solution for passaging cells available from many commercial sources.

2.2 siRNA-Mediated Gene Silencing

1. GeneFECTOR (Venn Nova, Inc.).

2. siRNA oligonucleotide duplexes: Dilute siRNA duplexes to a concentration of 20 µM in RNAse-free water.

2.3 Cell Staining

1. Anti-human PECAM-1 (CD31) primary antibody (e.g., R&D Systems, #BBA7).

2. Alkaline phosphatase-conjugated secondary antibody.

3. BCIP/NBT substrate solution: Add 1 tablet of SIGMAFAST BCIP/NBT (5-bromo-4-chloro-3-indolyl phosphate/nitro blue tetrazolium; Sigma) to 10 mL ultrapure water and 0.2 µm filter immediately prior to use.

4. 4 % paraformaldehyde (PFA): Add 4 g PFA to 100 mL of PBS and heat in a fume hood while stirring until dissolved. Prepare fresh and cool to room temperature prior to use.

5. 0.2 % Triton X-100: Add 0.2 % (v/v) Triton X-100 to PBS and mix.

6. 0.3 % hydrogen peroxide/methanol: Add 0.3 % (v/v) hydrogen peroxide to methanol. Prepare immediately prior to use.

7. 0.5 % Sodium borohydride solution: Dissolve 0.5 % (w/v) sodium borohydride in PBS. Make up fresh prior to use.

8. 1 % bovine serum albumin (BSA): Dissolve 1 % (w/v) BSA in PBS.

9. DAPI (4′,6-diamidino-2-phenylindole) nuclear counterstain: Prepare a stock solution of DAPI (1 mg/mL) in ultrapure water protect from light and store at –20 °C. Dilute to 1 μg/mL in PBS or water prior to use.

2.4 Lentiviral Transduction

1. Lentiviral vector and packing vectors. We use pLVX-Puro (Clontech) as it has a useful multicloning site, puromycin selection and expresses well in endothelial cells (ECs). For packing vectors, we use pMGD2 and p8.91, both of which can be obtained through Addgene.

2. Polyethylenimine (PEI) stock solution. Add 10 mL of PEI solution (Sigma, ~408,727) (weigh for accuracy) to 10 mol sterile ultrapure water and vortex. To neutralize the pH, add 12 M HCl, 1 mL at a time, and vortex after each addition using indicator paper to check the pH. It will take approximately 12 mL of HCl in total to reach pH 7.0. Make up the solution to a final volume of 41.2 mL with ultrapure water and store in aliquots at –70 °C.

3. Opti-MEM medium (Life Technologies).

4. Puromycin stock solution: Dissolve 100 μg in 1 mL of ultrapure water.

3 Methods

3.1 The Streamlined Coculture Assay

1. *Day 1*: Harvest NHDF using trypsin/EDTA solution. Once detached from the flask, dilute in DMEM + 10 % FCS.

2. Count the cells using a hemocytometer and adjust the cell density to 3×10^4 cells/mL with EGM-2 medium.

3. Seed the cells into either 12-well tissue culture dishes or onto glass coverslips, as appropriate. There is no need to coat the surface.

4. *Day 4*: Replace the medium with fresh EGM-2 medium.

5. *Day 5*: The NHDF should now be confluent. Harvest the HUVEC by trypsinization and dilute in Ham's F12/DMEM + 20 % FCS.

6. Collect the HUVEC by gentle centrifugation at $700 \times g$ for 5 min. Resuspend them in EGM-2 medium at a density of 3×10^4 cells/mL.

7. Seed the HUVEC onto the confluent fibroblasts at of 3×10^4 cells per well on the 12-well plate.

8. Refresh the medium with EGM-2 on *day 7* and *day 9*.

9. The assay is complete on *day 11* (*see* **Note 1**).

3.2 siRNA Gene Silencing in the Coculture Assay

1. *Day 1*: Prepare and seed the fibroblasts as above (*see* Subheading 3.1, **items 1–5**).

2. *Day 4*: Seed HUVEC at 6×10^4 cells/mL onto fibronectin-coated 6-well plates and incubate overnight.

3. *Day 5*: Transfect the HUVEC with siRNA oligonucleotide duplexes using GeneFECTOR (*see* **Note 2**). For transfection in a 6-well plate, dilute 3 µL of siRNA oligonucleotide stock with 97 µL Opti-MEM in a microfuge tube. Dilute 6 µL of GeneFECTOR with 94 µL Opti-MEM in a separate tube. Combine the two solutions and incubate for 5 min at room temperature.

4. Wash the HUVEC twice in Opti-MEM and then add 1 mL Opti-MEM.

5. Add the siRNA/lipid mix dropwise while swirling the plate gently.

6. Incubate for 3 h at 37 °C.

7. Harvest the HUVEC by trypsinization and seed the cells onto the NHDF monolayers as described above (*see* Subheading 3.1, **items 5–7**).

3.3 Staining ECs in Cocultures for the Quantification of Angiogenic Activity

1. Aspirate the medium and fix the coculture using 70 % ethanol at −20 °C for 30 min.

2. Incubate with 0.3 % hydrogen peroxide/methanol for 15 min to remove endogenous alkaline phosphatase activity.

3. Wash three times with PBS and then incubate with anti-CD31 antibody (0.25 µg/mL) in 1 % BSA for 1 h at 37 °C.

4. Wash three times with PBS and then incubate with 0.6 µg/mL alkaline phosphatase-conjugated secondary antibody in 1 % BSA for 1 h at 37 °C.

5. Wash six times in water and then add BCIP/NBT substrate solution.

6. Allow the stain to develop for 15–30 min at 37 °C and then wash the cells four times with water (prior to air-drying).

7. Image at low magnification without phase contrast (*see* **Notes 1, 3–5**).

3.4 Preparation of Cocultures for Immunofluorescence Microscopy

1. Wash the cocultures three times in PBS and fix in 4 % PFA in PBS for 15 min at room temperature.

2. Wash with PBS and permeabilize in 0.2 % Triton X-100 for 5 min.

3. Wash again and incubate with fresh 0.5 % sodium borohydride in PBS to reduce autofluorescence.

4. Wash three times in PBS and then incubate with an appropriate dilution(s) of primary antibodies for 1 h in 1 % BSA.

5. Wash three times in PBS and then incubate in fluorescent secondary antibody for 1 h in PBS.

6. Wash three times in PBS. A short incubation (~5 min) with DAPI can be used to counterstain the nuclei if required. Mount coverslips and image (*see* **Note 6**).

3.5 Lentiviral Transduction with GFP Constructs for Live Cell Imaging

1. Subclone your GFP (or similar fluorescent protein)-tagged gene of interest into the lentiviral vector using standard methods (*see* **Note 7**).

2. *Day 1*: Plate 12×10^6 HEK-293T cells in a T-150 tissue culture flask. The cells should be 80–90 % confluent the next day.

3. *Day 2*: Mix 40 µg lentiviral vector, 10 µg pMDG2, and 30 µg p8.91 in 5 mL of Opti-MEM. Add 1 µL of PEI stock to 5 mL Opti-MEM. Mix the two solutions together and leave for 20 min.

4. Wash the HEK-293T cells with PBS and then add the transfection mixture.

5. Incubate for 4 h in a tissue culture incubator.

6. Replace the transfection mixture with HEK-293T growth medium.

7. *Day 4*: Remove the medium from the cells and filter through a 0.4 µm syringe filter to remove any cells. This virus-containing medium can be aliquoted and frozen at –70 °C or used directly to infect cells.

8. Infect ECs with the lentivirus by replacing the EGM-2 growth medium with the equivalent volume of virus-containing HEK-293T medium and leave for 24 h.

9. Replace the medium with fresh EGM-2 growth medium. Or repeat the infection process (*see* Subheading 3.4, **item 5**) by adding fresh virus-containing HEK-293T medium to increase transduction if required.

10. If required, the transduced ECs can be selected by the addition of puromycin to the growth medium. In our experience, good selection lentiviral infected HUVEC is achieved with 2–4 µg/mL puromycin. Cell death should be visible within 24 h and full selection achieved within 4 days.

11. GFP expressing ECs are then plated into the coculture assay (*see* Subheading 3.1) and imaged by standard methods.

3.6 Fluorescent Cell Dyes

As an alternative to the infection of ECs with lentiviruses expressing fluorescent proteins, ECs can be labeled using fluorescent dyes (e.g., CellTracker Green CMFDA; *see* Fig. 1). ECs are incubated with 10 µM CellTracker Green CMFDA in EGM-2 growth medium for 45 min at 37 °C before replacing with fresh EGM-2 medium for 30 min at 37 °C. Other CellTracker dyes are available to allow co-imaging with GFP. ECs are then trypsinized and plated into the coculture assay (*see* Subheading 3.1, **items 5–7**).

4 Notes

1. Endothelial cells (ECs) in the coculture assay should form a network of capillary-like structures (Fig. 1). If the culture contains islands of cells, it is usually a sign that the fibroblasts were too old or not plated at a high enough density to ensure a confluent monolayer before plating the ECs.

2. Using siRNAs to transiently silence gene expression can be very effective when optimized. However, it is important to bear in mind that due to the length of assay, the gene knockdown in ECs will often not be maintained throughout the assay. This is not necessarily important when measuring effects on vessel growth, as even if the expression of the knockeddown gene is reestablished, the initial inhibition of EC outgrowth should be reflected in the end result. When measuring properties such as cell shape which may change/recover later in the assay, the use of stable gene silencing using lentiviral shRNAs may be preferable [5]. This can be achieved by following the protocol given for lentiviral transduction (*see* Subheading 3.2).

3. The stained coculture assay can be quantified by measuring the length of vessels per unit area. This is more sensible than attempting to quantify the length of individual vessels, as deciding what an individual vessel is becomes impossible in very dense cultures. A hemocytometer grid can be used to calibrate the area of the field of view. We use Adobe Photoshop and ImageJ [9] to trace and record these lengths. A typical density would be 3–6 mm/mm^2.

4. The assay also allows for the quantification of other parameters. The most common additional measurement is of the number of branches formed per unit length. It is important to remember that branches occur both by splitting of an individual vessel and by connection of two proximal vessels (anastomosis). As cultures become denser, the chances of two vessels meeting and forming a branch increases, and so it is important to remember that changes in the average length will have a causal, nonlinear relationship with branch number.

5. Other parameters that can be quantified include vessel thickness and the number, length, and position of filopodia.

6. The culture is deep enough that antibody staining may be weaker for ECs further from the surface. If you see a mixture of strongly stained and weakly stained vessels, you should increase the antibody incubation times to ensure equal labeling throughout the culture.

7. Lentiviral particles are highly infectious and local safety and containment rules will apply to their use. You should have a

thorough understanding of these procedures and obtain the relevant local permission and access to suitable laboratory facilities before commencing work.

Acknowledgment

This work was supported by project grants PG/10/014/28224 and PG/11/68/29074 from the British Heart Foundation.

References

1. Bishop ET, Bell GT, Bloor S et al (1999) An in vitro model of angiogenesis: basic features. Angiogenesis 3:335–344

2. Donovan D, Brown NJ, Bishop ET et al (2001) Comparison of three in vitro human 'angiogenesis' assays with capillaries formed in vivo. Angiogenesis 4:113–121

3. Sorrell JM, Baber MA, Caplan AI (2007) A self-assembled fibroblast-endothelial cell co-culture system that supports in vitro vasculogenesis by both human umbilical vein endothelial cells and human dermal microvascular endothelial cells. Cells Tissues Organs 186:157–168

4. Mavria G, Vercoulen Y, Yeo M et al (2006) ERK-MAPK signaling opposes Rho-kinase to promote endothelial cell survival and sprouting during angiogenesis. Cancer Cell 9:33–44

5. Hetheridge C, Mavria G, Mellor H (2012) Uses of the in vitro endothelial-fibroblast organotypic co-culture assay in angiogenesis research. Biochem Soc Trans 39:1597–1600

6. Hetheridge C, Scott AN, Swain RK et al (2012) The formin FMNL3 is a cytoskeletal regulator of angiogenesis. J Cell Sci 125:1420–1428

7. Jones MC, Caswell PT, Moran-Jones K et al (2009) VEGFR1 (Flt1) regulates Rab4 recycling to control fibronectin polymerization and endothelial vessel branching. Traffic 10:754–766

8. Van Hinsbergh VWM, Draijer R (1996) Culture and characterization of human endothelial cells. In: Shaw AJ (ed) Epithelial cell culture: a practical approach. Oxford University Press, Oxford, pp 87–110

9. Rasband WS (1997–2012) ImageJ. U S National Institutes of Health, Bethesda, MD. http://imagej.nih.gov/ij/

Chapter 11

Spheroid-Based In Vitro Angiogenesis Model

Larissa Pfisterer and Thomas Korff

Abstract

In vitro models mimicking capillary sprouting are important tools to investigate the tumor angiogenesis, developmental blood vessel formation, and pathophysiological remodeling processes of the capillary system in the adult. With this focus, in 1998 Korff et al. introduced endothelial cell (EC) spheroids as a three-dimensional in vitro model resembling angiogenic responses and sprouting behavior [1]. As such, EC spheroids are capable of giving rise to capillary-like sprouts which are relatively close to the physiologically and genetically programmed arrangement of endothelial cells in vessels. Co-culture spheroids consisting of endothelial cells and smooth muscle cells form a spheroidal core composed of smooth muscle cells and an outer monolayer of endothelial cells, similar to the physiological architecture of larger blood vessels. In practise, a defined number of endothelial cells are cultured in a round-bottom well plate or in "hanging drops" to allow the formation and arrangement of the spheroidal three-dimensional structure. Subsequently, they are harvested and embedded in a collagen gel to allow outgrowth of endothelial cell sprouts originating from each spheroid. To evaluate the pro- or antiangiogenic impact of a cytokine or compound, the number and length of sprouts is determined.

Key words Spheroids, Sprouting, Endothelial cells, Angiogenic stimuli, Angiogenesis, Vascular

1 Introduction

Endothelial cell (EC) heterogeneity is a common and known obstacle that researchers have to deal with when working with primary ECs. As with all primary cells, they alter their phenotype during cell culture, thereby losing their native quiescent and differentiated state [2]. Differentiated ECs express a typical set of markers characterizing their organ specificity and specialized functionality which is often altered upon cell culture. For instance, brain ECs show diminished tight-junction formation [3] and downregulation of CD34 and CD31 expression [4–7] in vitro. To allow functional and mechanistic in vitro studies using primary ECs, it is crucial to maintain their differentiated and native phenotype. Here, the EC spheroid culture model offers the possibility to analyze the organotypic differentiation and function in a defined experimental setup (Fig. 1).

Stewart G. Martin and Peter W. Hewett (eds.), *Angiogenesis Protocols*, Methods in Molecular Biology, vol. 1430, DOI 10.1007/978-1-4939-3628-1_11, © Springer Science+Business Media New York 2016

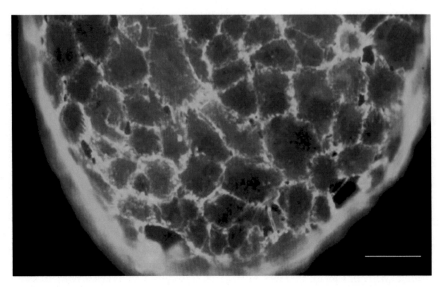

Fig. 1 The CD31-specific fluorescence (*green*) confirms the endothelial identity of the cobblestone-like cells located on the surface of the SMC core of SMC/EC co-culture spheroids (scale bar: 200 μm)

Korff et al. [1] characterized not only the spheroid formation process but also potential problems in this system. Interestingly, EC spheroids spontaneously organize in suspension culture and form a superficial monolayer surrounding the spheroidal core. ECs located within the core are not differentiated and undergo apoptosis. The surface monolayer, however, consists of differentiated ECs expressing a typical and characteristic subset of antigens, such as CD31, resembling the physiological arrangement of EC in the vascular system, and can be induced to re-express CD34 (Fig. 2).

As such, the spheroid culture model lends itself to a plethora of experimental approaches: (a) The analysis of the angiogenic potential and responses of vascular ECs, mimicking capillary sprouting upon transfer to a collagen gel [8]. (b) Based on this experimental setup, EC tip and stalk cell formation may be evaluated [9–11]. (c) The transfer of human EC spheroids into a complex organism as has been developed by Laib et al. [12]. Implantation of human EC spheroids into immunocompromised mice leads to fusion of the spheroid-based human capillary network with the host vasculature which is perfused and fully functional within 20 days. Using this approach, it is possible to evaluate sprouting of genetically manipulated or stimulated human ECs in vivo. (d) In addition to capillary network formation based solely on ECs, it is also possible to study interactions of EC and vascular smooth muscle cells (SMC) in co-culture spheroids [13]. Upon generating EC/SMC-co-culture spheroids composed of equal numbers of these cells, the ECs form a monolayer at the surface of the spheroid and the SMCs assemble in a multilayer core surrounded by ECs following their genetically programmed behavior and architecture of an inverted vessel wall. Under these conditions, ECs and SMCs maintain their differentiated

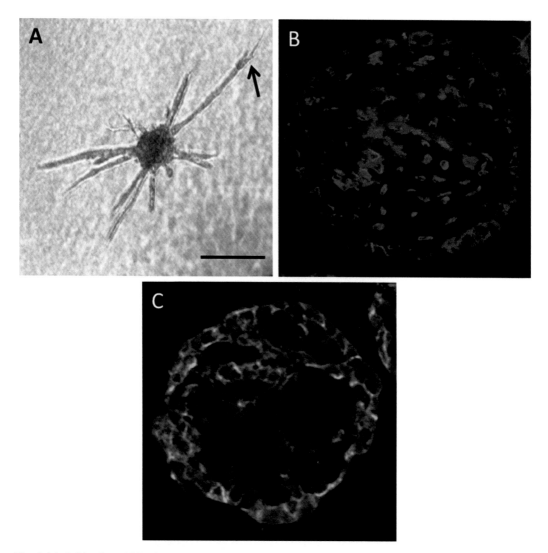

Fig. 2 (**a**), Cultivation of EC spheroids in collagen type-1 gels allows the analysis and documentation of tip cell formation (**a**, *arrow*, scale bar: 200 μm). (**b** and **c**) Cross sections of EC/SMC-co-culture spheroids reveals that that CD31-positive ECs are partially distributed in the spheroid center (**b**, *red fluorescence*) or below the surface (**c**, *green fluorescence*; *Red staining*: Evans Blue) during their formation

phenotype and do not proliferate and the ECs form more inter-endothelial junctions than on the surface of EC spheroids. The presence of SMCs appears to stabilize the quiescent phenotype of the EC monolayer. This is suggested by the stimulation of EC/SMC-spheroids with vascular endothelial growth factor (VEGF), a potent proangiogenic cytokine. While VEGF evokes sprouting of EC spheroids, it fails to do so in EC/SMC spheroids unless angiopoietin-2 (Ang-2) has been administered simultaneously. Ang-2 is known to have a destabilizing effect on the EC-SMC contacts thereby stimulating the VEGF responsiveness of ECs [14].

Aside from VEGF [15], fibroblast growth factor (FGF)-2 [16, 17] and platelet derived growth factor (PDGF) [18, 19] are also potent inducers of embryonic and adult angiogenesis. Stimulation of hemangioblastic cell lines with FGF-2 also stimulates the expression and abundance of flk1 (a.k.a. KDR, VEGF-R2) a high-affinity VEGF receptor. Consequently, simultaneous stimulation with VEGF and FGF-2 evokes potentiated proangiogenic effects. In line with this, VEGF treatment alone stimulates EC-derived tip-cell-rich sprout and tube formation, but does not lead to recruitment of mural cells or support the establishment of perfused microvessels. FGF-2 on the other hand facilitates endothelial sprouting, and stimulates the acquisition of mural cells and leads to lumen formation and finally to perfused microvessels [20].

Another way to analyze sprouting of EC spheroids is based on the use of Matrigel® as has been shown by Laib et al. [12] and Alajati et al. [20]. Matrigel® is a heterogeneous mixture of matrix proteins localized in the basal membrane such as collagen type IV, laminin, and heparan sulfate and is produced by the Engelbreth Holm Swarm (EHS) mouse sarcoma cell line. Matrigel® has a resilient structure, and is resistant to digestion by endogenous enzymes—an advantage for transplantation into the mouse. In contrast, the induction of EC spheroid sprouting usually takes longer and Matrigel® naturally contains a cocktail of cytokines which may affect pro- and antiangiogenic responses, or when transplanted can provoke immune responses. As an alternative, Matrigel® can be mixed with fibrin to optimize the hydrogel for outgrowth of capillary-like structures in vivo [21–23].

Furthermore, EC spheroid sprouting is also supported by pure fibrin or collagen type-I matrices. The method described in this chapter utilizes hydrogels based on the polymerization of collagen type-I extracted from rat tails.

2   Materials

All solutions should be prepared with sterile ultrapure water to ensure the absence of confounding factors, such as lipopolysaccharides (LPS). Store all reagents at 4 °C, unless indicated otherwise.

1. Prepare ECs from a suitable tissue, e.g., HUVEC from human umbilical cords using standard methods, or purchase from a commercial source (e.g., PromoCell, Lonza).

2. EC growth medium: An appropriate growth medium for the type of endothelial cells used. Endothelial cell basal medium supplemented with the recommended fetal bovine serum (FBS) and growth medium supplements (e.g., Promocell).

3. 50 mL conical tubes.

4. Non-adhesive 96-well round-bottom well plates.

5. Hank's balanced salt solution without calcium and magnesium (HBSS$^{-/-}$).

6. 1× Trypsin/EDTA solution.

7. 10× M199 medium.

8. Heat-inactivated FBS.

9. 0.2 % NaOH solution: Add an appropriate amount of NaOH to ultrapure water and filter (0.22 μm pore size) sterilize.

10. 24-well TC plates.

11. Cell counting chamber, e.g., Neubauer improved hemocytometer.

12. 4 % (w/v) Paraformaldehyde.

2.1 Methylcellulose Solution

1. Autoclave 6 g of methylcellulose in a 500 mL bottle containing a magnetic stirrer and a 250 mL bottle containing a magnetic stirrer.

2. Add 250 mL of basal EC growth medium without supplements to the autoclaved 250 mL bottle and heat to 60 °C while stirring (*see* **Note 1**).

3. Transfer the warm medium to the autoclaved bottle containing methylcellulose and stir on a magnetic stirrer for 20 min at 60 °C (*see* **Note 2**).

4. Add the remaining 250 ml of EC growth medium. It is recommended to avoid adding any supplements or FBS to the methylcellulose stock solution. In case any stimuli need to be administered, these should be resuspended at the doubled concentration in EC growth medium and added at this step to avoid overheating. Stir the viscous solution overnigt at 4 C (*see* **Note 3**).

5. Aliquot the methylcellulose solution into sterile 50 mL conical tubes and centrifuge for 2 h at $5000 \times g$ at room temperature (*see* **Note 4**).

2.2 Collagen Type-1 Solution

1. Prepare the following reagents and supplies: 70 % ethanol, 0.1 % acetic acid (resuspend acetic acid in ultrapure water and filter sterilize using a 0.22 μm pore size), an autoclaved 250 mL bottle, sterile cup or dish with a lid, a sterile scalpel, sterile forceps (1 course and 2 fine), sterile 50 mL centrifugation tubes. Rat tails can be stored at −20 °C prior to use.

2. Thaw 2–3 rat tails in 70 % (v/v) ethanol at room temperature for 20 min.

3. Working on a clean bench, straighten and twist the tails to loosen and soften the tissue and squeeze out any blood which may be left in the blood vessels.

4. Using a scalpel, cut the skin between the collagen fibers along the tail and pull the skin towards the tip either by using fingers or the serrated forceps. Any remaining blood vessels at the caudal vertebra and the connective tissue between ligaments should be removed carefully using fine forceps.

5. Cut the tendons between the most cranial vertebra and discard the vertebra. To isolate the tendons, break every second vertebra by pulling back and forth starting at the most caudal tip of the tail. Disconnect the loosened part from the rest and carefully extract the tendons. Cut the tendons at the vertebra (*see* **Note 5**) and collect the collagen fibers in the sterile dish (*see* **Note 6**). A total amount of 1.25 g of collagen fibres should be covered with 70 % ethanol and incubated for 30 minutes at room temperature.. Spread the tendons on a sterile surface (*see* **Note 7**) and allow to dry for 30 min in a laminar air flow hood.

6. Separate the individual collagen tendons using the fine forceps (*see* **Note 8**). Add 250 mL of sterile 0.1 % acetic acid and incubate for 48 h at 4 °C, inverting the mixture every 12 h (do not shake). To remove the insoluble matrix and tissue centrifuge for 1 h at $14,000 \times g$ at 4 °C. Carefully transfer the supernatant to sterile 50 mL conical bottom tubes without disturbing the sediment (*see* **Note 9**). The final collagen solution will have an approximate concentration of 2 mg/mL.

3   Methods

Carry out all steps at room temperature and under sterile conditions unless indicated otherwise.

3.1   Generation of EC Spheroids

1. Culture ECs (*see* Fig. 3) in a T75 flask to approximately 80 % confluence.

2. Remove the culture medium and wash the ECs with HBSS$^{-/-}$ to remove FBS.

3. Incubate the cells with approximately 2 mL of 1× Trypsin/EDTA at 37 °C, 5 % CO_2 for 1–2 min. Monitor the detachment of the ECs under a phase contrast microscope and knock the flask to aid detachment.

4. Once the cells have detached add 10 mL of EC growth medium immediately to inhibit the enzymatic activity. Break up any clumps of cells by pipetting and centrifuge for 5 min at $800 \times g$.

5. Resuspend the cell pellet in 10 mL EC growth medium. Add 500 ECs to each well of a round-bottom 96-well plate (*see* **Note 10**).

6. We routinely prepare four 96-well plates to ensure an adequate number of spheroids for experiments (*see* **Note 11**).

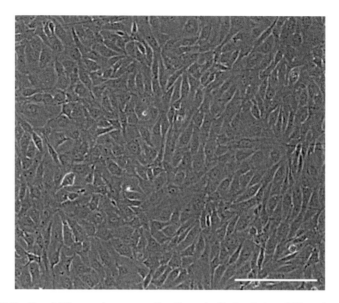

Fig. 3 Confluent EC monolayer on cell culture plastic (scale bar: 200 μm)

7. Resuspend 200,000 cells/mL in 32 mL of EC growth medium and then add 8 mL of methylcellulose solution (*see* Subheading 2.1) in a 50 mL tube (*see* **Note 12**) taking care to avoid introducing air bubbles and add 100 μL per well using a multichannel pipette (*see* Fig. 4) and incubate for 24 h at 37 °C and 5 % CO_2.

3.2 EC Spheroid-Based 3D Angiogenesis Assay

1. After controlling the spheroidal shape of the cell aggregates (*see* Fig. 5 and **Note 13**), harvest all spheroids and collect them in a 50 mL conical bottom tube (*see* **Note 14**). The spheroids are visible as small "cell balls" floating in the medium.

2. Centrifuge for 5 min at $1000 \times g$ and aspirate the supernatant with a vacuum pump taking care not to remove the spheroids at the bottom of the tube (*see* **Note 15**).

3. Prepare 20 % (v/v) FBS-methylcellulose (*see* **Notes 16** and **17**) and resuspend spheroids in 4.5 ml of the FBS-Methylcellulose (*see* **Note 16**) solution (*see* **Note 17**).

4. Prepare the collagen gels by mixing 4.5 mL of collagen solution (*see* Subheading 2.2) and 0.5 mL of 10× M199 medium and mix gently ensuring the equal distribution by monitoring the yellow-orange color of the pH indicator in the M199 medium.

5. To initiate the polymerization of the collagen, carefully titrate it with 0.2 % sodium hydroxide until a neutral pH is reached (around 250–400 μL in total, *see* **Note 18**). To ensure an optimal pH, add 0.2 % sodium hydroxide slowly mixing thoroughly

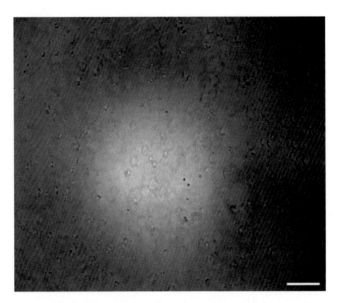

Fig. 4 ECs suspended in methylcellulose solution and distributed in a non-adhesive round-bottom-shaped 96-well plate (scale bar: 200 μm)

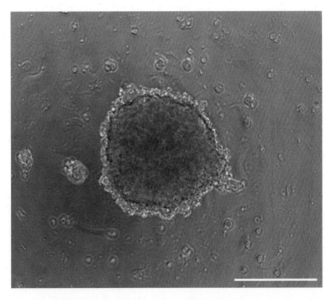

Fig. 5 EC spheroid after cultivating for 24 h in FBS-methylcellulose. The cell aggregate forms a compact structure with a well-defined edge (scale bar: 200 μm)

each time and place the solution on ice until the pH indicator (in M199) turns a reddish color.

6. Add 4.5 mL of the neutralized collagen-M199 mix to the spheroids resuspended in 4.5 mL FBS-methylcellulose (*see* **Note 19**). Mix immediately to achieve an even distribution of the spheroids. Avoid introducing air bubbles and ensure that mixing takes no longer than 20–30 s.

7. Transfer the spheroid suspension to a 24-well plate immediately using a 10 mL pipette. Add 1 mL per well to the eight center wells. Approximately 0.9 mL per well will be transferred to the wells due to the viscosity of the mixture. Fill the empty wells with 0.5 mL of HBSS$^{-/-}$ to avoid evaporation.

8. Place the plate in the incubator immediately and incubate for 30 min at 37 °C and 5 % CO_2 to allow polymerization. Add 100 μL mL of EC growth medium containing the stimuli of interest at 10× concentration on the top of the gel and incubate for another 24–48 h at 37 °C and 5 % CO_2. Sprout formation can be monitored under phase contrast microscopy.

9. *Optional step*: Prior to analysis fix the spheroids in 4 % paraformaldehyde. After fixation, spheroid sprouting can be analyzed immediately or stored at 4 °C.

3.3 Analysis of Sprout Formation

To evaluate angiogenic sprouting, analyze the sprouts originating from the spheroids by phase contrast microscopy. Parameters of interests are the number of sprouts originating and their mean and cumulative length (determined as the length from the spheroid core to the sprouting tip). Take images from 5 to 10 spheroids chosen at random (do not include those located near air bubbles, close to other spheroids, or on the bottom or at the edges of the well as these locations may influence the number, length, and orientation of spheroidal sprouts). Image J software may be utilized to determine the aforementioned parameters.

4 Notes

1. Do not increase the temperature above 60 °C; otherwise components of the medium will precipitate. Stop heating as soon as the required temperature is reached.

2. Ensure that the methylcellulose solution is mixed thoroughly.

3. The final solution has a viscous consistency which is why a continuous stirring using a magnetic stirrer is important.

4. Use a centrifuge with a rotor allowing a 90° angle to fully sediment the debris at the bottom tip and avoid its aspiration when removing the supernatant. Fragments of undissolved methylcellulose will impede the proper formation of spheroids once the cells are resuspended.

5. The extracted collagen fibers should be abscised 1 cm distance to the cranial vertebra. Both the loosened caudal vertebra and the remaining 1 cm fibers should be discarded.

6. The weight of the dish/vessel should be measured in advance and noted to allow for a precise determination of the weight of the tendons. The vessel should not be too flat and it should be possible to close it tightly.

7. Cut a small disposal bag open on two sides and unfold it on the bench to provide a clean surface.

8. To ensure an optimal homogenous collagen solution during the following steps, the tendons should be adequately separated: fix one end with both forceps and slide along the fiber with the one while holding the collagen with the other, repeat if necessary. By doing so, the dry and rigid collagen separates to thin fibers.

9. The pellet contains the remaining undissolved tissue pieces, which should not be transferred into the collecting tube. For that reason ~5 mL of the supernatant should be left behind.

10. For generation of co-culture spheroids consisting of ECs and SMCs, use 500 cells of each type.

11. The total volumes given are calculated for four 96-well plates. It is also possible to work with two 96-well plates and adapt the volume of methylcellulose and collagen accordingly. Using larger volumes offers easier handling and higher quality of gels.

12. The methylcellulose stock solution is too viscous to use a pipette to accurately measure the correct volume. For that reason, add methylcellulose until the meniscus reaches the 20 mL mark.

13. The medium should be clear and sterile and the spheroids themselves should have clear edges.

14. To avoid mechanical disruption of the spheroids during the harvesting process, cut 1–2 mm of a 1 mL pipette tip with sterilized scissors (use 70 % ethanol).

15. Sedimented spheroids do not appear as a clear cell pellet and may only be visible after suspension. Moreover the spheroids may easily be aspirated.

16. Due to the issue described in **Note 11**, add 20 % FBS first followed by methylcellulose solution when preparing the mixture prior to use.

17. When FBS is in the tip of the tube, it is easier to resuspend the mix equally, instead of first adding methylcellulose.

18. The amount of 0.2 % sodium hydroxide solution needed to neutralize the collagen solution may be determined in advance. It is crucial to place the M199/collagen solution and the 0.2 % sodium hydroxide solution on ice during the neutralization step.

19. Thorough mixing of the solutions is key critical for generating homogenous collagen gels. When mixing the solutions use a 10 mL pipette to give an even distribution of the spheroids by performing circular motions of the pipette tip while pipetting up and down.

References

1. Korff T, Augustin HG (1998) Integration of endothelial cells in multicellular spheroids prevents apoptosis and induces differentiation. J Cell Biol 143(5):1341–1352

2. Garlanda C, Dejana E (1997) Heterogeneity of endothelial cells. Specific markers. Arterioscler Thromb Vasc Biol 17(7):1193–1202

3. Wolburg H, Neuhaus J, Kniesel U et al (1994) Modulation of tight junction structure in blood-brain barrier endothelial cells. Effects of tissue culture, second messengers and cocultured astrocytes. J Cell Sci 107(Pt 5):1347–1357

4. Krause DS, Fackler MJ, Civin CI et al (1996) CD34: structure, biology, and clinical utility. Blood 87(1):1–13

5. Delia D, Lampugnani MG, Resnati M et al (1993) CD34 expression is regulated reciprocally with adhesion molecules in vascular endothelial cells in vitro. Blood 81(4):1001–1008

6. DeLisser HM, Newman PJ, Albelda SM (1994) Molecular and functional aspects of PECAM-1/CD31. Immunol Today 15(10):490–495

7. Baldwin HS, Shen HM, Yan HC et al (1994) Platelet endothelial cell adhesion molecule-1 (PECAM-1/CD31): alternatively spliced, functionally distinct isoforms expressed during mammalian cardiovascular development. Development 120(9):2539–2553

8. Korff T, Augustin HG (1999) Tensional forces in fibrillar extracellular matrices control directional capillary sprouting. J Cell Sci 112(Pt 19):3249–3258

9. Jakobsson L, Franco CA, Bentley K et al (2010) Endothelial cells dynamically compete for the tip cell position during angiogenic sprouting. Nat Cell Biol 12(10):943–953

10. Siemerink MJ, Klaassen I, Van Noorden CJ et al (2013) Endothelial tip cells in ocular angiogenesis: potential target for anti-angiogenesis therapy. J Histochem Cytochem 61(2):101–115

11. Siemerink MJ, Klaassen I, Vogels IM et al (2012) CD34 marks angiogenic tip cells in human vascular endothelial cell cultures. Angiogenesis 15(1):151–163

12. Laib AM, Bartol A, Alajati A et al (2009) Spheroid-based human endothelial cell microvessel formation in vivo. Nat Protoc 4(8):1202–1215. doi:10.1038/nprot.2009.96

13. Korff T, Kimmina S, Martiny-Baron G et al (2001) Blood vessel maturation in a 3-dimensional spheroidal coculture model: direct contact with smooth muscle cells regulates endothelial cell quiescence and abrogates VEGF responsiveness. FASEB J 15(2):447–457

14. Maisonpierre PC, Suri C, Jones PF et al (1997) Angiopoietin-2, a natural antagonist for Tie2 that disrupts in vivo angiogenesis. Science 277(5322):55–60

15. Hoeben A, Landuyt B, Highley MS et al (2004) Vascular endothelial growth factor and angiogenesis. Pharmacol Rev 56(4):549–580

16. Flamme I, Frolich T, Risau W (1997) Molecular mechanisms of vasculogenesis and embryonic angiogenesis. J Cell Physiol 173(2):206–210

17. Presta M, Dell'Era P, Mitola S et al (2005) Fibroblast growth factor/fibroblast growth factor receptor system in angiogenesis. Cytokine Growth Factor Rev 16(2):159–178

18. Battegay EJ, Rupp J, Iruela-Arispe L et al (1994) PDGF-BB modulates endothelial proliferation and angiogenesis in vitro via PDGF beta-receptors. J Cell Biol 125(4):917–928

19. Xue Y, Lim S, Yang Y et al (2012) PDGF-BB modulates hematopoiesis and tumor angiogenesis by inducing erythropoietin production in stromal cells. Nat Med 18(1):100–110

20. Alajati A, Laib AM, Weber H et al (2008) Spheroid-based engineering of a human vasculature in mice. Nat Methods 5(5):439–44521

21. Nakatsu MN, Sainson RC, Aoto JN et al (2003) Angiogenic sprouting and capillary lumen formation modeled by human umbilical vein endothelial cells (HUVEC) in fibrin gels: the role of fibroblasts and Angiopoietin-1. Microvasc Res 66(2):102–11222

22. Vernon RB, Angello JC, Iruela-Arispe ML et al (1992) Reorganization of basement membrane matrices by cellular traction promotes the formation of cellular networks in vitro. Lab Invest 66(5):536–547

23. Grant D, Cid M, Kibbey MC et al (1992) Extracellular matrix-cell interaction: matrigel and complex cellular pattern formation. Lab Invest 67(6):805–806

Chapter 12

Stem Cell Spheroid-Based Sprout Assay in Three-Dimensional Fibrin Scaffold: A Novel In Vitro Model for the Study of Angiogenesis

Fatemeh Sharifpanah and Heinrich Sauer

Abstract

Angiogenesis is a complex process of critical importance during development and in physiological and pathophysiological conditions. There is considerable research interest in studying the angiogenesis cascade and consequently a need for a physiologically valid, quantitative, and cost-effective assay. In this chapter, we describe the stem cell spheroid-based sprout assay in three-dimensional fibrin scaffold which allows fast and easy screening of pro- and anti-angiogenic effects of substances with a high degree of reproducibility.

Key words Sprout assay, Embryonic stem cells, In vitro model, Angiogenesis, Three-dimensional culture, Fibrin scaffold

1 Introduction

Angiogenesis refers to the development of new vessels from preexisting vessels [1] and is regulated by various pro- and anti-angiogenic factors. This complicated biological multi-step process is characterized by a complex cascade of events, during which quiescent endothelial cells become activated to degrade their surrounding extracellular matrix, directionally migrate toward the angiogenic stimulus, proliferate, and organize into new three-dimensional (3D) capillary networks [2–4]. Angiogenesis is a complex process which occurs during normal development and also under physiological and pathophysiological conditions. In normal physiological conditions, angiogenesis has a pivotal role in embryogenesis [5], the female reproductive cycle [6], wound healing [7], and bone formation [8]. Furthermore, angiogenesis is a key feature for pharmaceutical intervention in a number of diseases. In many pathological conditions such as tumor development [9–10], infantile hemangioma [11], rheumatoid arthritis [12, 13], various retinopathies [14], and psoriasis [15], angiogenesis is excessive,

Stewart G. Martin and Peter W. Hewett (eds.), *Angiogenesis Protocols*, Methods in Molecular Biology, vol. 1430,
DOI 10.1007/978-1-4939-3628-1_12, © Springer Science+Business Media New York 2016

and the aim of pharmaceutical intervention is to prevent this process. On the contrary, in wound healing [7] and some diseases such as coronary artery disease [16], the aim is to encourage angiogenesis. Since angiogenesis plays a pivotal role in a wide range of diseases, many research groups worldwide are focused on studying the molecular mechanisms by which the angiogenesis process can be either stimulated or inhibited.

One of the main considerations in angiogenesis research is the choice of an appropriate assay to evaluate the efficacy of new medications and to identify potential targets within the angiogenesis process. In accordance with the multi-step process of angiogenesis, it is essential to choose an assay which can mimic all these steps to investigate properties of new drugs for inhibiting or stimulating angiogenesis. There are several methods currently available both in vitro and in vivo, and while all of these methods have enhanced our understanding of angiogenesis and the underlying mechanisms, they also have limitations. Numerous in vivo models for studying angiogenesis have been already developed such as the Matrigel plug, sponge implantation, corneal pocket [17], and chick chorio-allantoic membrane (CAM) assays and the dorsal air sac model [2] with their own strengths and weaknesses [18–21]. Moreover, different in vitro models are also developed to study the process of angiogenesis including two different assay groups: (a) Endothelial cell-based assays [20, 22] such as the scratch wound, zymogen, Boyden chamber, tube formation, RAIN-Droplet assay [23], matrix invasion, and the microbead sprout assays; (b) Stem cell and organ culture-based assays such as the embryoid body (EB), mouse metatarsal, rat aortic ring, and the chick aortic arch assays [20, 21, 24]. With the exception of in vivo assays which can be more technically demanding to perform, only the endothelial cell-based microbead sprout assay plus all stem cell and organ culture-based assays provide the chance to assess all different steps of angiogenesis.

Embryonic stem (ES) cells have the capacity to self-renewal and differentiate into different types of cells [25]. For this reason, they are an interesting and suitable model for basic and clinical research in various fields such as developmental biology, cancer biology, infection biology, pharmacology, etc. Using this model gives an opportunity to investigate the biochemical process of differentiation/remodeling upon different conditions in more detail which is important for designing new drugs and also testing their toxicity effects. Mouse ES cells are maintained in an undifferentiated state using leukemia inhibitory factor (LIF) in the culture medium [26, 27]. In the absence of LIF, the ES cells are cultivated in a bioreactor spinner flask to form EBs and then differentiate into cell types of all three germ layers [27]. ES cells in form of EBs are used very commonly to investigate the angiogenesis process [28, 29]. For this propose, they are cultivated in suspension and/or on gelatin-coated tissue culture plates and fixed after distinct days to be stained against endothelial markers to assess properties of drug candidate on the angiogenesis process. Since two-dimensional culture does not cover all the steps of the angiogenesis process, it is essential to cultivate EBs in three-dimensional culture for mimicking the in vivo conditions of angiogenesis and investigating the complete angiogenesis function upon drug candidate treatment.

The special property of ES cells to differentiate toward various cell types is a major advantage of this assay [25, 27] since the formation of functional blood vessels requires the interaction of different cell types. Pericytes, smooth muscle cells, and leukocytes all play an important role during the angiogenesis/vasculogenesis process [30–33]. The close proximity of different cell types such as smooth muscle cells, leukocytes, and endothelial cells in the ES cell spheroids allows their interaction and more closely approximates the in vivo environment during the angiogenesis process. Therefore, the stem cell spheroid-based sprout assay allows the process of angiogenesis to be studied under the influence of a heterogeneous micro-milieu that mimics in vivo conditions. In this chapter, we describe the stem cell spheroid-based sprout assay in three-dimensional culture on fibrin scaffolds. This powerful technique is easily manageable, quantitative, and highly reproducible and can be used in various fields of angiogenesis research.

2 Materials

Prepare all media and solutions under sterile conditions in a class II cabinet. Store all reagents at +4 °C unless indicated otherwise. A humidified incubator at +37 °C, 5 % CO_2 is required with a stirring platform capable of 16 rotations per minute (e.g., Integra Biosciences CELLSPIN).

2.1 Cell Culture

Mouse fibroblast cells are used as a feeder layer for the ES cells and undifferentiated mouse ES cells.

1. Feeder culture medium: Supplement 500 mL of Dulbecco's Modified Eagle Medium (DMEM) with 10 % heat-inactivated (see **Note 1**) fetal calf serum (FCS), 5 mL of 100× nonessential amino acids (NEAA), 2 mM L-glutamine, 50 U/mL penicillin and 50 μg/mL streptomycin.

2. Mitomycin C solution: Dissolve 2 mg of mitomycin C powder in 4 mL of autoclaved double-distilled water and sterilize by passing through a sterile 0.22 μm filter. Aliquot and store at −20 °C (see **Note 2**). Dilute mitomycin C solution to 10 μg/mL in feeder culture medium immediately prior to use.

3. LIF culture medium: Supplement 500 mL of Iscove's Modified Dulbecco's Medium (IMDM) with 15 % heat-inactivated FCS (see **Note 1**), 100 μM β-mercaptoethanol (see **Note 3**), 5 mL NEAA, 2 mM L-glutamine, 10 mM sodium pyruvate, and 1000 U/mL ESGRO® LIF solution (see **Note 4**).

4. Differentiation culture medium: Supplement IMDM (500 mL) with 15 % heat-inactivated FCS (*see* **Note 1**), 100 µM β-mercaptoethanol (*see* **Note 3**), 5 mL NEAA, 2 mM L-glutamine, 10 mM sodium pyruvate, 50 U/mL penicillin and 50 µg/mL streptomycin.

5. 0.05% Trypsin/EDTA solution

6. Dulbecco's phosphate-buffered saline (PBS) without Ca^{2+} and Mg^{2+}. Sterilize by autoclaving for 15 min at +121 °C or passing through a sterile 0.22 µm filter.

7. PBS containing 0.9 mM calcium chloride and 0.5 mM magnesium chloride. Sterilize by autoclaving for 15 min at +121 °C or passing through a sterile 0.22 µm filter.

2.2 Sprout Assay

1. Collagenase solution: Dissolve 2 mg/mL collagenase B in prewarmed PBS (without Ca^{2+} and Mg^{2+}).

2. Methylcellulose solution: Add 250 mL pre-warmed (+37 °C) IMDM basal medium to 6 g of autoclaved methyl cellulose and place on a stirrer at +60 °C for 20 min. Then add a further 250 mL of pre-warmed IMDM basal medium and stir for 1 h at room temperature. Subsequently, stir overnight at +4 °C to further dissolve the methylcellulose (*see* **Note 5**). Dispense the solution into 50 mL culture tubes and centrifuge at $4000 \times g$ for 2 h at room temperature. Finally, transfer aliquots of the supernatant to sterile 50 mL tubes and store at +4 °C.

3. Fibrinogen: Dissolve fibrinogen in pre-warmed PBS (without Ca^{2+} and Mg^{2+}) to give a stock concentration of 5 mg/mL (*see* **Note 6**). Aliquot and store at −80 °C (*see* **Note 7**).

4. Aprotinin: Dissolve aprotinin in sterile PBS (without Ca^{2+} and Mg^{2+}) or 0.9 % saline solution (*see* **Note 6**) to give a stock solution of 10,000 KIU U/mL. Aliquot and store at +4 °C.

5. Thrombin: Dissolve thrombin (from bovine plasma) in sterile PBS (without Ca^{2+} and Mg^{2+}) to get stock of 390 NIH U/mL (*see* **Note 6**). Aliquot and store at −80 °C.

6. Sprouting culture medium: Use Opti-MEM® medium or basal growth medium without any additives, except for the substances to be studied (*see* **Note 8**).

7. Calcein AM stock solution: Dissolve 50 µg of calcein AM in 20 µL of dimethyl sulfoxide (DMSO) to give a 2.5 mM stock solution. Store at −20 °C.

2.3 Fixation and Staining

1. 4 % Paraformaldehyde (PFA): Dissolve 4 g PFA powder in 100 mL sterile 1× PBS without Ca^{2+} and Mg^{2+} in a chemical fume hood. Aliqout and freeze at −20 °C for long-term storage.

2. Immunostaining buffer: 0.01 % Triton X-100 in calcium and magnesium-free PBS at pH 7.4.

3 Methods

All the steps of this procedure should be performed in a class II biological safety cabinet with the exception of the steps under Subheading 3.5 and 3.6.

3.1 Cultivation and Maintenance of ES Cells

1. Thaw one vial of mouse fibroblasts, wash in feeder culture medium, and centrifuge at $400 \times g$ for 5 min.

2. Resuspend the cells in feeder culture medium (*see* Subheading 2.1, **item 1**) and seed them in 60 mm tissue culture plates in a total of 5 mL of medium and allow the cells to grow to subconfluence in a humidified incubator (+37 °C, 5 % CO_2).

3. Inactivate the subconfluent fibroblasts by incubating them in feeder culture medium with freshly added mitomycin C (10 μg/mL) (*see* Subheading 2.1, **item 2**) for 3 h (+37 °C, 5 % CO_2).

4. Wash the cells three times with pre-warmed feeder culture medium and maintain them in a humidified incubator (+37 °C, 5 % CO_2).

5. Thaw one vial of undifferentiated mouse ES cells, wash in LIF culture medium (*see* Subheading 2.1, **item 3**), and centrifuge at $400 \times g$ for 5 min.

6. Seed the ES cells on the inactivated fibroblast plate containing 5 mL of LIF culture medium.

7. Every day, exchange medium on the undifferentiated mouse ES cells plate with fresh pre-warmed LIF culture medium and passage them onto newly inactivated fibroblast plates when the mouse ES cell colonies on inactivated fibroblast cells reach a size of 214 ± 55 μm.

3.2 EB Preparation

1. Harvest confluent undifferentiated mouse ES cells with trypsin/EDTA solution at +37 °C for 2 min.

2. Transfer the cells into a sterile 50 mL tube containing 25 mL of pre-warmed differentiation culture medium.

3. Pellet the cells at $400 \times g$ for 5 min.

4. *Day 0*: Transfer 1×10^6 cells into the siliconized spinner flask bioreactor with 125 mL differentiation culture medium (*see* **Note 9**) and maintain in a humidified incubator (+37 °C, 5 % CO_2) on a stirring platform with speed of 16 rotations per minute.

5. *Day 1*: Add 125 mL of fresh pre-warmed differentiation culture medium to the flask to a final volume of 250 mL.

6. On each successive day, replace 125 mL with fresh differentiation culture medium until *day 4* of differentiation.

7. *Day 4*: Transfer the EBs from the spinner flask into a 100×20 mm bacteriological plate with 15 mL of differentiation culture medium and place it on a shaker in the humidified incubator (+37 °C, 5 % CO_2).

3.3 Spheroid
Preparation
with Methylcellulose

1. Wash the 4-day-old EBs twice with pre-warmed PBS without Ca^{2+} and Mg^{2+}.

2. Incubate the EBs with 2 mg/mL collagenase B solution for 5 min at +37 °C.

3. Transfer the dissociated cells into a sterile 50 mL culture tube with 25 mL pre-warmed PBS containing 0.9 mM Ca^{2+} and 0.5 mM Mg^{2+} (*see* **Note 10**).

4. Pellet the cells at $400 \times g$ for 5 min.

5. Discard the supernatant and resuspend the cell pellet in 1 mL of 20 % (v/v) methylcellulose (*see* Subheading 2.2, **item 2**) in differentiation culture medium and determine the cell number.

6. Dilute the cell suspension in a volume of 20 % methylcellulose in differentiation culture medium to give ~1000 cells per 40 μL (*see* **Note 11**).

7. Gently mix the cell suspension and dispense 40 μL of this cell suspension using a multichannel pipette into the lid of 100×20 mm bacteriological culture plates to make hanging drops.

8. Incubate the plates in the humidified incubator at +37 °C in an atmosphere of 5 % CO_2 for 24 h.

9. After 24 h incubation, collect the spheroids from each drop with wide-ended pipette and wash them once with pre-warmed $1\times$ PBS containing 0.9 mM Ca^{2+} and 0.5 mM Mg^{2+} and maintain them in the humidified incubator (+37 °C, 5 % CO_2). Spheroids should be used for the sprout assay on the day of collection (*see* **Note 10**).

3.4 Embedding
Spheroids in Fibrin
Scaffold

1. Prepare the embedding mixture by mixing 750 μL of fibrinogen stock solution (5 mg/mL) and 21 μL aprotinin stock solutions (10,000 KIU U/mL) and make up to 1 mL with $1\times$ PBS containing Ca^{2+} and Mg^{2+} (*see* **Note 10**). Final working concentration of fibrinogen and aprotinin will be 1.8 mg/mL and 200 KIU U/mL, respectively.

2. Gently mix the embedding mixture and add 250 μL of this mixture into each well of a 24-well plate.

3. Add 50 μL of spheroid suspension containing 8–10 spheroids into each well with a wide-ended pipette and mix them thoroughly with embedding mixture.

4. Dilute the thrombin stock solution 1:20 in PBS containing Ca^{2+} and Mg^{2+} and mix well (*see* **Note 10**).

5. Add 10 μL of diluted thrombin to each well and immediately mix to ensure that the spheroids are evenly distributed in the well. Final working concentration of thrombin will be 0.65 NIH U/mL.

6. Incubate the plates in the humidified incubator (+37 °C, 5 % CO_2) for 30 min to complete polymerization and formation of the fibrin scaffold.

7. After complete polymerization, add 400 μL sprouting culture medium (*see* Subheading 2.2, **item 6**) to equilibrate the scaffold for 1 h at +37 °C with 5 % CO_2.

8. Remove the sprouting culture medium and add 600 μL fresh pre-warmed sprouting culture medium containing growth factors, stimulants, substances, etc. as required. Incubate the plates for another 24–48 h in the humidified incubator (+37 °C, 5 % CO_2) to allow the formation of sprouts in the fibrin scaffold (*see* Fig. 1).

9. The sprouts can then be quantified without cell staining using a phase contrast microscope with an ocular micrometer and/or a digital imaging system in combination with an appropriate software analysis tool. The test substances may affect the number of sprouts and/or the average sprout length.

3.5 Quantification of Sprouts in the Fibrin Scaffold

1. After 24–48 h incubation in fibrin scaffold, vessels sprout from spheroids into the three-dimensional fibrin scaffold (*see* Fig. 1). Take transmission images from the sprouts of at least ten randomly selected spheroids per test condition using a phase contrast microscope (10× objective) equipped with an ocular micrometer.

2. The images can be analyzed using a suitable program such as MetaMorph® image analysis software. This allows an area of interest to be drawn around the outgrowths, and the sprouts

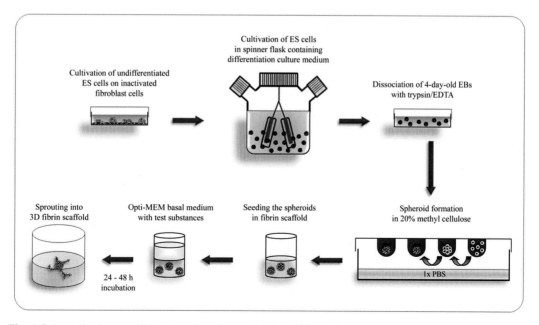

Fig. 1 Schematic diagram of the stepwise stem cell spheroid-based sprout assay in three-dimensional fibrin scaffold

are then highlighted by the software to allow the selection of light sprout areas against a dark background.

3. The number of sprouts and the average and cumulative sprout length for each treatment are then compared to the untreated control.

4. Sprouts may also be stained with a live cell marker such as calcein AM (*see* **Note 12**). Remove the medium with test substances very carefully and add calcein AM working solution 8 µg/mL (*see* Subheading 2.2, **item 7**) in Opti-MEM® medium and incubate for 30 min in the humidified incubator (+37 °C, 5 % CO_2). After 30 min incubation, remove the calcein AM working solution and wash the wells twice very carefully with pre-warmed Opti-MEM® medium (*see* **Note 8**). The plate is now ready for image acquisition using a confocal laser scanning microscope or a conventional fluorescence microscope. The excitation and emission wavelengths of calcein AM are 495 and 516 nm, respectively. We recommend recording images from the sprouts of at least ten randomly selected spheroids per each test condition. Process and analyze the acquired images using either the MetaMorph® image analysis software or equivalent software (*see* Fig. 2).

3.6 Fixation and Staining of the Endothelial Cells

The sprouts can be preserved for later analysis.

1. Remove the medium carefully and add 1 mL of +4 °C cold 4 % PFA to each well and incubate them at +4 °C for 30 min.

2. Remove the 4 % PFA, wash once with calcium and magnesium-free PBS, and store them in PBS at +4 °C for up to 3–4 days (*see* **Note 13**).

3. For staining, block the fixed sprouts with 10 % heat-inactivated FCS in immunostaining buffer (*see* Subheading 2.3, **item 2**) for 1 h at room temperature.

4. Carefully wash twice (*see* **Note 13**) with immunostaining buffer. Stain the sprout samples with primary antibodies against endothelial cell markers, e.g., platelet endothelial cell adhesion molecule (PECAM-1) and/or vascular endothelial cadherin (VE-cadherin) at +4 °C overnight.

5. Wash three times with immunostaining buffer for 10 min. Incubate with suitable fluorescence-conjugated secondary antibodies diluted in immunostaining buffer for 1 h in the dark at room temperature.

6. Wash three times with immunostaining buffer, 10 min each time. Store the plate with calcium and magnesium-free PBS in the dark at +4 °C for no more than 2 days prior to imaging.

7. Image the sprouts using a confocal laser scanning microscope or a conventional inverted fluorescence microscope.

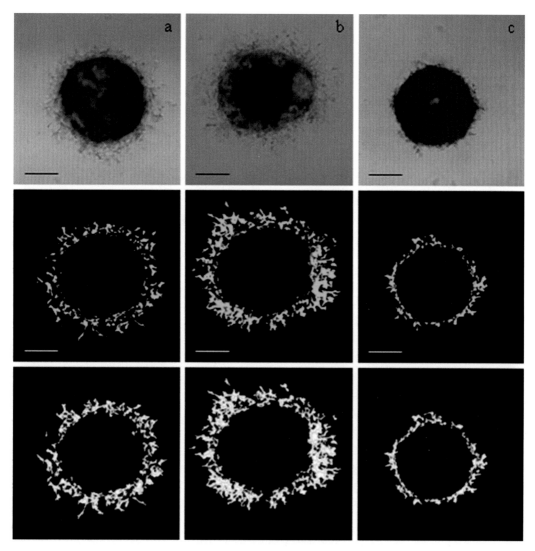

Fig. 2 Representative images of spheroid sprouts in the three-dimensional fibrin scaffold after 24 h of cultivation. The *upper panel* shows transmission images of the sprouts upon three different conditions: (**a**) unstimulated control, (**b**) stimulated with the pro-angiogenic substance VEGF (10 ng/mL), (**c**) incubated with the anti-angiogenic VEGFR2 inhibitor, SU5614 (1 µM). The *middle panel* shows fluorescent images of the sprouts after calcein AM staining. The *lower panel* represents the analyzed fluorescent images using MetaMorph® image analysis software. The bars on the images represent 200 µm

4   Notes

1. Gently mix FCS after thawing at +4 °C overnight. For inactivation of complement in FCS, incubate for 30 min at +56 °C in a water bath. Afterwards, FCS is immediately cooled on ice.

2. The mitomycin C solution is only stable for 2 weeks at +4 °C and should be aliquoted and stored at −20 °C for the long

term. The final working concentration is 10 μg/mL of mitomycin C in feeder culture medium and should be freshly prepared before use.

3. β-mercaptoethanol should only be handled in a chemical fume hood. Always wear gloves. Prepare a 10 mM stock solution in PBS without Ca^{2+} and Mg^{2+} and sterilize by passing through a sterile 0.22 μm filter. This can be stored at +4 °C for a maximum of 1 week.

4. The ESGRO® LIF stock solution should be thawed overnight at +4 °C. Afterwards, gently mix, aliquot, and store at –20 °C.

5. Sometimes the methylcellulose does not dissolve completely. Therefore, it should be centrifuged, and only clear supernatant should be stored at +4 °C for further use.

6. We find that the use of PBS (without Ca^{2+} and Mg^{2+}) gives the best results for this step, but the fibrinogen solution can also be made using 0.9 % saline solution. To prepare 0.9 % saline solution, 0.9 g NaCl should be dissolved in 100 mL double-distilled water and sterilized by autoclaving for 15 min at +121 °C or passing through a sterile 0.22 μm filter.

7. Do not store the dissolved fibrinogen at +4 °C because it will precipitate. The fibrinogen solution should always be stored at –80 °C.

8. We have observed the best results using Opti-MEM® medium without phenol red indicator.

9. Using the siliconized spinner flask bioreactor for EB preparation as a high-throughput technique is a fast and cost-effective method to prepare large numbers of EBs of uniform size as described in detail earlier by Wartenberg et al. [27].

10. It is absolutely essential to use 1× PBS containing 0.9 mM Ca^{2+} and 0.5 mM Mg^{2+} during this step of the protocol.

11. Using 1000 cells per drop for spheroid preparation is of critical importance for the technique as it gives the best results in the sprouting assay. Increasing the number of cells per drop leads to greater sprout density which complicates analysis.

12. The calcein AM working solution should be freshly prepared before use. It is critical that the final working concentration of calcein AM in Opti-MEM® medium is not greater than 8 μg/mL.

13. All washes, fixation, and staining steps are performed in the well and require careful pipetting and aspiration. If the samples are not treated gently, the sprouts may detach and be lost from the fibrin scaffold.

References

1. Folkman J, Klagsbrun M (1978) Angiogenic factors. Science 235:442–447
2. Auerbach R, Lewis R, Shinners BL et al (2003) Angiogenesis assay: a critical review. Clin Chem 49:32–40
3. Blacher S, Erpicum C, Lenoir B et al (2014) Cell invasion in the spheroid sprouting assay: a spatial organisation analysis adaptable to cell behaviour. Plos One 9:e97019
4. Sleeman JP (2010) Understanding the mechanisms of lymphangiogenesis: a hope for cancer therapy? Phlebolymphology 17:99–107
5. Breier G (2000) Angiogenesis in embryonic development: a review. Placenta 21:S11–S15
6. Reynolds LP, Grazul-Bilska AT, Redmer DA (2002) Angiogenesis in the female reproductive organs: pathological implications. Int J Exp Pathol 83:151–163
7. Sinno H, Prakash S (2013) Complements and the wound healing cascade: an updated review. Plast Surg Int. doi:10.1155/2013/146764
8. Portal-Nunez S, Lozano D, Esbrit P (2012) Role of angiogenesis on bone formation. Histol Histopathol 27:559–566
9. Auerbach R, Akhtar N, Lewis RL et al (2000) Angiogenesis assays: problems and pitfalls. Cancer Metastasis Rev 19:167–172
10. Weis AM, Cheresh DA (2011) Tumor angiogenesis: molecular pathways and therapeutic targets. Nat Med 17:1359–1370
11. Sharifpanah F, Saliu F, Bekhite M et al (2014).β-Adrenergic receptor antagonists inhibit vasculogenesis of embryonic stem cells by downregulation of nitric oxide generation and interference with VEGF signaling. Cell Tissue Res 358:443–452
12. Maruotti N, Annese T, Cantatore FP et al (2013) Macrophages and angiogenesis in rheumatic diseases. Vasc Cell 5:11–18
13. Maruotti N, Cantatore FP, Crivellato E et al (2006) Angiogenesis in rheumatoid arthiritis. Histol Histopathol 21:557–566
14. Crawford TN, Alfaro DV, Kerrison JB et al (2009) Diabetic retinopathy and angiogenesis. Curr Diabetes Rev 5:8–13
15. Guerard S, Pouliot R (2012) The role of angiogenesis in the pathogenesis of psoriasis: mechanisms and clinical implications. J Clin Exp Dermatol Res S2:007. doi:10.4172/2155-9554.S2-007
16. Kaminsky SM, Rosengart TK, Rosenberg J et al (2013) Gene therapy to stimulate angiogenesis to treat diffuse coronary artery disease. Hum Gene Ther 24:948–963
17. Norrby K (2006) In vivo models of angiogenesis. J Cell Mol Med 10:588–612
18. Nakatsu MN, Hughes CC (2008) An optimized three-dimensional in vitro model for the analysis of angiogenesis. Methods Enzymol 443:65–82
19. Nehls V, Drenckhahn D (1995) A novel microcarrier-based in vitro assay for rapid and reliable quantification of three-dimensional cell migration and angiogenesis. Microvasc Res 50:311–322
20. Staton CA, Reed MWR, Brown NJ (2009) A critical analysis of current in vitro and in vivo angiogenesis assays. Int J Exp Pathol 90:195–221
21. Staton CA, Stribbling SM, Tazzyman S et al (2004) Current methods for assaying angiogenesis in vitro and in vivo. Int J Exp Pathol 85:233–248
22. Koh W, Straman AN, Sacharidou A et al (2008) In vitro three-dimensional collagen matrix models of endothelial lumen formation during vasculogenesis and angiogenesis. Methods Enzymol 443:83–101
23. Zeitlin BD, Dong Z, Nör JE (2012) RAIN-Droplet: a novel 3D in vitro angiogenesis model. Lab Invest 92:988–998
24. Goodwin AM (2007) In vitro assays of angiogenesis for assessment of angiogenic and anti-angiogenic agents. Microvasc Res 74:172–183
25. Wobus AM (2001) Potential of embryonic stem cells. Mol Aspects Med 22:149–164
26. Graf U, Casanova EA, Cinelli P (2011) The role of the leukemia inhibitory factor-pathway in derivation and maintenance of murine pluripotent stem cells. Genes 2:280–297
27. Wartenberg M, Günther J, Hscheler J et al (1998) The embryoid body as a novel in vitro assay system for anti-angiogenic agents. Lab Invest 78:1301–1314
28. Li J, Stuhlmann H (2011) In vitro imaging of angiogenesis using embryonic stem cell-derived endothelial cells. Stem Cells Dev 21:331–342
29. Shinkaruk S, Bayle M, Lain G et al (2003) Vascular endothelial cell growth factor (VEGF), an emerging target for cancer chemotherapy. Curr Med Chem Anticancer Agents 3:95–117
30. Dalton HJ, Armaiz-Pena GN, Gonzalez-Villasana V et al (2014) Monocytes subpopulations in angiogenesis. Cancer Res 74:1287–1293
31. Lingen MW (2001) Role of leukocytes and endothelial cells in the development of angiogenesis in inflammation and wound healing. Arch Pathol Lab Med 125:67–71
32. Walter J, Sane DC (2000) The role of smooth muscle cells and pericytes in angiogenesis. In: Mousa SA (ed) Angiogenesis inhibitors and stimulators: potential therapeutic implications. Landes Bioscience, Georgetown, Texas, pp 25–30. ISBN 13: 978-158706022933
33. Sharifpanah F, De Silva S, Bekhite M et al (2015) Stimulation of vasculogenesis and leukopoiesis of embryonic stem cells by extracellular transfer RNA and ribosomal RNA. Free Radic Biol Med 89:1203–1217

Chapter 13

Human Arterial Ring Angiogenesis Assay

Giorgio Seano and Luca Primo

Abstract

In this chapter we describe a model of human angiogenesis where artery explants from umbilical cords are embedded in gel matrices and subsequently produce capillary-like structures. The human arterial ring (hAR) assay is an innovative system that enables three-dimensional (3D) and live studies of human angiogenesis. This ex vivo model has the advantage of recapitulating several steps of angiogenesis, including endothelial sprouting, migration, and differentiation into capillaries. Furthermore, it can be exploited for (1) identification of new genes regulating sprouting angiogenesis, (2) screening for pro- or anti-angiogenic drugs, (3) identification of biomarkers to monitor the efficacy of anti-angiogenic regimens, and (4) dynamic analysis of tumor microenvironmental effects on vessel formation.

Key words Human explants, Sprouting angiogenesis, 3D cell culture, Umbilical artery, Drug discovery, Lentiviral transduction, Tumor microenvironment, VEGF

1 Introduction

The intrinsic complexity of vessel sprouting limits the efficacy of cellular in vitro assays in elucidating molecular, cellular, and pharmacologic mechanisms [1]. A functional solution to bridge the gap between cultured endothelial cells (EC) and in vivo animal models is the use of ex vivo assays, in which explants of vascular tissues are embedded in extracellular matrix gels and produce new vascular sprouts that differentiate in capillary-like structures [2–6]. In this chapter we describe a reproducible system for investigating the angiogenic cascade by controlling experimental variables and precisely quantifying angiogenic outgrowth. In the human angiogenesis model human umbilical artery rings are embedded in basement membrane extract (BME), producing capillary-like structures, which recapitulate different steps of angiogenesis, including EC sprouting, migration, and differentiation into capillaries. In contrast to similar ex vivo assays [2–4, 7], the vessel outgrowth of the human arterial ring (hAR) assay is completely dependent on the addition of angiogenic factors in the culture medium [8]. The hAR assay offers the possibility for dynamic monitoring of the

Stewart G. Martin and Peter W. Hewett (eds.), *Angiogenesis Protocols*, Methods in Molecular Biology, vol. 1430, DOI 10.1007/978-1-4939-3628-1_13, © Springer Science+Business Media New York 2016

angiogenic process [8]. The system can be potentially used for the development of a human model for the screening of anti-angiogenic drugs, preventing failures caused by a lack of interspecies cross-reactivity [9]. This assay is relatively cheap, using biological material that would otherwise be discarded, and it requires fewer resources than in vivo assays in mice or other small animal models. However, this method has the advantage over many in vitro cell systems that it takes into account that blood vessels are a 3D network composed by endothelial cells and associated stromal cells [8]. Moreover, some modifications of the original assay open other interesting fields of research otherwise inaccessible to ex vivo models. Human arterial rings cultured with tumor spheroids give rise to massively branched structures, which mimic tortuous, abnormal tumor vessels that are more resistant to anti-angiogenic drugs [8]. Finally, the ease of genetic modification of hAR avoids the difficulties that scientists have encountered when knocking out genes that are required during the embryonic or adult life of the animal [9]. Here, we describe the preparation of explants from umbilical cords and their culture, along with all the modifications we have performed and validated and the analyses we implemented to dynamically investigate human sprouting angiogenesis in a 3D context (*see* Fig. 1a and *see* ref [8]).

2   Materials

2.1   Materials and Equipment

1. Umbilical cords obtained from healthy deliveries (*see* **Note 1**). These can be stored in sterile plastic bags at 4 °C (*see* **Note 2**).

2. A piece of polystyrene approximately 15 × 15 cm covered with aluminum foil and cleaned with 70 % ethanol.

3. Scalpel (#23 for arteries and #24 for umbilical cords), 0.2 mm forceps (extra or super fine points), dissecting scissors.

4. Stereomicroscope.

5. Sterile 48-well, 60-mm, and 30-mm culture dishes.

6. Sterile 200 μL and 1 mL pipette tips.

7. Basement membrane extract (BME), e.g., Matrigel (BD Biosciences). Aliquot and store at −20 °C. A maximum of 200 μL per hAR is required.

8. Dulbecco's phosphate-buffered saline (PBS), pH 7.4.

9. Endothelial Growth Medium (EGM)-2 (Clonetics) containing all the recommended supplements.

10. Endothelial Basal Medium (EBM)-2: endothelial cell basal medium (EBM)-2 containing 5 % fetal bovine serum (FBS), GA-1000 (gentamicin, amphotericin B) and heparin with the addition of user-specified growth factors.

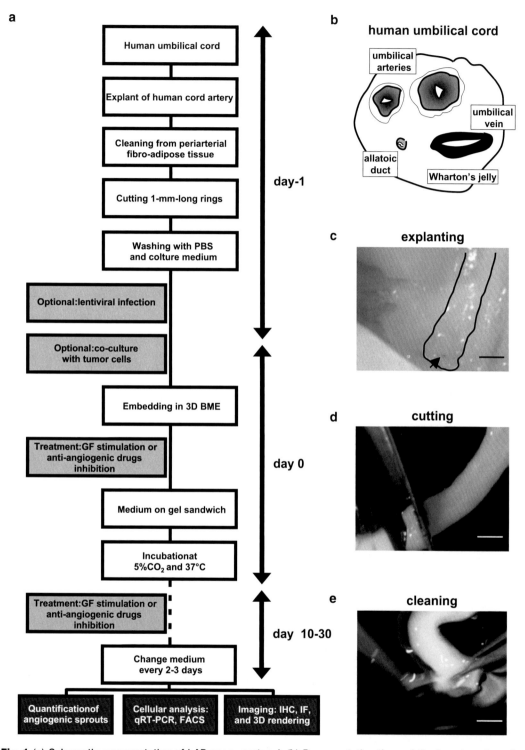

Fig. 1 (**a**) Schematic representation of hAR assay protocol. (**b**) Representation the umbilical cord anatomy in transverse section. (**c**) Stereomicroscope images of a cut margin of a human umbilical cord; lumen of one of the two arteries, *arrow*. The lumen of the artery is marked with an arrow and the edges of the artery delimitated by *black line*. (**d**) Explanted segment of artery from human umbilical cord. (**e**) Removal of peri-arterial fibro-adipose tissue and transverse cutting of explanted artery. Scale bar = 2 mm (for movies showing the dissection of the artery from the umbilical cord and preparation of the arterial rings, see supplementary movies of reference [8]). This research was originally published in Blood [8] © the American Society of Hematology

2.2 Stable Lentiviral-Mediated hAR Transduction	1. HEK-293T, SV40-transformed human embryonic kidney (ATCC® CRL-11268), and HeLa (ATCC® CCL-2) cell lines were obtained from the American Type Culture Collection (ATCC) and maintained as frozen stocks.

2. Growth medium: Dulbecco's Modified Eagle Medium (DMEM) supplemented with 10 % FBS, 2 mM L-glutamine, and antibiotics.

3. Polybrene stock solution: Dissolve polybrene in PBS at a concentration of 8 μg/μL 1 day before use and filter sterilize.

4. Lentiviral vectors carrying short hairpin RNA (shRNA) sequences against specific genes or scrambled sequences (used as control) from the RNAi Consortium library or similar resource.

5. Packaging (pCMVdR8.74) and envelope (pMD2.G-VSVG) DNA plasmids.

6. Ultracentrifuge for viral concentration.

2.3 Whole-Mount Live or Immunofluorescence Imaging

1. Glass-bottom dishes (e.g., WillCo wells).

2. Zinc fixative: 0.1 M Tris–HCl (pH 7.4), 3 mM calcium acetate, 27 mM zinc acetate, and 37 mM zinc chloride.

3. Cell permeabilizing solution: 0.2 % (v/v) Triton-X100 in PBS.

4. Antibody diluent: PBS containing 2.5 % (v/v) normal serum from the species in which the primary antibody was raised.

5. DAPI nuclear counterstain stock solution: Dissolve 1 mg/mL 4′,6-diamidin-2-fenilindolo (DAPI) in PBS.

6. Inverted fluorescence or confocal/multiphoton laser-scanning microscope.

2.4 Human Tumor Angiogenesis Assay

1. LNCaP prostate adenocarcinoma cell line obtained from ATCC (ATCC® CRL-1440) and maintained as frozen stocks.

2. RPMI-1640 medium: Roswell Park Memorial Institute (RPMI)-1640 supplemented with 10 % FBS, 2 mM L-glutamine and antibiotics.

3. Non-adherent, round-bottomed, 96-well plates.

4. Methylcellulose medium: DMEM containing 1 % (w/v) methylcellulose plus 10 % FBS.

2.5 Imaging/Quantification of Angiogenic Dynamics

1. Inverted bright-field microscope equipped with CCD camera.

2. Image Pro Plus (Media Cybernetics) and WinRhizo Pro (Regent Instruments Inc.) or ImageJ (NIH).

2.6 Sprouted Cell Extraction

1. BD Cell Recovery Solution (BD Biosciences).

2. 15 mL tubes.

3. Centrifuge.

2.7 Cytometric Analysis and Fluorescence-Activated Cell Sorting

1. Bovine Serum Albumin (BSA) solution; 1 % (w/v) BSA in PBS.

2. Cell fixative: 2 % (w/v) paraformaldehyde (PFA) in PBS, pH 7.4; heat on a stirrer at ~60 °C. This solution should be prepared when you are ready to fix or stored at 4 °C for up to 1 month. Formaldehyde is toxic and solutions should be prepared with care in a fume hood.

3. Flow cytometer and/or cell sorter.

2.8 RNA Extraction

1. TRIzol (Invitrogen) BD Cell Recovery Solution (Becton Dickinson).

3 Methods

An overview of the steps for the preparation of umbilical arteries, the explant culture, and the hAR assay is shown in Fig. 1a [8]. All procedures should be carried out working under sterile conditions in a class II biological cabinet.

3.1 Preparation of Human Arterial Rings

1. Obtain umbilical cords following normal obstetric procedures (*see* **Notes 1** and **2**).

2. Working under sterile conditions, remove clamps and cut away the end of cords directly exposed to air.

3. Wash twice with PBS by removing blood coagula.

4. Using dissecting scissors, cut the umbilical cords in 3–4 cm-long pieces and fix with needles on the foil covered piece of polystyrene, stretching the border of the cord to allow you to see both arteries and the vein (*see* Fig. 1b).

5. Using a scalpel, accurately cut the umbilical cord around one of the two arteries and progressively stretch the border of the cord using needles (*see* **Note 3**).

6. When it is possible to take one extremity of the artery by using micro-forceps, utilize the scalpel to dissect the artery from the fibrous tissue around it (*see* Fig. 1c).

7. When you dissect 1–2 cm of artery from the cord, cut it and place it in PBS (*see* **Note 4**) and proceed with the preparation of artery from other segments of cord.

3.2 Dissection and Removal of Fibro-adipose Tissue

1. Place the artery in a new, clean 60-mm dish and add 200–300 μL of PBS to avoid drying out (*see* **Note 5**).

2. Using a stereomicroscope and two micro-forceps, clean the fibro-adipose tissue from around the arteries. Hold one end of the artery with one set of forceps and clamp the external layer

using the other forceps by pulling away from it (Fig. 1d, *see* **Notes 4** and **6**).

3. Add PBS and proceed with the other arteries (*see* **Note 7**).

3.3 Cutting Arterial Rings

1. Place the artery in a new 30-mm plate without PBS.

2. Cut the arteries transversely into 1–2 mm-long arterial rings (ARs) using the #23 scalpel (Fig. 1e) (*see* **Notes 4** and **8**).

3. Add sterile EBM-2 and place in a humidified incubator.

3.4 Culture Conditions, Experimental Controls, and Sample Size

In our hands and with our experimental settings, the minimum sample size to be able to detect a difference of 50 % in sprouting length is 6 hAR per experimental point (confidence level of 0.05 and statistical power of 0.80). Ideal positive controls are hAR cultured in fresh EGM-2 where a rich cocktail of angiogenic growth factors is present. Ideal negative controls are hAR cultured in fresh EBM-2 plus 5 % FBS and heparin. The negative control hAR remain viable in culture for long periods, but no sprouting is ever detected.

3.5 Preparation of BME Gels and Embedding of hAR

1. Thaw BME on ice or overnight at 4 °C. At no time should BME be warmed to room temperature during handling.

2. Dilute BME gel with EBM-2 to obtain the final concentration of 8 mg/mL.

3. Cool 200 µL and 1 mL pipette tips in a –20 °C freezer.

4. Working under sterile conditions and on ice, pipette BME using pre-cooled tips and place (*see* **Note 9**).

5. Add 100 µL of BME to coat the bottom of 48-well culture dishes, on ice and using pre-cooled pipettes, and allow to gel in a humidified incubator to avoid dehydration (*see* **Note 10**).

6. Place hAR on 48-well culture dishes pre-coated with BME and seal them in position with an overlay of 100 µL of BME. Allow to gel in a humidified incubator to avoid dehydration (*see* Subheading 4, *see* **Note 11**).

7. Cover hAR cultures with 500 µL of EBM-2 plus 5 % FBS and heparin with addition of specific growth factors or EGM-2, with or without chemical inhibitors (Fig. 2c and *see* **Note 12**).

8. Change medium every 2–3 days (*see* **Note 13**).

9. Carefully remove medium by pipetting, do not use vacuum-driven aspirator.

10. Cover hAR cultures with fresh medium (*see* **Note 14**).

3.6 Variations on the Standard Technique

The use of this standard technique, or variations of it, allows the dynamic analysis of angiogenic growth (Fig. 2a), 3D vascular architecture (Fig. 2b), potential anti-angiogenic inhibitors (Fig. 2c), the role of specific genes (Fig. 3a, b), and the recapitulation of the tumor microenvironment (Fig. 3c).

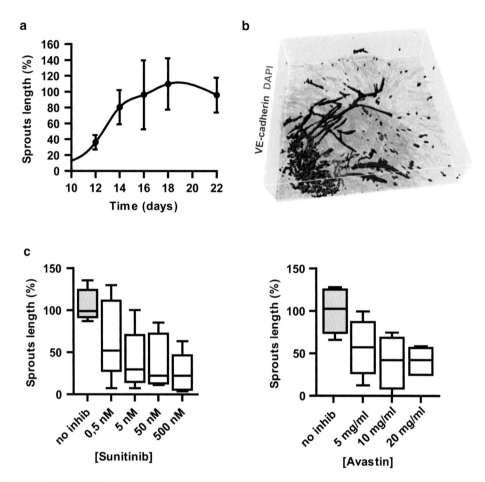

Fig. 2 (**a**) Time course of the angiogenic outgrowth of endothelial sprouts (normalized to day 16) from hARs cultured with complete medium (EGM-2). (**b**) 3D isosurface rendering of hAR sprouting of whole-mount immunofluorescence staining for VE-cadherin (*red*) and nuclei (*blue*) acquired through confocal microscopy. Tick marks on axis = 50 μm. (**c**) hARs treated with anti-angiogenic drugs. Quantification of the angiogenic outgrowth from hARs after 15 days of culture in EGM-2. The effect of various concentrations of Sunitinib or Avastin (VEGFRs RTKi or anti-VEGF monoclonal antibody, respectively) on angiogenic sprouting. This research was originally published in Blood [8] ©the American Society of Hematology

3.7 Stable Lentiviral-Mediated AR Transduction

Produce lentiviral particles by calcium phosphate-mediated transfection of the vector plasmid together with packaging and envelope plasmids in HEK-293T cells.

1. 24 and 48 h post-transfection, harvest the HEK-293T medium, pass through a 0.22 μm filter, and concentrate (19,000 $\times g$ for 2 h at 20 °C).

2. Freeze aliquots of concentrated and purified virus diluted in PBS.

3. Evaluate the multiplicity of infection (MOI) by infecting HeLa cells.

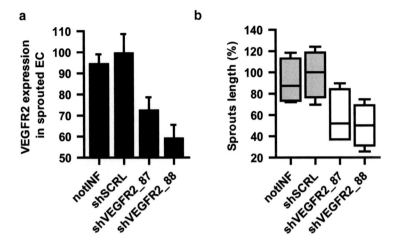

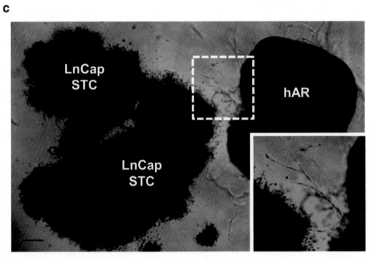

Fig. 3 (**a**) Cytofluorimetric analysis of VEGFR2 expression in hAR-sprouted cells extracted from BME. hARs transduced with lentiviruses carrying VEGFR2-targeted shRNA (shVEGFR2_87 and shVEGFR2_88) and scrambled shRNA control (shSCRL) or not infected (notINF). Values shown are mean ± SD, normalized to the mean of shSCRL hARs (*$P < 0.05$ vs. shSCRL hARs; **$P < 0.01$). (**b**) Quantification of sprouting angiogenesis in hARs transduced with lentiviruses after 15 days of culture in EGM-2 (*$P < 0.05$ vs. shSCRL mARs). (**c**) Representative image of hARs co-cultured with three-dimensional spheroidal aggregates of LnCap cells (STC) for 30 days in basal medium (EBM-2 plus 10 % FBS). *Inset*, higher magnification of the same photomicrograph. Scale bar: 400 μm. This research was originally published in Blood [8] ©the American Society of Hematology

4. Prepare hAR as described above (*see* Subheading 3.1).

5. Incubate hAR for 48 h in 100 µL of EGM-2 plus 1 µL of Polybrene stock solution (8 µg/µL) and 2–3×10^6 purified and concentrated virions (Fig. 3a, b).

3.8 Whole-Mount Live and/or Immunofluorescence (IF) Imaging

1. Prepare hAR as described above (*see* Subheading 3.1).

2. *Optional step*—Transduce hAR with lentiviruses carrying green fluorescent protein (GFP) or GFP-tagged protein (*see* Subheading 3.1).

3. Place glass-bottom dish on ice.

4. Place a 20 μL drop of BME on glass-bottom dishes and put in one hAR before BME is gelled (*see* Subheading 3.5).

5. Cover hAR cultures with EBM-2 plus 5 % FBS, antibiotics, and heparin with addition of specific growth factors or EGM-2, with or without chemical inhibitors.

6. *Optional step*—For live imaging, image hAR every 2–3 days with a fluorescent microscope.

7. Fix hAR with zinc fixative overnight and equilibrate them in PBS.

8. Permeabilize hAR with 0.2 % Triton-X100 in PBS for 2 h.

9. Saturate with 10 % normal serum for 2 h.

10. Incubate with primary antibodies diluted in PBS containing 2.5 % donkey serum and overnight at 4 °C in a humidified chamber.

11. Wash hAR with PBS three times for 10 min.

12. Incubate them for 1 h at room temperature with secondary antibodies.

13. Wash hAR with PBS three times for 10 min.

14. Stain nuclei with a 1/1000 dilution of DAPI nuclear counter-stain stock solution in PBS for 10 min.

15. Observe hAR using an inverted fluorescence or confocal/multiphoton laser-scanning microscope (Fig. 2b).

3.9 Tumor Human Angiogenesis Assay

1. Culture LNCaP cells in RPMI-1640 medium.

2. The day before gel embedding, induce formation of autonomous 3D spheroidal aggregates of cells (STC) by seeding 6×10^4 LNCaP cells into non-adherent, round-bottomed, 96-well plates overnight at 37 °C in methylcellulose medium.

3. Prepare hAR as described above (*see* Subheading 3.1).

4. The following day, gently wash 2 LNCaP STC once in PBS and twice in methylcellulose-free DMEM medium (*see* **Notes 11** and **15**).

5. Embed STC with 20 μL BME in the proximity of hAR.

6. Seal in place with an overlay of 50 μL BME and allow to gel in a humidified incubator to avoid dehydration (*see* **Note 11**).

7. Cover with 500 μL of 10 % FBS EBM-2 or complete EGM-2.

8. Change the medium every 2–3 days. EBM-2 or EGM-2 medium may be used depending on the aim of the assay: EBM-2 medium allows a solely tumor-driven stimulation (Fig. 3c), while culture in EGM-2 medium is faster but less specific (*see* **Note 13**).

9. It is also possible to perform this human tumor angiogenesis assay for whole-mount live or IF imaging (*see* Subheading 3.8).

3.10 Imaging/ Quantification of Angiogenic Dynamics

1. Image tubular structures with an inverted bright-field microscope and photograph them every 2–3 days (1024×1024, $4\times$, or $5\times$ magnification; *see* **Note 16**).

2. To quantify capillary-like structures, process bright-field photomicrographs using Image Pro Plus, as shown in Fig. 4.

3. Process images with the Kernel well 7×7 (strength: 7 pixels) filter. It makes binary-like modifications of the images without the requirement for user-defined threshold selection.

4. Select capillary-like structures using WinRhizo Pro image analysis software to recognize elongated particles (*see* **Note 14**).

5. Quantify angiogenic sprouting as total sprout length (sum of every sprout from each hAR), normalized with the positive control (Fig. 2c).

6. It is also possible to quantify the branching index of the capillary-like structures. For branching analysis, count all branching points manually and divide them by the WinRhizo-quantified sprout length.

Original Brightfield image	Kernel7x7 graphical filter	Analyzed by WinRhizo

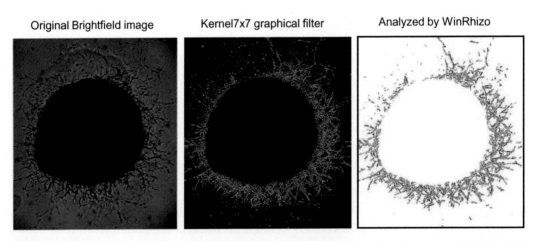

Fig. 4 Computer-assisted image analysis strategy for quantification of angiogenic outgrowth. Bright-field photomicrographs ($4\times$ magnification) of hAR outgrowth are digitally modified with 7×7 Kernel graphical filter and finally processed and quantified with threshold and noise removal methods through WinRhizo software. This research was originally published in Blood [8] ©the American Society of Hematology

3.11 Extraction of Sprouted Cells

1. Remove the medium and wash three times with cold PBS.

2. Add 1 mL of BD Cell Recovery Solution per well and after 10 min, pool into a 15 mL tube by adding 2 mL BD Cell Recovery Solution.

3. After 1 h on ice, centrifuge the cells to a pellet at the bottom of the tube for 5 min at 4 °C.

4. Wash cell pellet by gentle resuspension in cold PBS and pellet the cells by centrifugation at 4 °C.

3.12 Cytometric Analysis and Fluorescence-Activated Cell Sorting

1. *Optional Step* Permeabilize sprouted cells with cell permeabilizing solution.

2. Incubate with primary antibody or control IgG for 30 min.

3. After three washes with PBS containing 1 % BSA, incubate cells with secondary antibody for 30 min.

4. After final washes with PBS, fix cells with cell fixative. Acquire using cytometer or sort them by cell sorter (Fig. 3a).

3.13 RNA Extraction

Extract RNA from (sorted or unsorted) sprouted cells using TRIzol (Invitrogen) according to the manufacturer's instructions.

4 Notes

1. Umbilical cords must be obtained under the appropriate local ethical approval in compliance with national laws and policies governing the use of human tissue.

2. Umbilical cords can be preserved at 4 °C for several days after collection. However, a reduction in sprouting efficiency (~40 %) was observed when arteries were processed from umbilical cords more than 2 days after clamping.

3. Umbilical cord arteries are highly tortuous. If you are not careful in following the path of the artery with the scalpel, you may inadvertently cut the artery.

4. A high level of operator skill is required for explanting, dissecting, and cleaning the arteries. Skilled operators can produce rings from around six arteries in 2–3 h.

5. If you use too much PBS, the artery will float and may make any further microprocedure more complicated.

6. Take care to remove all fibro-adipose tissue from arteries as we observed that failure removing periarterial fibro-adipose tissue results in a lack of EC outgrowth.

7. Arteries can remain at room temperature in PBS for maximum 1 h.

8. To be sure to have efficiently cut the segment, make sure that the whole length of the scalpel blade is pressing on the plastic of the dish and then displace the hARs from their positions. Arteries cannot be left dry for more than 5 min without affecting their viability.

9. Bubbles may form when BME is pipetted. Try to minimize bubble formation as it may cause problems with the setting of the BME gels. Also, allowing the tubes to stand on ice for an additional 10 min will reduce the presence of bubbles.

10. Variation in color of BME is normal and will not affect efficacy. These differences will disappear upon equilibration with 5 % CO_2 in the incubator.

11. Sometimes, a drop of medium can be transported with the hAR when placing on the BME. Take care to blot the hAR by placing hAR on a dry sterile plastic well to prevent the transfer of medium which may dilute the BME immediately around hAR.

12. Take care to add the medium down the side of the tissue culture well and not directly on the gel.

13. Maintenance of EBM-2 at pH 7.4 is of utmost importance for optimal growth.

14. Image analysis can also be performed using ImageJ (NIH) or MATLAB (Mathworks), but this requires the in-house design of segmentation algorithms.

15. STC are very delicate. To avoid the dismantling of formed aggregates, very carefully pipette on the border of the well and avoid aspirating STC. If necessary, use a stereomicroscope to pipette as close as possible.

16. For optimal contrast and depth of field, reduce the aperture as far as possible.

Acknowledgments

This work was supported by Associazione Italiana pe la Ricerca sul Cancro IG 14635; FPRC-ONLUS (Intramural Grant 5x1000 2008); MIUR-Fondo Investimenti per la Ricerca di Base RBAP11BYNP (Newton); Susan G. Komen Foundation Fellowship (PDF14301739).

References

1. Folkman J (2007) Angiogenesis: an organizing principle for drug discovery? Nat Rev Drug Discov 6:273–286

2. Nicosia RF, Ottinetti A (1990) Growth of microvessels in serum-free matrix culture of rat aorta. A quantitative assay of angiogenesis in vitro. Lab Invest 63:115–122

3. Zhu WH, Iurlaro M, MacIntyre A, Fogel E, Nicosia RF (2003) The mouse aorta model: influence of genetic background and aging on bFGF- and VEGF-induced angiogenic sprouting. Angiogenesis 6:193–199

4. Primo L, Seano G, Roca C, Maione F, Gagliardi PA, Sessa R, Martinelli M, Giraudo E, di Blasio L, Bussolino F (2010) Increased expression of alpha6 integrin in endothelial cells unveils a proangiogenic role for basement membrane. Cancer Res 70:5759–5769

5. Sessa R, Seano G, di Blasio L, Gagliardi PA, Isella C, Medico E, Cotelli F, Bussolino F, Primo L (2012) The miR-126 regulates angio-poietin-1 signaling and vessel maturation by targeting p85beta. Biochim Biophys Acta 1823:1925–1935

6. Seano G, Chiaverina G, Gagliardi PA, di Blasio L, Puliafito A, Bouvard C, Sessa R, Tarone G, Sorokin L, Helley D, Jain RK, Serini G, Bussolino F, Primo L (2014) Endothelial podosome rosettes regulate vascular branching in tumour angiogenesis. Nat Cell Biol 16(10):931–41. doi:10.1038/ncb3036, PMID: 25218639, 1–8; Epub 2014 Sep 14

7. Kruger EA, Duray PH, Price DK, Pluda JM, Figg WD (2001) Approaches to preclinical screening of antiangiogenic agents. Semin Oncol 28:570–576

8. Seano G, Chiaverina G, Gagliardi PA, di Blasio L, Sessa R, Bussolino F, Primo L (2013) Modeling human tumor angiogenesis in a three-dimensional culture system. Blood 121:e129–e137

9. Cheon DJ, Orsulic S (2011) Mouse models of cancer. Annu Rev Pathol 6:95–119

Chapter 14

A Modified Aortic Ring Assay to Assess Angiogenic Potential In Vitro

Nina Zippel, Yindi Ding, and Ingrid Fleming

Abstract

Angiogenesis, an integral part of many physiological and pathological processes, is a tightly regulated multistep process. Angiogenesis assays are used to clarify the molecular mechanisms and screen for pharmacological inhibitors. However, most in vitro angiogenesis models measure only one aspect of this process, whereas in vivo assays are complex and difficult to interpret. The ex vivo aortic ring model allows the study of many key features of angiogenesis, such as endothelial activation, branching, and remodeling as well as later steps such as pericyte acquisition. This model can be modified to include genetic manipulation and can be used to assess the pro- or anti-angiogenic effects of compounds in a relatively controlled system.

Key words Angiogenesis, Endothelium, Mouse aorta, Organ culture, Pericyte recruitment, Sprouting

1 Introduction

Angiogenesis, the sprouting of new blood vessels from pre-existing ones, is an essential process of normal development, tissue remodeling, wound healing, as well as the reproductive cycle and pregnancy. At the same time, angiogenesis is an integral part of pathological processes, such as tumor growth and metastasis, proliferative diabetic retinopathy, rheumatoid arthritis, and psoriasis [1]. Angiogenesis is a complex event comprised of several distinct steps, including endothelial activation, basement membrane disruption, and invasion of the extracellular matrix by endothelial sprouts which develop from the parent vessel followed by their elongation, branching, and structural remodeling [2]. These processes are tightly regulated by a fine equilibrium of pro- and anti- angiogenic modulators, such as vascular endothelial growth factor (VEGF), angiopoietins, basic fibroblast growth factor (bFGF), and others [2–4]. However, the interplay between these regulatory factors at different stages of the angiogenic process is not completely understood. The importance of angiogenesis in pathology has led to a

Stewart G. Martin and Peter W. Hewett (eds.), *Angiogenesis Protocols*, Methods in Molecular Biology, vol. 1430, DOI 10.1007/978-1-4939-3628-1_14, © Springer Science+Business Media New York 2016

general interest in clarifying the cellular and molecular mechanisms that are required for the formation of new blood vessels in order to identify novel therapeutic targets. Currently, a number of in vitro and in vivo angiogenesis assays are in general use, each with their own advantages and limitations [5, 6].

Well-established in vivo assays of angiogenesis [7] are the chick embryo chorioallantoic membrane assay (CAM assay) [8, 9], tumor cell injection and implantation [10, 11], the matrix plug assay [12], as well as retina angiogenesis [13, 14] and the ischemic hindlimb [15, 16]. In vivo assays simulate the natural, sequential process of angiogenesis and have the benefit of potentially involving all of the relevant cell types and growth factors. However, despite their relevance, in vivo assays are extremely complex and time consuming and thus not suitable for high throughput analysis, can be difficult to interpret, and results are often confounded by inflammatory responses that have a direct impact on angiogenesis.

While it is clear that in vivo studies cannot be completely avoided, it is important that the most be made of available in vitro studies that closely mimic the in vivo situation. Thus, while complete replacement may not be possible, angiogenesis assays that can assess responses in native endothelial cells can certainly help to reduce and refine in vivo studies.

Several in vitro angiogenesis assays are used widely to investigate and quantify angiogenesis [17, 18]. Perhaps the most frequently used are the endothelial tube (or cord) formation assay, network formation by fibroblast and endothelial co-cultures, or 3D endothelial spheroid sprouting assays [19–22]. However, these cell-based assays are limited by the fact that they require the isolation and culture of the cells of interest and are only able to mimic selective phases of angiogenesis. These models also exclude important contributions from supporting cells such as pericytes as well as the processes involved in sprouting from the parent vessel, invasion and maturation, as well as later remodeling steps.

It follows, therefore, that more biologically relevant information can be obtained by using native endothelial cells, i.e., avoiding the dedifferentiating culture step and organ culture models such as the aortic ring assay. The assay was originally described for the rat aorta by Nicosia and Ottinetti [23] and bridges the gap between in vitro and in vivo methods and is equally applicable to other species [24, 25], including the mouse [26, 27]. This means that it is also useful for the study of a broad spectrum of genetically modified mouse lines [16, 28, 29].

As the name suggests, this model involves the embedding of aortic rings in an extracellular matrix (usually fibrin or collagen gels) and culture in a defined medium. Over a short period of time, i.e., 6–7 days (Fig. 1), hollow capillaries sprout spontaneously from the cut surfaces of the aortic sections, which are thought to be initiated by the wound of the dissection procedure.

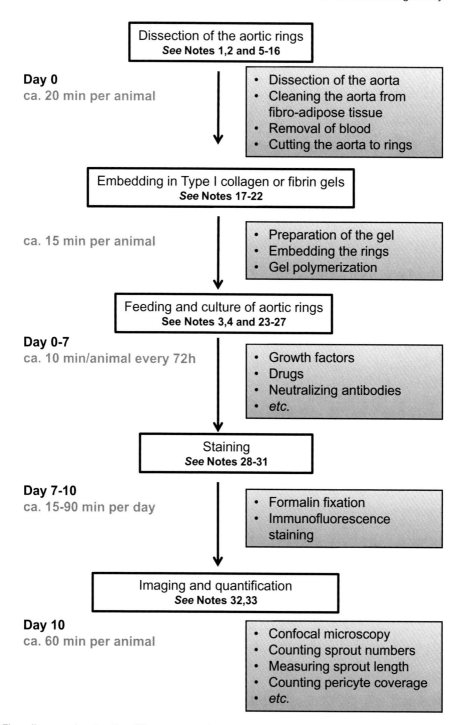

Fig. 1 Flow diagram showing the different steps of the aortic ring assay. The timing is indicated on the left with the main stages highlighted in solid boxes and single steps in *gray* boxes

The capillary-like tubes formed are made up of endothelial cells as well as supporting cells involved in vessel maturation, such as pericytes. Endothelial sprouts grow out from the aortic sections over a time course comparable to that seen in vivo [30, 31] in an exponential growth phase that can result in a relatively complex network before being followed by a regression phase.

This sensitive assay allows assessment of the cellular and molecular steps in angiogenesis as well as the identification, characterization, or screening of novel angiogenic modulators [32], such as cytokines, peptides [11, 33], anti-angiogenic compounds [34], and neutralizing antibodies. It is important to realize that the vast majority of cells that grow out of the aortic ring are of adventitial origin, i.e., fibroblasts. This means that any conclusions about angiogenesis can only be made after the endothelial cells and/or pericytes have been visualized.

2   Materials

2.1   Preparation of Aortic Rings

1. Rats or mice of appropriate age, sex, and genetic background (*see* **Note 1**).

2. Microdissection scissors and forceps: bone scissors, a pair of skin forceps, a pair of straight fine forceps (i.e., Dumont #5), and iris scissors (Vannas spring scissors with a blade size of 3 mm for mice, or 4 mm for rat aortae).

3. Dulbecco's modified Eagle's medium/Ham's F12 nutrient mix (DMEM/F12) containing 100 U/mL penicillin/100 µg/mL streptomycin (*see* **Note 2**).

4. Scalpel.

5. Culture dishes (10 cm).

6. Syringe (1 mL).

7. Needles (27-G and 30-G).

8. Dissecting microscope.

9. A culture dish half filled with polymerized silicone (SYLGARD 184 silicone elastomer). Prepare several days in advance to allow full polymerization of the SYLGARD 184.

10. 70 % (v/v) ethanol.

2.2   Collagen Gel

1. Rat tail collagen type 1 (Becton Dickenson).

2. Sterile water, 4 °C.

3. Medium 199 (M199) 10× solution.

4. Sterile aqueous 7.5 % sodium bicarbonate ($NaHCO_3$).

5. Sterile aqueous 0.5 N sodium hydroxide (NaOH).

2.3 Embedding

1. 48-Well plates for mouse aortae, or 96-well plates for rat aortae.

2. A pair of straight fine forceps (i.e., Dumont #5).

3. Laminar flow tissue culture hood.

2.4 Culture Medium

1. Culture medium: Endothelial basal medium (EBM) containing 2.5 % (v/v) autologous (mouse or rat) serum and 100 U/mL penicillin/100 μg/mL streptomycin (*see* **Notes 3** and **4**).

2. Humidified incubator with a 5 % CO_2 atmosphere at 37 °C.

2.5 Staining,
Imaging,
and Quantification

1. Dulbecco's phosphate-buffered saline (PBS) solution.

2. Blocking solution: PBS containing 0.5 % (v/v) Triton X-100 and 1 % (w/v) bovine serum albumin (BSA).

3. Anti-CD31 antibodies (e.g., Becton Dickinson, BD 550274).

4. Anti-rat secondary antibodies (e.g., 488 Alexa Fluor).

5. 4 % paraformaldehyde (PFA), pH 7.4.

6. Confocal microscope.

3 Methods

The protocol for aortic ring assay involves different steps including the dissection of the aorta, embedding the aortic rings in collagen or fibrin gels, culture of the aortic rings, staining, as well as image capture and the quantification of the capillary sprouts.

3.1 Preparation
of the Aortic Rings

1. Sacrifice the mouse or rat according to the relevant local ethical guidelines for animal care and experimentation (*see* **Note 5**).

2. Sterilize the surface of the carcass with 70 % ethanol.

3. Lay the carcass on its back and open the ventral skin.

4. Cut through the sternum, and open the rib cage using a bone scissor and skin forceps.

5. Remove the lungs and the esophagus (Fig. 2a).

6. Remove aorta, now visible as a fat-covered vessel tracking down along the spine, rapidly from the sacrificed mouse or rat.

7. Gently detach the thoracic aorta from the spine using closed microdissection forceps starting from the diaphragm toward the heart (Fig. 2b, c).

8. Cut once at the posterior end and once at the anterior end (Fig. 2d–f) (*see* **Notes 6** and **7**).

9. Transfer the dissected aortae immediately to ice-cold serum-free DMEM/F12 with antibiotics. Keep the aortae on ice until dissection.

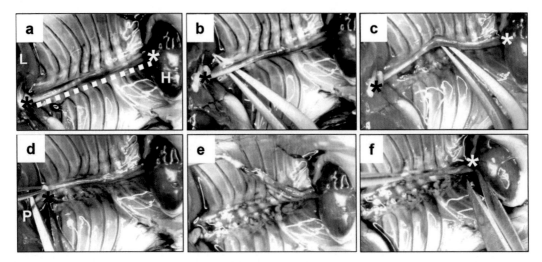

Fig. 2 Dissection of the thoracic mouse aorta. (**a**) After opening the rib cage and removing the lungs and esophagus, the aorta (*white dotted line*) is visible as it runs along the spine from the heart (*white asterisk*) to the diaphragm (*black asterisk*). The heart (H) and liver (L) are visible. (**b–c**) The aorta is detached from the spine by blunt dissection using closed forceps, beginning at the posterior end moving gently toward the anterior end until the aorta is mostly detached. (**d**) The aorta is cut once at the posterior end (P) and (**e**) lifted gently without applying tension. (**f**) Cut the aorta again at the anterior end (*white asterisk*)

10. For the dissection, add ice-cold DMEM/F12 plus antibiotics to a dish, which is half filled with polymerized silicone.

11. Transfer the aorta to the silicone-filled dissection dish.

12. Pin the ends of the aorta to the silicone with fine 30-G needles without applying tension to the aorta (Fig. 3a) (*see* **Notes 8** and **9**).

13. Carefully remove the surrounding periaortic fibroadipose tissues and branching vessels with fine microdissection forceps and iridectomy scissors under a microscope (Fig. 3b, c), paying special attention not to damage the aortic wall (*see* **Note 10**).

14. Gently flush out blood from the lumen of the aorta with DMEM/F12 using a 1 mL syringe fitted with a 27-G blind needle (Fig. 3d) until the aorta is free of clotted blood (Fig. 3e) (*see* **Notes 11** and **12**).

15. Place a piece of millimeter paper underneath the dish.

16. Using a scalpel blade, cut away the proximal and distal 1 mm segments of the aorta.

17. With help of the scale paper, cut the aorta into equal sections ~1 mm long (Fig. 3f) (*see* **Notes 13** and **14**).

18. Store the aortic rings on ice until the gel has been prepared (*see* **Notes 15** and **16**).

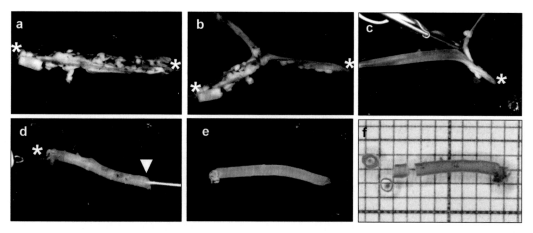

Fig. 3 Clean and cut the extracted aorta. (**a**) The fat-covered aorta should be pinned at its ends (*white asterisks*) to the silicone bottom of a dissection dish. (**b–c**) Clean the aorta of the surrounding fibroadipose tissues by running the scissors parallel to the aorta, thus removing whole strips of fat. (**d**) After removing one pin, the syringe should be partially inserted into the fat-free aorta (*white arrow*) and flushed gently with medium until any remaining blood has been removed (**e**). (**f**) Place a piece of graph paper underneath a culture dish—this helps to cut the aorta into sections of equal length

3.2 Preparation of the Collagen Gel

The use of a collagen matrix is recommended for the aortic ring assay to support the outgrowth of large microvessel sprouts, which are easily distinguishable from fibroblast outgrowth. In addition, a collagen matrix supports the response to growth factors, such as VEGF [27]. However, specific conditions may require the use of a different matrix, as the type of matrix can be important for the angiogenic response to different growth factors [27, 35, 36], for example, sprouting in response to bFGF is stronger in a fibrin matrix [27]. The preparation of a fibrin gel is described as an option in Subheading 4 (*see* **Note 17**).

Prepare the collagen gel on ice to avoid premature polymerization of the matrix. Note that all pipette tips and plates should be stored at 4 °C until use to avoid the premature coagulation of the gel. Work under sterile conditions using a laminar flow.

1. Add rat tail collagen type-1 to a final concentration of 1.5 mg/mL to a 1.5 mL Eppendorf tube.

2. Adjust with sterile water to a volume of 871 μL/mL and invert immediately.

3. Add 100 μL/mL of 10× Medium 199.

4. Add 34 μL/mL of 7.5 % NaHCO$_3$ solution

5. Mix well for 10 s to prevent uneven polymerization. Avoid the formation of air bubbles.

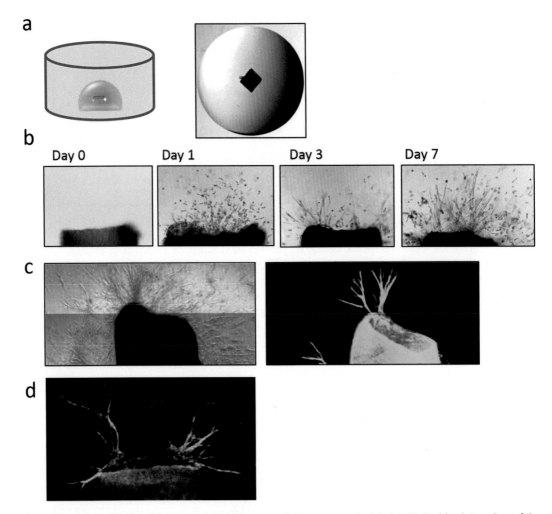

Fig. 4 Generating and visualizing sprouts. (**a**) Mouse aortic rings are embedded on their sides into a drop of the collagen matrix. (**b**) Phase-contrast images of the aortic rings embedded in collagen at days 0, 1, 3, and 7. (**c**) Comparison of a phase-contrast image and CD31 immunofluorescent staining of the same aortic ring. (**d**) Co-staining with CD31 (*green*) and α-actin (*red*) allows for the simultaneous visualization of endothelial cells and pericytes

6. Adjust the pH with a few drops of 0.5 N NaOH to approximately 7.4. The gel color should turn from orange to cherry red. Put the gel on ice for approximately 1 min until the color no longer changes (*see* **Note 18**).

7. Keep on ice and use within 30 min.

3.3 Embedding the Aortic Rings

3.3.1 Mouse Aortae

1. Place a 25 μL drop of the gel in the middle of each well of a pre-cooled 48 well plate (*see* **Note 19**).

2. Embed one aortic ring into the middle of each collagen drop using fine forceps (Fig. 4a). The gel should remain its drop-like shape (*see* **Note 20**).

3.3.2 Rat Aortae

1. Add 100 µL of the gel to each well of a pre-cooled 96 well plate (*see* **Note 21**).

2. Embed one aortic ring into the middle of each well using fine forceps (*see* **Note 22**).

3.4 Culturing the Aortic Rings

1. Carefully place the plates in an incubator at 37 °C with 5 % CO_2 for approximately 45–60 min until the gel has solidified (*see* **Note 23**).

2. Add endothelial basal medium supplemented with 100 U/mL penicillin/100 µg/mL streptomycin, optionally supplemented with 2.5 % autologous serum and/or growth factors (*see* **Notes 24** and **25**).

3. Change the culture medium every 72 h (*see* **Note 26**).

4. Cultivate the aortic sections at 37 °C in a humidified environment for 7 days (*see* **Note 27**).

3.5 Fluorescence Staining and Imaging

Although phase-contrast microscopy may provide an overview of microvessel outgrowth (Fig. 4b), it does not allow for discrimination between endothelial cells and other cell types, such as fibroblasts, which grow out from the aortic explants in large numbers. Immunohistochemical staining can be used to identify the interacting cell types, and endothelial sprouts can be quantified following staining with antibodies against CD31 (Fig. 4c). The interaction with supporting pericytes can also be visualized, e.g., by staining against α-actin (Fig. 4d) or αNG2 chondroitin sulfate proteoglycan. Sprout length can be quantified by measuring the number and length of the sprouts (Fig. 5), and coverage by pericytes can be quantified by counting the number of associated pericytes per unit length.

1. Remove the culture medium and wash with PBS (*see* **Note 28**).

2. Fix with 4 % PFA for 60 min at room temperature (*see* **Note 29**).

3. Remove the fixative and wash three times with PBS, each washing step should be at least 5 min.

4. Permeabilize and incubate with blocking solution overnight at 4 °C.

5. Prepare the required primary antibodies, e.g., anti-CD31 for endothelial cells in PBS at an appropriate dilution (*see* **Note 30**).

6. Incubate overnight at 4 °C.

7. Wash five times in PBS for 10 min.

8. Prepare secondary antibodies in PBS 1/500 and incubate overnight at 4 °C or for 2 h at room temperature (*see* **Note 31**).

9. Remove the antibody solution and wash three times for 10 min in PBS.

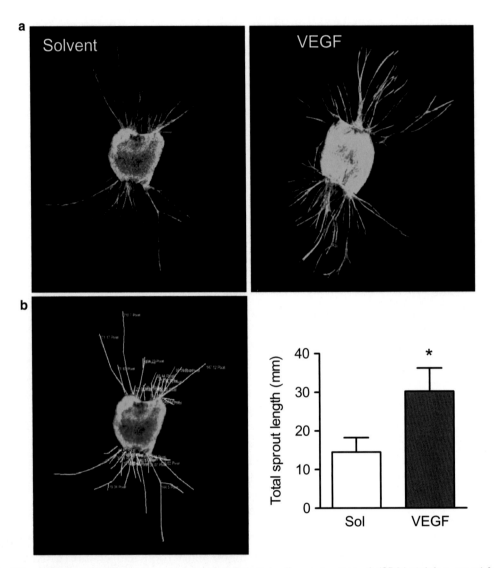

Fig. 5 Quantification. (**a**) Fluorescence images of endothelial cell sprout outgrowth (CD31 staining, *green*) from solvent- or VEGF-treated rings after 7 days. (**b**) A skeletonized representation of the aortic ring shown in (**a**) generated with the help of quantification software. Total sprout length calculated following the conversion of pixels to mm

10. Postfix for 5 min with 4 % PFA at room temperature.

11. Wash two times for 10 min in PBS.

12. Cover with PBS or mounting medium.

13. Store at 4 °C in the dark.

14. Image aortic sections using a laser scanning microscope with 10 or 20× magnification (*see* **Note 32**).

3.6 Quantification   1. Count the number of sprouts that emerge from the main ring as well as the individual branches arising from it as separate vessels [37].

2. Sprout length can be quantified by manually measuring and converting pixels into μm (Fig. 4d) (*see* **Note 33**).

---

4 Notes

1. Outgrowth is optimal in aortic rings from 6- to 12-week-old animals, and with older animals, i.e., 6–10 months of age, sprouting is minimal [35, 38]. If using such animals, the use of collagen gels and medium containing autologous serum is recommended. Fibrin gels, which poorly support sprouting and vascular endothelial growth factor (VEGF) response, or serum-free conditions would not produce meaningful results with these animals. In addition, it has been shown that gender influences outgrowth from aortic explants, and angiogenic response in female rats or mice tends to be impaired compared to males [38]. For these reasons, the age and sex of the animals used within one experiment should be kept as consistent as possible to limit variability. In view of the significant impact that aging and genetic background have on sprouting, aortic rings from littermates should be used.

2. When transfection of the aortic sections is planned, the use of Opti-MEM medium instead of DMEM/F12 is recommended.

3. Do not add endothelial growth medium (EGM) supplements to the basal endothelial medium (EBM).

4. To isolate the autologous serum collect blood (without adding anticoagulants) prior to sacrificing the mouse. This must be performed by trained and licensed individuals according to the relevant ethical guidelines for animal care and experimentation. To obtain the serum, let the blood coagulate at room temperature and centrifuge at $800 \times g$ for 10 min. The serum is the supernatant. Transfer the supernatant to a fresh tube and pass through a sterile filter before use. Heat inactivate for 10 min at 54 °C.

5. This must be performed by trained and licensed individuals.

6. Avoid overstretching which could damage the smooth muscle and endothelial cell layers.

7. The abdominal aorta can also be used. However, results do tend to vary between sections from the thoracic and abdominal aorta.

8. Prepare SYLGARD 184-filled dishes according to the manufacturer's protocol in advance, since complete polymerization takes several days.

9. Avoid overstretching which damages the smooth muscle and endothelial cell layers.

10. Handle the aorta only at its ends and avoid stretching the vessel.

11. Make sure all remaining blood is removed since clotted blood impairs sprouting.

12. Insert the syringe only into the tip of the aorta in order to avoid damage to the tissue during syringing. Take care that no air bubbles are introduced into the aorta as this is a great way of removing the endothelial cells.

13. Keep ring width as consistent as possible within the experiment.

14. Approximately 30 rings can be cut from each rat aorta, or ten rings from each mouse aorta. At least three rings per aorta and three different aortae should be used per condition. It is recommended to repeat each experiment several times, since animal experiments contain an inherent variance.

15. If the assay cannot be continued immediately, store the aorta at 37 °C in an incubator with 5 % CO_2.

16. If transfection of the aortae is planned, starve the aortic rings in serum-free medium in an incubator at 37 °C with 5 % CO_2 overnight prior to transfection, or viral transduction.

17. To prepare a fibrin gel, dissolve lyophilized bovine fibrinogen in endothelial basal medium to a final concentration of 3 mg/mL and add 5.0 mg/mL aprotinin, which prevents the degradation of the fibrin matrix by proteases. Incubate at 37 °C for 1 h to dissolve and pass through a sterile filter (0.45 μm) to remove unpolymerized clumps of fibrinogen. Prepare aliquots of the fibrinogen solution and keep them on ice to prevent premature polymerization when thrombin is added. Add 0.5 U thrombin to each 1 mL aliquot of fibrinogen immediately before use. Prepare only one aliquot at a time. Mix well but avoid the formation of air bubbles. Keep on ice and use within 1 min.

18. A pH of 7.4 is critical for collagen gel polymerization, as well as optimal angiogenic responses in the aortic ring assay [39]; sprouting is markedly delayed at a pH of 6.9 [40].

19. Do not pipette all of the drops at once, but only a few wells at a time to avoid polymerization before there is time to embed the aortic rings.

20. To avoid the dilution of the gel with medium, carefully remove excess medium from the aortic sections by tapping them on an empty dish or sterile filter paper.

21. Rat aortic rings are larger in diameter and thus require a larger amount of matrix, which can no longer be applied in a drop. Although it is also possible to embed mouse aortic sections in

50 μL gel in 96 well plates, this increases the amount of gel needed and the response to growth factors would be impaired.

22. The rings can be embedded in the three-dimensional matrix either with the luminal axis perpendicular or parallel to the bottom of the well. It is preferable to embed the aortic rings on their sides rather than as open rings. This allows for optimal imaging of microvascular sprouting, as the sprouts mostly grow from the cut surfaces along the axis of the lumen. When the rings are placed on their end, the sprouts grow toward the observer and become more difficult to visualize and quantify.

23. The addition of medium before the gel has polymerized can damage the gel. If the gel takes longer than 120 min to set, it will begin to dry out.

24. Add the medium carefully to the side of the well to prevent the fragile gels from lifting or being damaged.

25. Under serum-free conditions, spontaneous angiogenesis in murine aortic rings is usually minimal, whereas rat aortic sections sprout in serum-free conditions. The addition of 2.5 % autologous serum is recommended for murine aortic rings; indeed, the assay rarely functions in its absence. Sprouting can be increased by supplementing small amounts of growth factor (e.g., 10–30 ng/mL VEGF, 10–30 ng/mL bFGF) (Fig. 4d, e); this is particularly useful when anti-angiogenic agents are being tested.

26. If the medium is replaced more often, sprouting is impaired, probably due to the depletion or dilution of endogenous growth factors.

27. Aortic explants can be cultured for up to 14 days before breakdown of the collagen or fibrin matrix occurs [39]. During the first 3 days, fibroblasts grow out of the rings, followed by endothelial cell sprouts. Initially there is an exponential phase of sprouting and growth which peaks approximately at day 7, depending on age and genetic background of the mice and choice of extracellular matrix. In the second week of culture, explants become quiescent and the sprouts begin to regress, probably due to the lack of flow and the depletion of growth factors and other soluble components and the reduction in integrity of the surrounding matrix [30, 41].

28. The PBS buffer should be at room temperature.

29. The 4 % PFA solution should be at room temperature.

30. To co-stain for pericytes, antibodies against α-NG2 chondroitin sulfate proteoglycan (i.e., Millipore, AB5320, 1/200 dilution) are recommended.

31. The addition of 4′,6-diamidino-2-phenylindole (DAPI) or Hoechst 1/5000 to counter stain nuclei can be performed during this step.

32. Imaging can be performed directly in the culture wells if appropriate hardware is available. Alternatively, the gels can be mounted on microscope slides; fill the wells with water and detach the gel from the wells using needles or forceps and a spatula. Carefully arrange the gel on microscope slides and add mounting medium.

33. Quantification of sprout outgrowth from immunostained rings can be performed either by manually counting the number of sprouts and measuring their length as described above or by using computer-assisted image analysis programs, as described elsewhere (*see* refs. [32, 42]). However, if sprouts grow out in large numbers, computer-assisted automatic image analysis is error-prone since it can no longer distinguish between single sprouts, which visually fuse together. In order to obtain meaningful results, manual counting is recommended.

Acknowledgment

This work was supported by the Deutsche Forschungsgemeinschaft (SFB834/A9 and Exzellenzcluster 147 "Cardio-Pulmonary Systems").

References

1. Carmeliet P (2005) Angiogenesis in life, disease and medicine. Nature 438:932–936

2. Kumar R, Yoneda J, Bucana CD et al (1998) Regulation of distinct steps of angiogenesis by different angiogenic molecules. Int J Oncol 12:749–757

3. Ferrara N, Kerbel RS (2005) Angiogenesis as a therapeutic target. Nature 438:967–974

4. Nicosia RF, Lin YJ, Hazelton D et al (1997) Endogenous regulation of angiogenesis in the rat aorta model. Role of vascular endothelial growth factor. Am J Pathol 151:1379–1386

5. Auerbach R, Lewis R, Shinners B et al (2003) Angiogenesis assays: a critical overview. Clin Chem 49:32–40

6. Staton CA, Stribbling SM, Tazzyman S et al (2004) Current methods for assaying angiogenesis in vitro and in vivo. Int J Exp Pathol 85:233–248

7. Norrby K (2006) In vivo models of angiogenesis. J Cell Mol Med 10:588–612

8. Ribatti D (2008) Chick embryo chorioallantoic membrane as a useful tool to study angiogenesis. Int Rev Cell Mol Biol 270:181–224

9. Tay SL, Heng PW, Chan LW (2012) The chick chorioallantoic membrane imaging method as a platform to evaluate vasoactivity and assess irritancy of compounds. J Pharm Pharmacol 64:1128–1137

10. Falcon BL, Pietras K, Chou J et al (2011) Increased vascular delivery and efficacy of chemotherapy after inhibition of platelet-derived growth factor-B. Am J Pathol 178:2920–2930

11. Reynolds AR, Hart IR, Watson AR et al (2009) Stimulation of tumor growth and angiogenesis by low concentrations of RGD-mimetic integrin inhibitors. Nat Med 15:392–400

12. Passaniti A, Taylor RM, Pili R et al (1992) A simple, quantitative method for assessing angiogenesis and antiangiogenic agents using reconstituted basement membrane, heparin, and fibroblast growth factor. Lab Invest 67:519–528

13. Pitulescu ME, Schmidt I, Benedito R et al (2010) Inducible gene targeting in the neonatal vasculature and analysis of retinal angiogenesis in mice. Nat Protoc 5:1518–1534

14. Hu J, Popp R, Fromel T et al (2014) Muller glia cells regulate Notch signaling and retinal angiogenesis via the generation of 19,20-dihydroxydocosapentaenoic acid. J Exp Med 211:281–295

15. Silvestre JS, Mallat Z, Duriez M et al (2000) Antiangiogenic effect of interleukin-10 in

ischemia-induced angiogenesis in mice hindlimb. Circ Res 87:448–452

16. Shi L, Fisslthaler B, Zippel N et al (2013) MicroRNA-223 antagonizes angiogenesis by targeting β1 integrin and preventing growth factor signaling in endothelial cells. Circ Res 113:1320–1330

17. Montesano R, Pepper MS, Vassalli JD et al (1992) Modulation of angiogenesis in vitro. EXS 61:129–136

18. Bishop ET, Bell GT, Bloor S et al (1999) An in vitro model of angiogenesis: basic features. Angiogenesis 3:335–344

19. Schor AM, Ellis I, Schor SL (2001) Collagen gel assay for angiogenesis: induction of endothelial cell sprouting. Methods Mol Med 46:145–162

20. Nakatsu MN, Hughes CC (2008) An optimized three-dimensional in vitro model for the analysis of angiogenesis. Methods Enzymol 443:65–82

21. Nakatsu MN, Sainson RC, Aoto JN et al (2003) Angiogenic sprouting and capillary lumen formation modeled by human umbilical vein endothelial cells (HUVEC) in fibrin gels: the role of fibroblasts and Angiopoietin-1. Microvasc Res 66:102–112

22. Bell SE, Mavila A, Salazar R et al (2001) Differential gene expression during capillary morphogenesis in 3D collagen matrices: regulated expression of genes involved in basement membrane matrix assembly, cell cycle progression, cellular differentiation and G-protein signaling. J Cell Sci 114:2755–2773

23. Nicosia RF, Ottinetti A (1990) Growth of microvessels in serum-free matrix culture of rat aorta. A quantitative assay of angiogenesis in vitro. Lab Invest 63:115–122

24. Stiffey-Wilusz J, Boice JA, Ronan J et al (2001) An ex vivo angiogenesis assay utilizing commercial porcine carotid artery: modification of the rat aortic ring assay. Angiogenesis 4:3–9

25. Chen CH, Cartwright J Jr, Li Z et al (1997) Inhibitory effects of hypercholesterolemia and ox-LDL on angiogenesis-like endothelial growth in rabbit aortic explants. Essential role of basic fibroblast growth factor. Arterioscler Thromb Vasc Biol 17:1303–1312

26. Masson VV, Devy L, Grignet-Debrus C et al (2002) Mouse aortic ring assay: a new approach of the molecular genetics of angiogenesis. Biol Proced Online 4:24–31

27. Baker M, Robinson SD, Lechertier T et al (2012) Use of the mouse aortic ring assay to study angiogenesis. Nat Protoc 7:89–104

28. Carmeliet P, Collen D (2000) Transgenic mouse models in angiogenesis and cardiovascular disease. J Pathol 190:387–405

29. Zippel N, Malik RA, Fromel T et al (2013) Transforming growth factor-β-activated kinase 1 regulates angiogenesis via AMP-activated protein kinase-alpha1 and redox balance in endothelial cells. Arterioscler Thromb Vasc Biol 33:2792–2799

30. Nicosia RF, Zorzi P, Ligresti G et al (2011) Paracrine regulation of angiogenesis by different cell types in the aorta ring model. Int J Dev Biol 55:447–453

31. Nicosia RF, Bonanno E, Villaschi S (1992) Large-vessel endothelium switches to a microvascular phenotype during angiogenesis in collagen gel culture of rat aorta. Atherosclerosis 95:191–199

32. Go RS, Owen WG (2003) The rat aortic ring assay for in vitro study of angiogenesis. Methods Mol Med 85:59–64

33. Nicosia RF, Bonanno E (1991) Inhibition of angiogenesis in vitro by Arg-Gly-Asp-containing synthetic peptide. Am J Pathol 138:829–833

34. Bocci G, Danesi R, Benelli U et al (1999) Inhibitory effect of suramin in rat models of angiogenesis in vitro and in vivo. Cancer Chemother Pharmacol 43:205–212

35. Zhu WH, Iurlaro M, MacIntyre A et al (2003) The mouse aorta model: influence of genetic background and aging on bFGF- and VEGF-induced angiogenic sprouting. Angiogenesis 6:193–199

36. Madri JA, Williams SK (1983) Capillary endothelial cell cultures: phenotypic modulation by matrix components. J Cell Biol 97:153–165

37. Aplin AC, Fogel E, Zorzi P et al (2008) The aortic ring model of angiogenesis. Methods Enzymol 443:119–136

38. De Rossi G, Scotland R, Whiteford J (2013) Critical factors in measuring angiogenesis using the aortic ring model. J Genet Syndr Gene Ther 4(5):pii: 1000147

39. West DC, Burbridge MF (2009) Three-dimensional in vitro angiogenesis in the rat aortic ring model. Methods Mol Biol 467:189–210

40. Burbridge MF, West DC, Atassi G et al (1999) The effect of extracellular pH on angiogenesis in vitro. Angiogenesis 3:281–288

41. Aplin AC, Zhu WH, Fogel E et al (2009) Vascular regression and survival are differentially regulated by MT1-MMP and TIMPs in the aortic ring model of angiogenesis. Am J Physiol Cell Physiol 297:C471–C480

42. Blacher S, Devy L, Burbridge MF et al (2001) Improved quantification of angiogenesis in the rat aortic ring assay. Angiogenesis 4:133–142

Chapter 15

Quantitative Imaging-Based Examination of Pericytes Controlling Endothelial Growth Dynamics and Angiogenesis

Anthony R. Sheets, Jennifer T. Durham, and Ira M. Herman

Abstract

Microvascular endothelial cell-mural cell interactions are instrumental in modulating both physiological and pathologic angiogenesis. Pericyte-endothelial cell communication through direct physical associations and secreted effectors comprises a bidirectional signal array that regulates vascular maturation and integrity. As endothelial cell proliferation, migration, and morphogenesis are key elements of vascular growth and remodeling during angiogenesis, we have developed novel preclinical systems for studying the roles of endothelial-mural cell dynamics on cell cycle entry and angiogenic activity in vitro. These coculture models not only enable evaluation of endothelial cell-pericyte "cross talk" but also allow for the quantitative analysis of both heterotypic contact-dependent and contact-independent cell cycle progression in either cell population, as well as angiogenic sprouting in three-dimensional vascular networks. Cells actively proliferating in two-dimensional assays can be labeled via incorporation of 5-ethynyl-2′-deoxyuridine (EdU) into their DNA. Additionally, each cell population can be vitally labeled with a variety of cell-specific and/or membrane-permeant lipophilic dyes prior to coculture, such as DiO, or through immunofluorescence of mural or endothelial cell-specific markers after cellular fixation and/or permeabilization. Ultimately, this experimental approach can be used to investigate cellular contact-dependent and soluble mechanisms mediating mural-endothelial cell interactions, which may be instrumental in microvascular development and remodeling in vivo.

Key words Mural cell, Microvasculature, Morphogenesis, Proliferation, Diabetic retinopathy, Diabetic ulcer, Chronic wound, Tumor vessels, Vascular stabilization

1 Introduction

Microvascular morphogenesis occurs through two parallel yet distinct processes: vasculogenesis, the creation of new blood vessels from endothelial precursor cells, and angiogenesis, the sprouting of capillaries from existing vascular structures. As microvascular remodeling and maturation proceeds, blood vessels are primarily composed of endothelial cells in direct contact with blood and mural cells that invest the abluminal endothelial surfaces [1]. Mural

Stewart G. Martin and Peter W. Hewett (eds.), *Angiogenesis Protocols*, Methods in Molecular Biology, vol. 1430,
DOI 10.1007/978-1-4939-3628-1_15, © Springer Science+Business Media New York 2016

cell populations, including microvascular pericytes, maintain vascular stability, aid in barrier function, and regulate vascular tone [2–4]. In order for remodeling during angiogenesis to occur, pericytes must divest from nascent vessels, allowing the endothelial cells to divide, migrate, and undergo morphogenesis into a new vascular conduit. Indeed, these processes must occur throughout life, as tissues and organs continue to grow, and capillary sprouting and remodeling are indispensible for cutaneous wound healing in children and adults. Several studies have revealed the bidirectional nature of microvascular pericyte-endothelial cell interactions, shedding light on the regulatory contributions of soluble and extracellular matrix-associated growth factors [5–8] as well as heterotypic intercellular contacts [9–11]. Importantly, whereas the cellular and molecular mechanisms governing blood vessel growth and remodeling are tightly regulated in development and healing, perturbations in these signaling cascades and mural-endothelial cell associations underlie the pathogenesis of multiple disease states, including proliferative diabetic retinopathy, tumor angiogenesis, and chronic, nonhealing wounds [12–14].

Given the multiple factors regulating vascular stability in health and disease, we present a method of coculture to study the roles of endothelial-mural cell dynamics on cell cycle entry in vitro. Here, we describe a system that allows the user to simultaneously determine the roles of pericyte and/or endothelial cell-specific mechanisms that may control heterotypic cell contact-dependent and contact-independent cell cycle progression and angiogenic reactivation. Additionally, we present a protocol for quantitative analysis of mural-endothelial cell interactions in vascular morphogenesis in 3D, as extensions of our previously published methods [15, 16]. Thus, these experimental paradigms can be used to test the effects of altered microvascular cellular parameters on angiogenesis, as achieved through methods of forced expression and gene silencing, as well as the use of neutralizing antibodies against cell-specific secreted entities. Notably, these methods can be applied to studies of endothelial cell interactions with several types of mural cells, such as pericyte-like adipose-derived stem cells as well [17]. The outcomes of such experiments will provide information that may have critical implications for microvascular angiogenesis in health and disease.

2   Materials

In order to study the contributions of heterotypic cellular contacts and soluble growth factors on microvascular cell proliferative status in vitro, pericytes and endothelial cells can be plated at varying densities, depending on the vascular bed being modeled. For our assays of microvascular cell cycle progression in coculture (*see* schematic, Fig. 1), we typically culture retinal capillary pericytes, labeled

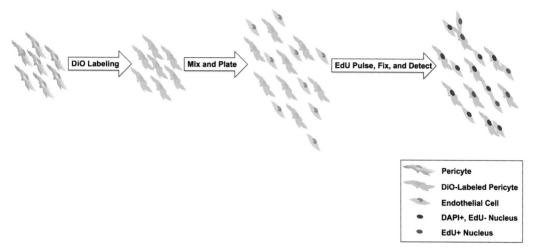

Fig. 1 Schematic of pericyte-endothelial coculture proliferation assay

with vital lipophilic tracers, and endothelial cells together at a 1:1 ratio, thereby recapitulating the cellular distribution of the in vivo retinal microvasculature [18]. For our three-dimensional coculture model of vascular morphogenesis and angiogenic sprouting, we culture DiO-labeled pericytes and capillary endothelial cells together at a ratio of 1:25, embedded between layers of an extra-cellular matrix solution (Fig. 3).

2.1 Coculture Proliferation Assay Components

1. Primary pericyte and capillary endothelial cell populations.
2. 15 mL conical tubes (VWR Scientific).
3. 12 mm-diameter circular glass coverslips (Thermo Fisher Scientific).
4. 24-well tissue culture plates (BD).
5. Vybrant DiO Cell Labeling Solution (Life Technologies).
6. Serum-free Dulbecco's Modified Eagle's Medium (DMEM).
7. Coculture media: DMEM containing 2 % bovine calf serum (BCS), 1 % L-glutamine, and 1 % antibiotic-antimycotic solution.
8. Click-iT EdU Alexa Fluor 594 imaging kit (Life Technologies).
9. 4 % formaldehyde solution, pH 7.2.
10. PBS pH 7.2 without sodium azide.
11. PermaFluor (Thermo Fisher Scientific), or other aqueous mounting media.

2.2 Three-Dimensional Angiogenesis Assay Materials

1. Primary pericyte and capillary endothelial cell populations.
2. Vybrant DiO cell-labeling solution (Life Technologies).
3. 15 mL conical tubes.
4. Serum-free Dulbecco's Modified Eagle's Medium (DMEM).

5. Growth factor-reduced (GFR) matrigel.

6. Type I collagen from rat tail tendon.

7. Coculture media containing 25 mM HEPES pH 7.1.

8. 1 M NaOH.

9. 8-well chamber slides.

10. Sterile 18G needles.

11. Sterile, 1″ long, 22G blunt-end needles, connected to a vacuum source.

3   Methods

3.1   Ultraviolet Sterilization of 12 mm Cover Glass

1. One cover glass is placed into each well of a 24-well plastic tissue culture plate.

2. The lid is removed from the plastic tissue culture plate, and the plate containing coverslips is placed in a biosafety cabinet.

3. The open plate is exposed to UV light for 20 min at room temperature.

4. Following sterilization, wells are filled with 400 μL coculture media and placed in a humidified tissue culture incubator until ready to be seeded with cells.

3.2   Vital Labeling of Pericytes

1. Pericytes are suspended in serum-free DMEM at a density of 1×10^6 cells per mL in 15 mL conical tubes and mixed with 10 μL of the lipophilic, membrane-permeant tracer DiO for every mL of cell suspension. This suspension is then incubated for 20 min at 37 °C.

2. Following this incubation, the cells are pelleted by centrifugation at $600 \times g$ for 5 min. The media is then aspirated, and the cells are washed and are resuspended in PBS, followed by another centrifugation at $600 \times g$ for 5 min.

3. The PBS wash is aspirated, and the procedure is repeated two additional times.

4. After the final PBS wash, pericytes are then suspended in coculture media at a density of 1.5×10^5 cells/mL.

3.3   Coculture of DiO-Labeled Pericytes and Unlabeled Endothelial Cells

1. Capillary endothelial cells are mixed at a 1:1 ratio with vitally labeled pericytes in coculture media.

2. 50 μL of coculture cell suspension, containing 7500 of each cell type (a total of 15,000 cells), are added to UV-sterilized glass coverslips, prepared as described above.

3. Cocultures are maintained for a total of 24 h in a tissue culture incubator at 37 °C and 5 % CO_2.

3.4 Determining Mural/Endothelial Cell Cycle Progression In Vitro

1. At 20 h after plating, cocultures are pulsed with 50 μL of coculture media containing 100 μM EdU, for a final concentration of 10 μM EdU in solution.

2. Four hours later, at a total of 24 h after plating, cocultures are fixed with 4 % formaldehyde for 5 min at room temperature.

3. After fixation, cocultures are washed with azide-free PBS (*see* **Note 1**) three times for 5 min each.

4. Cocultures are then incubated in Click-iT EdU reaction mixture, prepared according to the manufacturer's instructions. Coverslips are placed cell side down on 25 μL drops of Click-iT EdU reaction buffer on parafilm in a humidified chamber for 1 h at room temperature, protected from light.

5. Subsequently, nuclear counterstaining is performed using Hoechst 33342 (1:1000, Life Technologies) under the same conditions described in the previous step (*see* **Note 2**).

6. Finally, coverslips are mounted on glass slides using PermaFluor or other aqueous mounting media, and allowed to dry prior to phase-contrast and fluorescence microscopy.

7. Quantitative image analysis of pericyte/endothelial cell cycle progression is performed by calculating ratios of EdU-positive nuclei to total nuclei for the cells of interest; this methodology can be applied to studies of heterotypic cellular contact-dependent and soluble growth-factor-driven S-phase entry. A representative image of pericyte-endothelial cell coculture is shown in Fig. 2.

3.5 Preparing Three-Dimensional Angiogenesis Arrays

1. Equilibrate 8-well chamber slides, 15 mL conical tubes, and coculture media at 4 °C for 1 h.

2. Once reagents are equilibrated, prepare extracellular matrix solution by mixing GFR matrigel, collagen I isolated from rat tail tendon, and coculture media in a 15 mL conical tube, at a ratio of 2:1:1 (*see* **Note 3**). Keep this mixture on ice.

3. Pipet 250 μL of matrix mixture into the bottom of each well of an 8-well chamber slide. Spread this mixture evenly across the well bottoms using a sterile, 18G needle.

4. Allow the matrix mixture to polymerize, by placing chamber slides containing the bottom matrix layer in a 37 °C, 5 % CO_2 tissue culture incubator for 1 h.

5. While the bottom matrix layer is polymerizing, prepare the pericyte-endothelial cell coculture as follows: to achieve a pericyte-endothelial cell ratio of 1:25, combine DiO-labeled pericytes (prepared as above) and unlabeled endothelial cells at densities of 2.4×10^4 and 4×10^5 cells/mL, respectively. Mix this suspension thoroughly, and store on ice until use.

6. Following polymerization, pipet the pericyte-endothelial cell suspension until the mixture is homogeneous, and add 250 μL

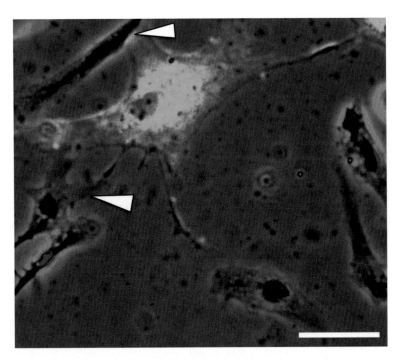

Fig. 2 Representative image of pericyte-endothelial coculture proliferation assay. Bovine retinal endothelial cells and DiO-labeled bovine retinal pericytes were cultured together at a 1:1 ratio on UV-sterilized glass coverslips in DMEM supplemented with 2 % BCS and antibiotics for 24 h. After 20 h of coculture, wells were pulsed with 10 μM EdU for the remaining 4 h. At the experimental end point, cocultures were fixed with 4 % formaldehyde, and the EdU "click" reaction was performed according to the manufacturer's instructions. In this image, a DiO-labeled pericyte is present in *green*, and *white* arrowheads denote areas of physical contact with endothelial cells. Hoechst 33342-labeled nuclei are depicted in *blue*, and an EdU-positive nucleus of an actively proliferating endothelial cell is shown in *red*

of coculture mixture to each well of an 8-well chamber slide. Allow cells to settle for 1–2 h in a tissue culture incubator at 37 °C and 5 % CO_2.

7. After visually confirming that cells are adherent on the bottom matrix layer under a microscope, carefully aspirate the supernatant, and add an additional 250 μL of matrix mixture to form the top layer. Allow this layer to polymerize as above, and incubate 3D arrays overnight to allow microvascular network formation (see schematic, Fig. 3).

3.6 In Vitro "Wounding" and Quantifying Angiogenic Activity

1. Prepare matrix mixture as described above, and store on ice until ready to use.

2. Visually confirm capillary network formation using phase-contrast and fluorescent microscopy.

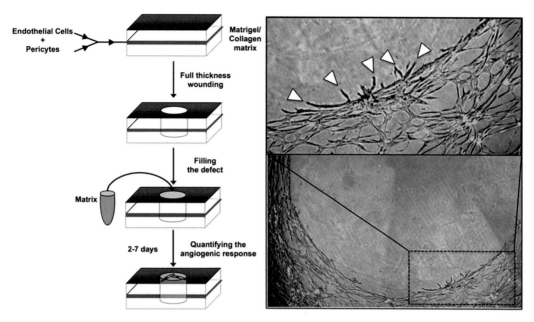

Fig. 3 Schematic and representative image of three-dimensional pericyte-endothelial cell coculture angiogenesis assay. DiO-labeled bovine retinal pericytes and bovine retinal endothelial cells were cultured together at a 1:25 ratio and embedded between two layers of an extracellular matrix solution. Tube networks were allowed to form overnight, and full-thickness wounds were created in the gels the following day. These defects were immediately filled with fresh matrix solution, and live fluorescent and phase-contrast images were captured immediately after injury and at 24-h intervals after injury. In this image, endothelial cells are unlabeled, DiO-labeled pericytes are present in *green*, and arrowheads within the inset denote areas of angiogenic sprouting into the defect

3. Attach a sterile, 22G blunt-end needle to a vacuum supply. Using a gentle suction, create a full-thickness defect in the microvascular network by placing the needle directly in the center of each well of the 8-well chamber slide.

4. Immediately fill this defect with 100 μL of matrix mixture (*see* **Note 4**), and capture images of each well using live phase-contrast and fluorescent microscopy on a temperature-controlled stage heated to 37 °C.

5. Additional images are captured daily over a period of 7 days. Quantitative analysis of angiogenic activity can be performed using NIH Image J, by subtracting any cells/structures inside the wound immediately after injury (day 0) from those structures present within the defect on subsequent days, thereby normalizing quantification for each wound and ensuring that only new angiogenic sprouts are measured (Fig. 3). In this way, average tube length, total tube length, and total number of sprouts can be calculated.

4 Notes

1. EdU staining can be performed immediately following fixation, or coverslips can be stored in 1× PBS, without sodium azide, for later use. Sodium azide may react with the EdU alkyne.

2. Postfixation immunostaining for cell-specific markers can be performed in between Click-iT EdU incubation and nuclear counterstaining.

3. Type I collagen is supplied as a solution in 0.02 N acetic acid. After combining the matrix components, neutralize the acidic solution with 10 μL 1 M NaOH for each mL matrigel-collagen I-media mixture.

4. Matrix mixture can be combined with growth factors, such as FGF2 or VEGF, or other substances of interest, to determine their impact on angiogenic sprouting in this model of microvascular injury and repair.

References

1. Conway EM, Collen D et al (2001) Molecular mechanisms of blood vessel growth. http://cardiovascres.oxfordjournals.org/content/49/3/507.short

2. Orlidge A, D'Amore PA (1987) Inhibition of capillary endothelial cell growth by pericytes and smooth muscle cells. J Cell Biol 105:1455–1462

3. Armulik A, Genové G, Mäe M, Nisancioglu M, Wallgard E, Niaudet C, He L, Norlin J, Lindblom P, Strittmatter K, Johansson B, Betsholtz C (2010) Pericytes regulate the blood-brain barrier. Nature 468:557–561

4. Hall CN, Reynell C, Gesslein B, Hamilton NB, Mishra A, Sutherland BA, O'Farrell FM, Buchan AM, Lauritzen M, Attwell D (2014) Capillary pericytes regulate cerebral blood flow in health and disease. Nature 508:55–60

5. Papetti M, Shujath J, Riley KN, Herman IM (2003) FGF-2 antagonizes the TGF-beta1-mediated induction of pericyte alpha-smooth muscle actin expression: a role for myf-5 and Smad-mediated signaling pathways. Invest Ophthalmol Vis Sci 44:4994–5005

6. Antonelli-Orlidge A, Saunders K, Smith S, D'Amore P (1989) An activated form of transforming growth factor beta is produced by cocultures of endothelial cells and pericytes. Proc Natl Acad Sci 86:4544–4548

7. Hellstrom M, Lindahl P, Abramsson A, Betsholtz C (1999) Role of PDGF-B and PDGFR-beta in recruitment of vascular smooth muscle cells and pericytes during embryonic blood vessel formation in the mouse. http://dev.biologists.org/content/126/14/3047.short

8. Greenberg JI, Shields DJ, Barillas SG, Acevedo LM, Murphy E, Huang J, Scheppke L, Stockmann C, Johnson RS, Angle N, Cheresh DA (2008) A role for VEGF as a negative regulator of pericyte function and vessel maturation. Nature 456:809–813

9. Wu D, Minami M, Kawamura H, Puro D (2006) Electrotonic transmission within pericyte-containing retinal microvessels. Microcirculation 13:353–363

10. Gerhardt H, Wolburg H, Redies C (2000) N-cadherin mediates pericytic-endothelial interaction during brain angiogenesis in the chicken. Dev Dyn 218:472–479

11. Sainson RC, Harris AL (2008) Regulation of angiogenesis by homotypic and heterotypic notch signalling in endothelial cells and pericytes: from basic research to potential therapies. doi: 10.1007/s10456-008-9098-0

12. Geevarghese A, Herman IM (2014) Pericyte-endothelial crosstalk: implications and opportunities for advanced cellular therapies. Transl Res 163:296–306

13. Dulmovits BM, Herman IM (2012) Microvascular remodeling and wound healing: a role for pericytes. Int J Biochem Cell Biol 44:1800–1812

14. Kutcher ME, Herman IM (2009) The pericyte: cellular regulator of microvascular blood flow. Microvasc Res 77:235–246

15. Nayak RC, Herman IM (2001) Bovine retinal microvascular pericytes. http://link.springer.com/10.1385/1-59259-143-4:247

16. Herman IM, Leung A (2009) Creation of human skin equivalents for the in vitro study of angiogenesis in wound healing. doi: 10.1007/978-1-59745-241-0_14

17. Mendel TA, Clabough EB, Kao DS, Demidova-Rice TN, Durham JT, Zotter BC, Seaman SA, Cronk SM, Rakoczy EP, Katz AJ, Herman IM, Peirce SM, Yates PA (2013) Pericytes derived from adipose-derived stem cells protect against retinal vasculopathy. PLoS One 8:e65691

18. Frank RN, Dutta S, Mancini MA (1987) Pericyte coverage is greater in the retinal than in the cerebral capillaries of the rat. http://www.iovs.org/content/28/7/1086.short

Chapter 16

Static and Dynamic Assays of Cell Adhesion Relevant to the Vasculature

Lynn M. Butler, Helen M. McGettrick, and Gerard B. Nash

Abstract

Methods are described for analyzing adhesion of isolated cells (such as leukocytes, tumor cells, or precursor cells) to purified adhesion receptors or cultured endothelial cells. "Static" assays (where cells are allowed to settle on the adhesive substrates) and flow-based assays (where cells are perfused over the substrates) are compared. Direct observations of the time course of adhesion and migration can be made when purified proteins or endothelial cells are cultured in plates, after cells are allowed to settle onto them for a desired period. In the flow-based assay, cells are perfused through coated glass capillaries, flow-channels incorporating coated plates, or commercially available preformed channels. Again, direct video-microscopic observations are made. In this assay various stages of capture, immobilization, and migration can be followed. In general, the static systems have higher throughput and greatest ease of use, but yield less detailed information, while the flow-based assay is most difficult to set up but is most physiologically relevant if one is interested in the dynamics of adhesion in the vasculature.

Key words Adhesion, Leukocyte, Endothelial cells, Tumor cells, Flow, Blood vessel

1 Introduction

Adhesion of several types of cells in the vascular system may be relevant to the process of angiogenesis: leukocytes, platelets, metastatic tumor cells, endothelial cell progenitors, and stem cells. We have recently described methods for studying dynamics of leukocyte migration [1] and attachment of flowing platelets [2] on endothelial monolayers. Here we concentrate on "generic" methods, both static and with imposed flow, for studying cell adhesion to purified proteins or to cultured endothelial cells. These may be adapted for a variety of adhesion proteins or endothelial stimuli, or for different adherent cells, relevant to specific scenarios chosen for investigation.

In general, flow-based assays are required to evaluate capture processes occurring in the vasculature, where cells are initially traveling rapidly on a scale of the receptors used. Microscopic observation of such models allows the different steps in the adhesion process to be dissected (e.g., capture, rolling, activation-dependent stabilization and migration, observed for leukocytes) [3–5]. While these stages have been well-described for flowing leukocytes, their existence for other cells types remains unclear and worthy of examination. Static assays are easier to set up and utilize, but results relating to the roles of specific receptors may be ambiguous when several are available, and one cannot be certain whether an interaction described could occur under more realistic flow conditions in the vasculature. Specificity in assays (static or flow-based) can be improved by choosing individual adhesion receptors to coat surfaces, while endothelial cells may be used to generate more "physiological" data, provided one realizes that the definition of the operative adhesive mechanisms may require subsequent detailed evaluation. Either type of surface can be used to assess motility of adherent cells, with endothelial monolayers offering the potential to study transendothelial diapedesis.

In this chapter, we describe static assays on purified protein and on cultured human umbilical vein endothelial cells (HUVEC). The proteins most relevant for these assays would be basement membrane constituents such as collagen, laminin, or fibronectin. HUVEC are a widely used primary cell capable of expressing a range of adhesion molecules. Alternatively, endothelial cell lines or EC whose derivation is described earlier (cf. Chapter-) might also be used. In the following description of flow-based assays, relevant purified proteins are those adapted for "capture" such as E-selectin, P-selectin, or vascular cell adhesion molecules-1 (VCAM-1). EC would typically be stimulated with agonist such as inflammatory cytokines (tumor necrosis factor-α, TNF, or interleukin-1, IL-1) so that such capture receptors are expressed [4, 6].

2 Materials

2.1 Culture of Endothelial Cells

1. Medium 199 with glutamine (M199) supplemented with gentamycin sulfate (35 μg/ml), human epidermal growth factor (10 ng/ml), and fetal calf serum (20 % v/v heat-inactivated). Adding hydrocortisone (1 μg/ml, from 10 mg/ml stock in ethanol) improves growth if going beyond first passage.

2. Phosphate buffered saline with 1 mM Ca^{2+} and 0.5 mM Mg^{2+} (PBS).

3. Bovine skin gelatin (Type B, 2 % solution, culture tested).

4. Collagenase (type IA) stored at –20 °C at 10 mg/ml in PBS . Thawed and diluted to 1 mg/ml with M199 for use.

5. Autoclaved cannulae and plastic electrical ties.

6. EDTA solution (0.02 %, culture tested).

7. Trypsin (culture tested).

8. 70 % (v/v) ethanol or industrial methylated spirits.

9. Tumor necrosis factor-α (TNF) and interleukin-1β (IL-1), stored in aliquots at 80 °C.

2.2 Purified Proteins *(See* **Note 1***)*

All dissolved in PBS at ~mg/ml and stored at –20 °C unless stated otherwise.

1. Basement membrane or extracellular matrix proteins:
   (a) Collagen type IV (Human placenta).
   (b) Fibronectin (Human foreskin fibroblasts).
   (c) Laminin type 1 (Human placenta).
   (d) Human Laminin type 10 (Human placenta) (Chemicon).

2. Capture receptors for flowing cells, e.g.:
   (a) Recombinant human VCAM-1 (R&D systems).
   (b) Recombinant human P-selectin stop protein (R&D systems).

3. Bovine serum albumin (BSA) (diluted in PBS from 7.5 % (w/v) culture-tested solution, typically to 1 % (w/v)).

2.3 Surfaces for Protein Coating or Endothelial Culture

1. *Multi-well plates*; we typically use 6-well plates but other formats may be used (*see* Subheadings 3.1.1 and 3.2.3).

2. *Chamber slides*: Plastic chamber with cover, mounted on glass slide.

3. *Microslides*: Glass capillaries with rectangular cross-section (0.3 mm × 3 mm; length 50 mm) (Vitro Dynamics Inc., New Jersey, USA; available through Camlab Ltd., Cambridge, UK).

4. Aminopropyltriethoxysilane 4 % (v/v) in acetone with molecular sieve (BDH Laboratory Supplies) added to ensure anhydrous.

5. A special culture dish for use with microslides (Fig. 1). Constructed by fusing either three or six glass tubing side arms into the wall of a Pyrex glass petri dish 100 mm in diameter. Made to order by the Glassblowing Workshop of the School of Chemistry, The University of Birmingham, UK.

6. Coatable preformed channels with a range of dimensions are commercially available:
   (a) Ibidi μ-Slides (www.ibidi.com).
   (b) Cellix Biochips (www.cellixltd.com).
   (c) BioFlux Plates (www.fluxionbio.com/bioflux/).
   (d) GlycoTech parallel plate flow chambers (http://www.glycotech.com/apparatus/parallel.html)

2.4 Flow-Based Adhesion Assay

1. *Flow system*: Syringe pump with smooth flow (e.g., PHD2000 infusion/withdrawal, Harvard Apparatus, South Natic, MA, USA).

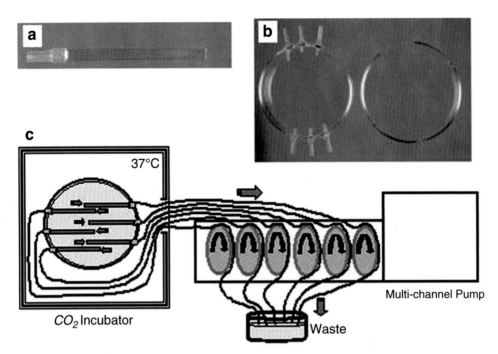

Fig. 1 Apparatus for culture of endothelial cells in microslides. (**a**) Microslide with silicone rubber tubing adapter attached. (**b**) Glass culture dish with six ports fused into the wall; diameter 90 mm; volume of culture ~50 ml. (**c**) Schematic diagram of apparatus for culture of endothelial cells in six microslides. The glass culture dish with six microslides attached is placed in a CO_2-incubator. Tubing attached to each outlet passes through a port in the incubator wall. Perfusion is supplied for 30 s in each hour by a multichannel roller pump which pumps to waste

Electronic 3-way microvalve with zero dead volume (LFYA1226032H Lee Products Ltd., Gerrards Cross, Buckinghamshire, UK.) with 12 V DC power supply for valve. Silicon rubber tubing, internal diameter/external diameter (ID/OD) of 1/3 and 2/4 mm (Fisher Scientific). Scotch double-sided adhesive tape, ~1 cm wide (3M Ltd.). Three-way stopcocks (BOC Ohmeda AB). Sterile, disposable syringes (2, 5, 10 ml) and glass 50 ml syringe for pump (Popper Micromate; Popper and Sons Inc., New York, USA).

2. For chamber slides: Custom made flow-channel assembly, made to order by Wolfson Applied Technology, University of Birmingham, UK, or by suitable workshop (Fig. 2). This includes a gasket, cut to the size of the chamber slide, with a slot that forms the flow channel when the assembly is clamped together (Fig. 2). The gasket can be cut fresh each time using lengths of Parafilm sealing film and a scalpel (typical depth ~130 μm), or cut from silicone rubber sheet (e.g., 250 μm thick; ESCO, Bibby Sterilin Ltd., Staffordshire, UK) and reused. In either case, the depth of the channel should be checked after the system is clamped together, by focussing alternately on the upper and lower surfaces of the channel on a microscope, using a calibrated microscope focus control.

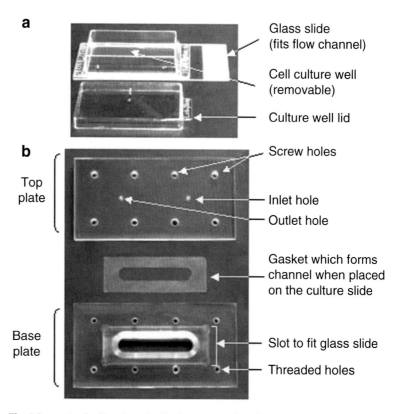

Fig. 2 Apparatus for flow-based adhesion assay using chamber slides. (a) Photograph of chamber slide incorporated in flow-channel. (b) Photograph of parallel plate flow-channel to hold chamber slide

3. Commercially available flow channels were listed above (Subheading 2.3, **item 6**). We have used the Ibidi μ-slide VI$^{0.4}$ in studies of leukocyte recruitment [7]. This slide contains six individual channels, and is suitable to perform analysis of cell recruitment as described in Subheading 3.3, **step 3**. We have no experience with the other systems. The Glycotech chambers have been incorporated in flow-based adhesion assays by others, e.g., ref. 8. This is also the case for Cellix [9] and Fluxion [10] devices which can be supplied with their own custom flow systems.

4. *Video-microscope*: Microscope with heated stage or preferably, with stage and attached flow apparatus enclosed in a temperature controlled chamber at 37 °C, and phase-contrast optics. Fluorescence capability is desirable for some variants of assay. Video camera (e.g., analog Cohu 4912 monochrome camera with remote gain control), monitor and video recorder (e.g., time lapse, Panasonic AG-6730), or digital camera for direct capture to computer.

5. *Image analysis*: Computer with video capture card (if recording to video) and specialist software for counting cells, measuring

motion etc. There are a range of commercial packages available, as well as image analysis software available free over the Internet (NIH Image http://rsb.info.nih.gov/nih-image/). We currently use Image Pro software.

3   Methods

This chapter aims to provide detailed protocols for adhesion assays that can be used for various cell types (such as leukocytes, stem cells or tumor cell lines). Cell isolation protocols are given in the chapters in Section 2 of this volume or in previous works on leukocyte adhesion [1]. Isolation of HUVEC is described below. We primarily use microscopy to directly examine the adhesive behavior of different leukocyte subsets to purified proteins or endothelial cells (but *see* **Note 2**). For flow-based assays various systems can be used. We have used glass capillaries (microslides) coated with either purified proteins or cultured endothelial cells [11]. Microslides are economical (content only ~50 µl) and easy to use with purified proteins, but endothelial cells require a specialized culture system (*see* Subheading 3.2.5). For endothelial cells, it may be easier initially to culture in commercially available chamber slides and incorporate them in a custom-made flow channel (Fig. 2), or to use commercially designed flow channels (*see* Subheading 2.4, **item 3**) or 35 mm dishes with the GlycoTech system. Flow systems have been used mainly with leukocytes, and they will need to be optimized (e.g., with respect to flow rates, cell concentrations etc.) for other cells types. In addition, pretreatment of the endothelial cells (e.g., with cytokines) may be necessary to stimulate them to present adhesion receptors for the cells under study.

3.1   Preparing Surfaces Coated with Purified Protein for Adhesion Assays

3.1.1   Coating Multi-well Plates (See **Note 3***) with Purified Proteins*

1. Make up the purified protein (e.g., a basement membrane or extracellular matrix constituent) to the desired concentration (*see* **Note 4**).

2. Add 1 ml to each well in a 6-well, or proportionately less in smaller-well formats.

3. Incubate for 2 h or overnight at 37 °C.

4. Aspirate off residual protein solution and add 2 ml 1 % (w/v) BSA to each well for 2 h to block nonspecific binding sites.

3.1.2   Coating Microslides with Purified Proteins

3.1.2.1   Pretreatment with APES

1. Immerse microslides in nitric acid (50 % (v/v) in distilled water) for 24 h (e.g., in batch ~100–300).

2. Wash thoroughly in beaker using running tap water and rinse through with deionized distilled water.

3. Blot water on tissue paper, and dry microslides at 37 °C.

4. Place in polystyrene tubes and rinse twice with anhydrous acetone by gently inverting the tubes over 30 s.

5. Immerse in a freshly prepared solution of APES (4 % v/v in anhydrous acetone) for 1 min, ensuring all capillaries are filled (*see* **Note 5**).

6. Remove microslides from the APES and blot out onto tissue, ensuring all capillaries are emptied.

7. Reinsert into a fresh aliquot of the APES solution for a further 1 min.

8. Remove APES by blotting and rinse the microslides once with anhydrous acetone, followed by three washes with deionized distilled water and then dry at 37 °C. Between each change, care must be taken to remove all the liquid from the microslides.

9. Attach a short length of 2 mm ID silicon rubber tubing to APES-coated microslide. The tubing assists in handling and filling of microslides, and is required for adaptation to the endothelial culture system.

10. Autoclave the microslides at 121 °C for 11 min and store aseptically, indefinitely.

3.1.2.2 Coating
with Purified Proteins

1. Using a pipettor inserted in the silicon tubing adaptor, aspirate ~50 µl of adhesion receptor at desired concentration into a microslide (*see* **Note 6**).

2. Incubate at 37 °C for 2 h.

3. Aspirate ~100 µl of PBS with 1 % (w/v) bovine albumin into a microslide as above and incubate at 37 °C for 2 h to block nonspecific binding sites.

4. In our experience, similar coating can be carried out for Ibidi µ-slides. For other commercial chambers, refer to manufacturer's instructions.

**3.2 Culture
of Endothelial Cells
on Chosen Surfaces**

There are various methods for culture of endothelial cells from different sources, and for the novice, it is probably best to start by buying cells and media from commercial suppliers. Our current method for isolating and culturing human umbilical vein endothelial cells is given below, adapted from Cooke et al. [11], which also described methods for coating microslides (*see* "Pretreatment with APES") and culturing HUVEC in them (Subheading 3.2.5).

3.2.1 Isolation
and Primary Culture
of HUVEC

1. Place the cord on paper toweling in a tray and spray liberally with the 70 % ethanol. Choose sections of about 3–4 in. that do not have any clamp damage. Each 3–4 in. piece of cord equates to 25 cm² flask of primary cells.

2. Locate the two arteries and one vein at one end of the cord.

3. Cannulate the vein and secure the cannula with an electrical tie.

4. Carefully wash through the vein with PBS using a syringe and blow air through to remove the PBS.

5. Cannulate the opposite end of the vein and tie off.

6. Inject collagenase (~10 ml per 3–4 in.) into vein until both cannulae bulbs have the mixture in them.

7. Place the cord into an incubator for 15 min at 37 °C.

8. Remove from the incubator and tighten the ties. Massage the cord for ~1 min.

9. Flush the cord through using a syringe and 10 ml PBS into a 50 ml centrifuge tube.

10. Push air through to remove any PBS, repeat this twice more (3×10 ml).

11. Centrifuge at $400 \times g$ for 5 min. Discard supernatant.

12. Resuspend the cells in ~1 ml of culture medium and mix well with pipette

13. Make up to 4 ml in complete medium.

14. Add cell suspension to a 25 cm² culture flask.

15. Culture at 37 °C in a 5 % CO_2 incubator.

16. Change medium after 2 h, again the next day, and every 2 days thereafter. Cells should be confluent in 3–7 days.

3.2.2 Dispersal of Endothelial Monolayers for Seeding New Surfaces

1. Rinse a flask containing a confluent primary monolayer of HUVEC with 2 ml EDTA solution.

2. Add 2 ml of trypsin solution and 1 ml of EDTA for 1–2 min at room temperature, until the cells become detached. Tap on bench to loosen.

3. Add 8 ml of culture medium to the flask, to neutralize the trypsin, and remove the resulting suspension and centrifuge at $400 \times g$ for 5 min.

4. Remove supernatant and resuspend the cell pellet in 0.5 ml of culture medium and disperse by sucking them in and out of a 1 ml pipette tip.

5. Make up to desired volume of culture medium for seeding onto the chosen surface.

3.2.3 Seeding HUVEC in Multi-well Plates (See **Note 3***)*

1. Add 1 ml of 1 % gelatin (in PBS) to each well in a 6-well, or proportionately less in smaller-well formats, for 15 min.

2. Trypsinize a single flask of HUVEC as in Subheading 3.2.2.

3. Make up to 8 ml with culture medium (*see* **Note 7**) and add 2 ml of HUVEC suspension to each of 4 wells, or proportionately less in smaller-well formats.

4. Culture at 37 °C in a 5 % CO_2 incubator.

5. Replace the medium 24 h later and culture for 1–3 days (*see* **Note 7**).

6. If required add cytokine stimulant, e.g., TNF (100 U/ml) or IL-1 (5×10^{-11} g/ml), to cultures for desired period before assay (e.g., typically 4–24 h when studying neutrophils or lymphocytes).

3.2.4 Seeding HUVEC in Chamber Slides

1. Trypsinize a single flask of HUVEC as in Subheading 3.2.1.
2. Make up to 6 ml with culture medium (*see* **Note 7**).
3. Remove chamber cover and add 2 ml to each Lab-Tek Chamber Slide (Nalge Nunc International, Naperville, IL) (Fig. 2).
4. Replace plastic cover.
5. Use the tray in which the chamber slides come as an incubation rack and incubate at 37 °C in a 5 % CO_2 incubator.
6. Replace the medium after 24 h and culture for 1–3 days.
7. Add cytokine stimulant (as suggested above) to culture for desired period before migration assay.

3.2.5 Culturing HUVEC in Microslides (See **Note 8***)*

1. Prepare the special culture dish (Subheading 2.3, **item 5**) by attaching a length of silicon rubber tubing (40 cm long) onto each external arm.
2. Autoclave the dish with tubing attached at 121 °C for 11 min before use.
3. Draw in 1 % gelatin (in PBS) using a pipettor with tip inserted into the adaptor tubing on microslides. Allow to coat for 30 min. Microslides hold ~50 μl.
4. Wash microslides with PBS followed by air to remove excess gelatin using a pipettor with tip inserted into the adaptor tubing.
5. Trypsinize a single flask of HUVEC as in Subheading 3.2.2.
6. Resuspend cell pellet to ~400 μl with culture medium and transfer to one corner of a tilted 35 mm petri dish.
7. Aspirate ~50 μl of cell suspension into each of the six microslides using a pipettor inserted in the silicon tubing adaptor.
8. Place a sterile, glass microscope slide inside a 100 mm petri dish, rest the filled microslides across it (to keep them horizontal), and incubate in the dish at 37 °C for 1 h.
9. Add 50 ml of culture medium to the special culture dish, prime the tubing and clamp ends.
10. Connect the microslides aseptically to the internal side-arms of the special culture dish, via adaptor tubing, using sterile forceps.
11. Placed the dish in a humidified CO_2 incubator and pass the silicon tubing through a service port located in the incubator wall (manufactured to order; e.g., either model GA2000, LEEC Ltd., Nottingham, UK or Nuaire DH; Triple Red, Thame, UK).

12. Attach the tubing to individual channels of a multichannel, roller pump (e.g., Watson Marlow 500 series pump with 308MC pumpheads; Watson-Marlow Bredel Pumps Ltd., Falmouth, UK), itself linked to a timed power supply (e.g., RS Components Ltd., Corby, UK.).

13. Pump medium through each microslide to waste (*see* Fig. 1 for layout) at a flow rate ~0.2 ml/min for 30 s in each hour to change medium contained in the microslides.

14. The original seeding is designed to yield confluent monolayers within 24 h.

15. After 24 h, treat HUVEC with TNF (100 U/ml) or IL-1 (5×10^{-11} g/ml) if desired. Cytokines can be added to the dish to treat all microslides equally. Or, detach the microslides with tubing adaptor and place them in a separate disposable plastic culture dish. Aspirate differently diluted cytokines into the separate microslides as desired and repeat aspiration at hourly intervals.

3.2.6 Culturing HUVEC in Ibidi μ-Slides VI$^{0.4}$ or Other Commercial Chambers

1. Trypsinize a single flask of HUVEC as in Subheading 3.2.2.

2. Resuspend cell pellet to ~600 μl with culture medium.

3. Aspirate ~30 μl of cell suspension into each of the six channels, by placing the pipette at the entrance of the channel and gently releasing the cells.

4. Incubate the slide at 37 °C for 1 h.

5. Add 100 μl of medium to one end of the channel and remove from the other to remove non-adherent endothelial cells.

6. Add 100 μl of medium and return the slide to the incubator for 24 h prior to use in the assay.

7. Other commercial chambers may also be coated with endothelial cells; for methods, refer to manufacturer's instructions.

3.3 Static Cell Adhesion Assays

*3.3.1 Cell Adhesion to Purified Proteins or Endothelial Cell Monolayers (See **Note 2**)*

1. Pre-warm the microscope and PBSA wash buffer to 37 °C.

2. Pretreat the endothelial cells with cytokine if desired.

3. Rinse the surface of the plate/monolayer (6-well format) with 2 ml PBSA, using a plastic pipette to remove any residual protein.

4. Add 2 ml of cell suspension to each well.

5. Leave to settle for desired time (*see* **Note 9**).

6. Aspirate off the cellular suspension and gently rinse twice with PBS.

7. Add 2 ml PBSA and view the well under phase-contrast video-microscope with an objective magnification of ×20.

8. Make video/digital recordings immediately after the wash stage, choosing at least five different fields at random and record

them for 5 s each, to allow counting of adherent cells. If migratory behavior is of interest, choose one field at random and record continuously for 5 min (for leukocytes) or longer for analysis of movement (*see* **Note 10**).

3.3.2 Data Analysis

1. Make video/digital recordings of a microscope stage micrometer to calibrate the size of the field observed on the monitor and to calibrate scale of the image analysis system.

2. To measure cell *adhesion*: Count all cells visible in each video field (*see* e.g., Fig. 3) (*see* **Note 11**). Take the average for the fields and convert to number per mm² using known dimensions of field. Multiply this by the area of the wells (9.6 cm² for 6-wells) and divide by the number of cells added. Multiply by 100 to obtain percentage of cells adherent.

3. To measure cell *migration*: Take images at intervals from the prolonged video/digital sequence into a program such as Image Pro (DataCell, Finchampstead) in a sequence (e.g., at 1 min intervals for leukocytes). Outline cells using a pointer and record the positions of their centroids. Calculate the distances and directions of cell migration in each interval. Calculate the average migration velocity over time and the direction of migration as required.

4. To measure cell *transmigration in endothelial cell assays* (*see* **Note 11**): Express the count of phase-dark cells in each video field as a proportion of the total number of adherent cells. Take the average for the repeated fields. For leukocytes, phase dark cells are under the monolayer. This can be verified, e.g., by removing the endothelial layer with trypsin, to leave the transmigrated cells [5].

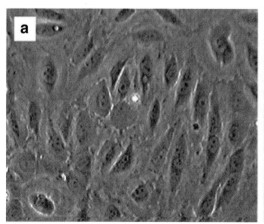 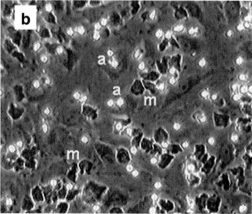

Fig. 3 Phase-contrast micrographs of confluent monolayers of (**a**) untreated HUVEC or (**b**) HUVEC treated with100 U/ml TNF, after completion of a neutrophil adhesion assay. Neutrophils were settled for 5 min on HUVEC, and the non-adherent cells were washed off, and allowed to migrate for a further 15 min. In (**b**) phase bright neutrophils are adherent to the surface of the HUVEC (a) and phase-dark spread cells (m) are migrated underneath the monolayer

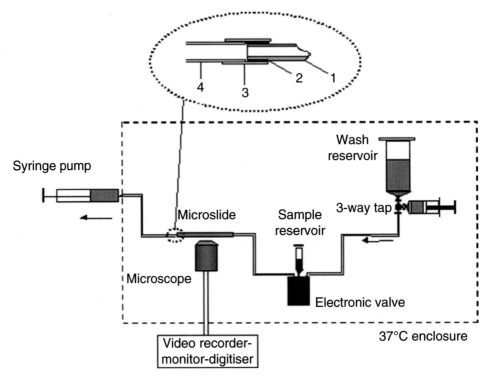

Fig. 4 Schematic of flow-based adhesion assay flow system. In expanded section of C. 1 = microslide, 2 = double-sided tape, 3 = silicon rubber tubing with ID/OD 2/4 mm, 4 = silicon rubber tubing with ID/OD 1/2 mm

3.4 Flow-Based Assay of Cell Adhesion

3.4.1 Setting up the Flow Assay

1. Assemble flow system shown in Fig. 4 but without chamber slide or microslide attached. The electronic valve has a common output, and two inputs, from "Wash reservoir" and "Sample reservoir," which can be selected by turning the electronic valve on or off.

2. Fill wash reservoir with PBSA and rinse through all tubing, valves, and connectors with PBSA, ensuring bubbles are displaced (e.g., using syringe attached to 3-way tap for positive ejection). Fill sample reservoir with PBSA and rinse through valve and attached tubing. Prime downstream syringe and tubing with PBSA and load into syringe pump. All tubing must be liquid-filled.

3.4.2 Connecting the Chamber Slides (Endothelial Cell Coated)

1. Remove the medium from the chamber slide.

2. To detach slide from the chamber, grip one end of slide and gently squeeze both ends of the chamber toward the center, lifting the chamber as its sealing gasket releases.

3. To remove the gasket: insert the tip of a thin bladed spatula or similar tool under gasket at one corner. Without stretching or tearing the gasket, smoothly lift it away from the slide (*see* **Note 12**).

4. Use dampened cotton bud to remove any gasket residue.

5. Put the glass slide in the bottom Perspex plate of the flow channel and place a pre-cut gasket over the slide. Put the top Perspex plate onto the gasket (Fig. 2).

6. Screw the top and bottom Perspex plates together evenly, tightening diagonally opposite screws by turn, a bit at a time.

7. Put the tubing into the inlet and outlet holes in the top Perspex plate of flow chamber.

3.4.3 Connecting the Microslides (Purified Protein or Endothelial Cell Coated)

1. Glue a coated microslide across the middle of a glass microscope slide using two spots of cyanoacrylate adhesive (Superglue Locktite UK., Welwyn Garden City, UK.) applied to the edges of the slide. Discard tubing adaptor used for filling.

2. Wrap double-sided adhesive tape around each end of the microslide, without obstructing lumen.

3. Connect microslide to silicon rubber tubing by pushing over each of the taped ends. Start at the upstream (sample) end to avoid injection of air. Squeeze the 2 mm ID silicone rubber tubing to flatten and ease over rectangular end of microslide, one corner first.

3.4.4 Connecting the Coated Ibidi μ-Slides or Other Commercial Chambers

1. Mount the slide on the microscope stage, using a slide holder or some adhesive tape.

2. Connect the supplied tubing, with Luer fittings, to the system and connect to the slide through the ports. Start at the upstream (sample) end to avoid the introduction of air bubbles.

3. Cellix and Fluxion systems are supplied with custom-made flow systems specific to these devices.

3.4.5 Perfusing Cellular Suspension and Recording Behavior

1. Place chamber slide or microslide onto microscope stage and start flow by turning on syringe pump in withdrawal mode, with electronic valve and 3-way tap in position to allow delivery of PBSA from wash reservoir.

2. Wash out culture medium and observe surface using phase-contrast microscopy.

3. Adjust flow rate to that required for assay (*see* **Note 13**).

4. Perfusion of leukocytes through microslides in our group is typically carried out at a flow rate ($Q=0.37$ ml/min) equivalent to a wall shear rate of 140 s^{-1} (which is comparable to that found in post-capillary venules) and wall shear stress of 0.1 Pa (=1 dyn/cm^2). Descriptions of flow rates and corresponding shear stresses for the commercially available chambers are available from the manufacturers.

5. Load isolated cells into sample reservoir (Fig. 4) and allow to warm for 5 min.

6. Switch the electronic valve so cell suspension is drawn through microslide.

7. Deliver timed bolus (e.g., 4 min). Typically, flowing cells will be visible after about 30 s, the time required to displace dead volume in valve and tubing.

8. Switch electronic valve so that PBSA from wash reservoir is perfused. Again, 30–60 s will be required before all cells have been washed through the chamber slide or microslide.

9. Video recordings can be made as desired during inflow and washout of cells. Typically, a series of fields should be recorded along the centreline of the microslide during inflow (e.g., six fields recorded for 10 s each during the last minute of the bolus), for offline analysis of the behavior (e.g., rolling or stationary adhesion) of the cells. Another series should be made after 1 min washout (when the bolus is complete) for analysis of the numbers of adherent cells. Fields are recorded over a desired period, or at later times (e.g., after a further 5 and 10 min) to assess cellular movement or transmigration through the monolayer when endothelial cells are used. All analyses are made offline.

10. If a defined timing protocol is developed, digital images, or sequences of digital images could be recorded instead of video images. The continual recording of the latter gives flexibility in analysis.

3.4.6 Analysis of Cell Behavior from Video Recordings

1. Calibrate size of video fields and image analysis system as in Subheading 3.3, **step 3**.

2. Digitize a sequence of images at 1 s intervals from recordings made at desired times.

3. Count cells present on a stop-frame video field at the start of a sequence. Repeat and average counts for the series of sequences recorded, e.g., after washout of non-adherent cells. Convert this to count of adherent cells/mm^2.

4. Divide this by the numbers of cells perfused (in units of 10^6 cells) to obtain number adherent/mm^2/10^6 perfused. The number perfused is simply calculated by multiplying the concentration of the suspension (typically 10^6/ml) by the flow rate (e.g., 0.37 ml/min) by the duration of the bolus (e.g., 4 min). This normalization allows correction for changes in conditions (bolus duration, cell concentration, flow rate) between experiments, and effectively calculates an efficiency of adhesion.

5. The count includes cells that are rolling (circular phase bright cells tumbling slowly at ~1–10 μm/s over the surface) or stably adherent on the endothelial surface (phase bright cells typically with distorted outline and migrating slowly on the surface at <10 μm/min) or transmigrated cells (phase-dark spread cells

migrating under the HUVEC at >10 μm/min). Non-adherent cells will only be visible as blurred streaks.

6. To assist in obtaining data for proportion of adherent cells behaving in the different ways, play the digitized sequence as a loop. Observing cells in turn it is easy to classify them as rolling, stably adherent or transmigrated.

7. Repeat the analysis at different times (e.g., after 1, 5, or 10 min of washout) to quantify the progress of migration through the endothelium or any changes in behavior such as transmigration through the endothelial monolayer.

8. To measure rolling velocity, mark the leading edges of a series of cells to be followed and move to second captured frame. Re-mark the leading edges and record the distance moved. Repeat through the 10 s sequence. This will yield data for position versus time. Velocity for each cell can be averaged over the observation time, and estimates of variation in velocity made if desired.

4 Notes

1. The proteins suggested are typical of those used in static or flow-based assays for leukocytes. Previous studies of leukocyte adhesion have used laminin type I from Elizabeth-Horm-Sarcoma, but this is not strictly appropriate as it is only found in the prenatal human basement membrane. The alternative laminin mixture, containing predominantly LN10, along with LN1, LN2, LN6, and LN8, may be more physiologically relevant.

2. Microscopy gives unequivocal, direct evaluation of cell adhesion and allows the state of the cells and surface, and uniformity of adhesion to be checked. There are alternative methods, for example, a radioactive or fluorescent label, or a dye can be preloaded into the target cells. The label is released upon lysis of the cells adherent to the surface and quantitated by measurement of radioactivity, fluorimetry or densitometry respectively. There can be problems of label leakage and also uptake by the endothelial cells. Other methods include total lysis of adherent cells and measurement of total protein., or analysis of a released enzyme (e.g., myeloperoxidase from neutrophils) or cell-specific marker. All of these approaches should be calibrated using known numbers of cells to obtain quantitative results.

3. Six well plates are useful for adhesion assays as the large well format allows for efficient washing and provides a large, optically clear, area to view under the microscope. However, if limited purified proteins or cells are available, a smaller well format can

be used. 12- or 24-well plates give reasonable washing and viewing. Others have used 96-well plates, but visual quantification of adhesion is problematic (due to the limited area and poor optical properties). In our experience the problem with well-based assays lies in difficulty in achieving efficient washing and the tendency of cells to collect around the edges of the well. This is worse the smaller the well, and higher nonspecific background adhesion occurs which decreases sensitivity. Direct visual observation alleviates this problem to a degree, but is again better in larger wells. With the smaller well plates, washing can be improved using a swinging-bucket, plate centrifuge, with a sealed plate, to "spin" cells off the surface before analysis.

4. The main constituents of the endothelial basement membrane are laminin, collagen type IV, and fibronectin, which we have used at coating concentrations ranging between 0.02 and 20 μg/ml. A 0.2 μg/ml laminin solution deposited protein onto the surface equivalent to that found in the basement membrane deposited from HUVEC which had been cultured for 20 days (measured by ELISA) [12]. The same was true for a collagen solution of around 2 μg/ml. It is advisable to titrate proteins for particular cells under study.

5. Successful coating with APES requires that the reagent be anhydrous. We buy small volumes adequate to coat a batch of microslides, use a fresh bottle for each batch, and discard any unused reagent. It is important to efficiently remove all liquid from each microslide between changes, and to ensure all bubbles are displaced on refilling with agents.

6. In general, microslides are coated with receptors involved in capture of flowing cells, such as P- or E-selectin, or VCAM-1 (–), where concentrations ~1 μg/ml have been effective. Leukocytes typically form unstable rolling attachments on these receptors.

   In our hands, titration of P-selectin has shown that as the concentration used increases, attachment quickly saturates, but velocity at which cells decreases steadily. If capture receptors are combined with other receptors for firm adhesion, then cells may become stably attached and migrate. For leukocytes this typically requires addition of an activating agent, such as a chemokine, to activate integrin receptors.

7. One confluent 25 cm² flask of HUVEC, resuspended in 6 ml, will seed 3 wells (2 ml per well), to produce a confluent monolayer within 24 h. Alternatively, one 25 cm² flask can be resuspended in 8 ml and used to seed four wells, which will be confluent in 2–3 days. We have found no difference in neutrophil adhesion or migration on 1 or 3 day cultures [12]. Chamber slides can be seeded as if equivalent to a single well in a 6-well plate.

8. The microslide culture system is one we have developed for culture of endothelial cells under flow as well as for testing of adhesion and migration [13]. The culture dishes can have three ports or six ports. Three port dishes are ideal for setting up different cytokine treatments or culture regimes, whilst six port dishes are ideal when those parameters are fixed but the endothelial cells or perfused cells are different. A system is available commercially (GlycoTech, Rockville, MD) based on culture of endothelial cells in 35 mm diameter culture dishes. An adaptor and gasket are inserted on top of the cells and held in place by vacuum. The device allows perfusion of fluid (± suspended cells of interest) over a central region of the endothelial surface.

9. When studying neutrophil adhesion we found 5 min to be an appropriate settling time. This short duration and the effective wash procedure in 6-well plates results in little or no background neutrophil adhesion on albumin coated proteins or unstimulated HUVEC controls (*see* Fig. 3a). However, the settling time needs to be tested depending on the type of cell being studied, and may need to be longer for poorly adherent cells.

10. We have found that 5–10 min is sufficient to visualize neutrophils migrating on proteins or over, through and underneath endothelial monolayers. However, the rate of migration varies between cell types. Longer recordings maybe required allowing analysis of migration over hours rather than minutes e.g., for endothelial cells or fibroblasts.

11. Adherent leukocytes can be classified into two groups: (1) phase-bright cells adherent to the upper surface of the endothelial cells; (2) phase-dark, spread cells that are transmigrated under the endothelial monolayer (*see* Fig. 3b). Leukocyte adhesion levels should not change between the two sets of recordings—only the proportion that have transmigrated.

12. Take care with this procedure, so as not to disrupt the endothelial monolayer.

13. The flow rate (Q) required to give a desired wall shear rate (γ_w in s^{-1}) or wall shear stress (τ_w in Pascal, Pa) is calculated from the internal width (w) and internal depth (h) of the microslide (or flow channel) and the viscosity (n) of the flowing medium using the formulae,

$$\gamma = \left(6Q\right)/\left(wh^2\right); \tau = n\gamma$$

For microslides, since $w = 3$ mm, $h = 0.3$ mm, $n = 0.7$ mPa s for simple cell-suspension buffers at 37 °C, this can be manipulated to give,

$$Q\left(\text{ml}/\text{min}\right) = 0.0027\gamma_w\left(\text{s}^{-1}\right) \text{ or } Q\left(\text{ml}/\text{min}\right) = 3.95\tau_w\left(\text{Pa}\right)$$

When using flow channels which are formed by slots cut in the gasket, the length and width of the slot, and the thickness of the gasket (i.e., depth) are used in the calculation.

References

1. McGettrick HM, Butler LM, Nash GB (2005) Analysis of leukocyte migration through monolayers of cultured endothelial cells. Methods Mol Biol 370:37–54

2. Nash GB (2004) Adhesion between platelets and leukocytes or endothelial cells. Methods Mol Biol 272:199–214

3. Springer TA (1995) Traffic signals on endothelium for lymphocyte recirculation and leukocyte emigration. Annu Rev Physiol 57:827–872

4. Bahra P, Rainger GE, Wautier JL, Luu N-T, Nash GB (1998) Each step during transendothelial migration of flowing neutrophils is regulated by the stimulatory concentration of tumour necrosis factor-alpha. Cell Ad Commun 6:491–501

5. Luu NT, Rainger GE, Nash GB (1999) Kinetics of the different steps during neutrophil migration through cultured endothelial monolayers treated with tumour necrosis factor-alpha. J Vasc Res 36:477–485

6. Smith CW, Kishimoto TK, Abbassi O, Hughes B, Rothlein R, McIntire LV, Butcher E, Anderson DC, Abbass O (1991) Chemotactic factors regulate lectin adhesion molecule 1 (LECAM-1)-dependent neutrophil adhesion to cytokine-stimulated endothelial cells in vitro. J Clin Invest 87:609–618

7. Butler LM, Jeffery HC, Wheat RL, Rae PC, Townsend K, Alkharsah KR, Schulz TF, Nash GB, Blackbourn DJ (2011) Kaposi's sarcoma-associated herpesvirus infection of endothelial cells inhibits neutrophil recruitment through an interleukin-6-dependent mechanism: a new paradigm for viral immune evasion. J Virol 85:7321–7332

8. Cinamon G, Shinder V, Alon R (2001) Shear forces promote lymphocyte migration across vascular endothelium bearing apical chemokines. Nature Immunol 2:515–522

9. Konya V, Ullen A, Kampitsch N, Theiler A, Philipose S, Parzmair GP, Marsche G, Peskar BA, Schuligoi R, Sattler W, Heinemann A (2013) Endothelial E-type prostanoid 4 receptors promote barrier function and inhibit neutrophil trafficking. J Allergy Clin Immunol 131:532–540

10. Warren KJ, Iwami D, Harris DG, Bromberg JS, Burrell BE (2014) Laminins affect T cell trafficking and allograft fate. J Clin Invest 124:2204–2218

11. Cooke BM, Usami S, Perry I, Nash GB (1993) A simplified method for culture of endothelial cells and analysis of adhesion of blood cells under conditions of flow. Microvasc Res 45:33–45

12. Butler LM, Rainger GE, Rahman M, Nash GB (2005) Prolonged culture of endothelial cells and deposition of basement membrane modify the recruitment of neutrophils. Exp Cell Res 310:22–32

13. Sheikh S, Gale Z, Rainger GE, Nash GB (2004) Methods for exposing multiple cultures of endothelial cells to different fluid shear stresses and to cytokines, for subsequent analysis of inflammatory function. J Immunol Methods 288:35–46

Part IV

In Vivo Techniques

Chapter 17

Dorsal Skinfold Chamber Preparation in Mice: Studying Angiogenesis by Intravital Microscopy

Axel Sckell and Michael Leunig

Abstract

Intravital microscopy represents an internationally accepted and sophisticated experimental method to study angiogenesis, microcirculation, and many other parameters in a wide variety of neoplastic and non-neoplastic tissues. Since 1924, when the first transparent chamber model in animals was introduced, many other chamber models have been described in the literature for studying angiogenesis and microcirculation. Because angiogenesis is an active and dynamic process, one of the major strengths of chamber models is the possibility of monitoring angiogenesis in vivo continuously for up to several weeks with high spatial and temporal resolution. In addition, after the termination of experiments, tissue samples can be excised easily and further examined by various ex vivo methods such as histology, immunohistochemistry, and molecular biology. This chapter describes the protocol for the surgical preparation of a dorsal skinfold chamber in mice as well as the method to implant tumors in this chamber for further investigations of angiogenesis and other microcirculatory parameters. However, the application of the dorsal skinfold chamber model is not limited to the investigation of neoplastic tissues. To this end, the investigation of angiogenesis and other microcirculatory parameters of nonneoplastic tissues such as tendons, osteochondral grafts, or pancreatic islets has been an object of interest.

Key words Angiogenesis, Intravital microscopy, Microcirculation, Skinfold chamber, Human tumor, Mouse model

1 Introduction

Since 1924, when the first transparent chamber model in animals was introduced by Sandison [1], many other chamber models have been described in the literature for studying angiogenesis and microcirculation in a wide variety of neoplastic and nonneoplastic tissues by means of intravital microscopy (for reviews *see* refs. [2–4]). Because angiogenesis is an active and dynamic process, one of the major strengths of chamber models is the possibility of monitoring angiogenesis in vivo continuously up to several weeks with high spatial and temporal resolution. In addition, after the termination of experiments, tissue samples can be excised easily and further

Stewart G. Martin and Peter W. Hewett (eds.), *Angiogenesis Protocols*, Methods in Molecular Biology, vol. 1430,
DOI 10.1007/978-1-4939-3628-1_17, © Springer Science+Business Media New York 2016

examined by various ex vivo methods such as histology, immuno-histochemistry, and molecular biology.

The advantages of using mice as experimental animals are, for instance, the availability of a large number of different well-defined mouse strains, including transgenic or knockout mice, and the wide variety of commercially generated agents suitable for mice, such as monoclonal antibodies, nanoparticles, and single-gene products.

This chapter describes the protocol for the surgical preparation of the dorsal skinfold chamber in mice as well as the method to implant tumors in this chamber for further investigations of angiogenesis and other microcirculatory parameters. The model [5, 6] presented here is the development of a similar model in hamsters [7]. In brief, take a fold of the depilated dorsal skin of an anesthetized mouse and cut out surgically a circular area of one skin layer (consisting of epidermis, dermis, subcutis, cutaneous muscle, and subcutaneous fatty tissue) completely. Then, fix the skinfold like a sandwich between the two titanium frames of the chamber and close the operation field with a sterile coverslip to avoid drying, infection, or mechanical damage of the inner layer (i.e., the cutaneous muscle) of the unprotected side of the opposite skin. For tissue implantation or other local treatments, the chamber can easily be opened again by removing the coverslip and be reclosed with a new sterile coverslip. The cutaneous muscle serves as a site for implantation of tissues such as little chunks of solid tumors (*see* Subheading 3.2). From now on, intravital microscopy can be performed for monitoring angiogenesis and other parameters such as tumor growth, microvascular perfusion index, microcirculation, and leukocyte endothelium interaction. However, the application of the dorsal skinfold chamber model is not limited to the investigation of neoplastic tissues. To this end, the investigation of angiogenesis and other microcirculatory parameters of nonneoplastic tissues such as tendons [8], osteochondral grafts [6, 9], or pancreatic islets [10] has been an object of interest. In principle, the implantation of these tissues follows the same rules like the implantation of tumors as described in this chapter (*see* Subheading 3.2).

2 Materials

2.1 Dorsal Skinfold Chamber Preparation

2.1.1 Facilities and Apparatus

1. Laminar flow hood.
2. Dry sterilizer.
3. Dissecting microscope.
4. Two flat custom-made thermal pads.
5. Halogen lamp with two flexible swan-neck light transmission tubes.

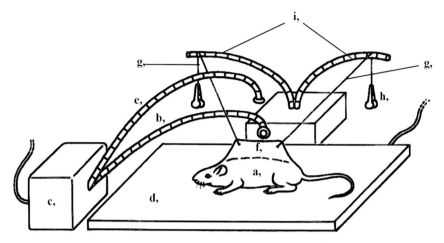

Fig. 1 Setup for the surgical preparation inside the hood: *a* mouse, *b* swan-neck light transmission tube 2 (for transillumination from behind), *c* halogen lamp, *d* thermal pad, *e* swan-neck light transmission tube 1 (for epiillumination from above), *f* skinfold, *g* holding thread, *h* baby mosquito, *i* flexible swan-necks of the skin-spreading device

6. Custom-made skin-spreading device (Workshop, Department of Experimental Surgery, University of Heidelberg, Heidelberg, Germany; Fig. 1). This consists of a heavy metal base and two flexible swan-neck tubes for applying tension to spread out the mouse skinfold, prior to the fixation of the back titanium frame of the dorsal skinfold chamber to the skinfold of the mouse.

2.1.2 Drugs

1. Isotonic sodium chloride (0.9 % NaCl solution injectable).

2. Anesthesia: Mixture consisting of isotonic sodium chloride, ketamine hydrochloride, xylazine, and acepromazine.

3. Depilatory cream.

2.1.3 Dorsal Skinfold Chambers

1. Custom-made dorsal skinfold chambers (Workshop, Department of Experimental Surgery, University of Heidelberg, Heidelberg, Germany; Fig. 2) consisting of 2 titanium frames, 3 screws (M2 × 6), 6 nuts (size 4), and one tension ring to keep the sterile coverslip in position after closing of the chamber preparation.

2. Sterile coverslips (0.13 to 0.16 mm thick, 11.75 mm diameter, circular, Assistent, Sontheim, Germany).

3. Special pair of pliers (Garant®, Germany) to bring the tension ring in position to keep the chamber closed and to remove it again, respectively.

4. Wrench (CHR-VAN, size 4, SKG, Germany).

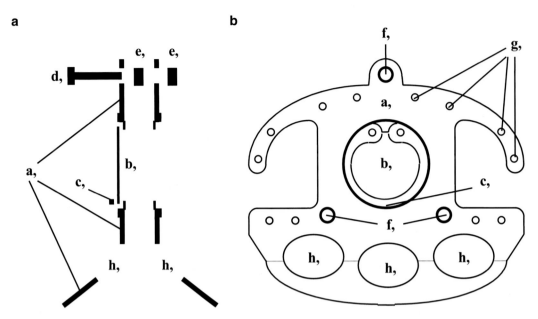

Fig. 2 Construction plan of the dorsal skinfold chamber: (**a**) Cross-section; (**b**) lateral view: *a* titanium frame, *b* coverslip, *c* tension ring, *d* screw, *e* nut, *f* screw holes, *g* bore holes for holding sutures, *h* holes for weight reduction of the chamber

2.1.4 Surgical Instruments

1. Electric hair clipper equipped with a 1/20 mm cutting head (GH 700; Aesculap®).

2. Two delicate hemostatic forceps (baby mosquito, BH 115, Aesculap®).

3. One needle holder (Castroviejo, BM 2, Aesculap®).

4. One delicate dissecting forceps (Micro-Adson, BD 220, Aesculap®).

5. Two microforceps (BD 331, Aesculap®).

6. One pair of dissecting scissors, fine patterns (Cottle-Masing, sharp, OK 365, Aesculap®).

7. One pair of microscissors (spring type) with round handles (FD 103, Aesculap®).

8. One pair of microscissors (spring type) with flat handles and cross-serration (Vannas, FD 15, Aesculap®).

9. Sterile scalpel blades.

10. Polypropylene sutures.

2.1.5 Other Materials

1. Mouse (25–30 g body weight, 6–12 weeks); depending on the research goal, inbreed, outbreed, immune-competent, immune-deficient, and so on.

2. One cage per animal (*see* **Note 1**).

3. Syringes (1 ml).

4. 26-gauge needles (26G3/8, 0.45×10).

5. Sterile nonwoven swabs (5×5 cm).

6. Sterile Q-tips (cotton pads on wooden sticks).

7. Fine black waterproof permanent pen.

8. Surgical masks.

9. Rubber gloves.

10. 70 % alcohol to disinfect skin of the mouse, surgical instruments, and rubber gloves.

2.1.6 Additional Equipment Necessary for Tissue Implantation

1. Custom-made device consisting of a slitted polycarbon tube (24 mm internal diameter, 120 mm long) and a special mounting stage to fix the tube with the animal in it (Workshop, Department of Experimental Surgery, University of Heidelberg).

2. Adhesive tape (hypoallergenic, ≈1.2 cm×9.1 m).

3. Hank's balanced salt solution (100 ml) stored at 6 °C.

4. Sterile petri dishes (diameter ≈ 100×20 mm).

3 Methods

3.1 Surgical Preparation of the Dorsal Skinfold Chamber

A scheme of the setup for the surgical preparation of the dorsal skinfold chamber is shown in Fig. 1. All surgical procedures should be performed under aseptic conditions (*see* **Note 2**).

1. Anesthetize the mouse by an injection of a mixture of ketamine (65 mg/kg body weight), xylazine (13 mg/kg body weight), and acepromazine (2 mg/kg body weight) intramuscularly into the limb (*see* **Note 3**).

2. For depilation of the entire dorsum of the mouse, carefully shave the anesthetized mouse with the electric hair clipper. Put the mouse on a thermal pad outside the hood and apply a thick layer (1–2 mm) of depilatory cream on the shaved skin area. After the cream was allowed to take effect for 5–10 min (*see* **Note 4**), it can be easily removed using a nonwoven swab soaked in hand-warmed, germ-free water, wiping in caudal-to-cranial direction (*see* **Notes 5** and **6**).

3. Dry the wet depilated skin with a dry sterile swab and disinfect the skin with a 70 % alcohol-soaked swab.

4. Place the mouse between the light transmission tube (tube 2) of the halogen lamp (Fig. 1) and the surgeon, in a prone position on an opened sterile nonwoven swab lying on the thermal pad inside the hood. The longitudinal axis of the mouse (with its head lined up to the left-hand side of the surgeon) should be parallel to the frontal plane of the surgeon. Illuminate the

mouse from top with the light transmission tube (tube 1) of the halogen lamp. Adjust tube 2 with its central light about 2 cm parallel to the surface of the thermal pad and perpendicular to the longitudinal axis of the mouse (Fig. 1).

5. Lift up a fold of the depilated dorsal skin. This fold, running from the sacrum to the neck of the mouse, should be located directly over, and parallel to the spine. Adjust the fold under transillumination (tube 2) in such a way that both sides of the skinfold become congruent (*see* **Note 7**). Then, spread the skin in the upright position by fastening two holding threads (polypropylene 4-0; no knots but one baby mosquito each fixed at their ends) at the edge of the skinfold (*see* **Note 8**) and hanging them over two flexible swan-necks of the custom-made skin spreading device (Fig. 1; *see* **Note 9**).

6. Fix the first titanium frame of the chamber to the side of the skinfold facing away from the surgeon with two temporary holding sutures (polypropylene 4-0) through both of the bore-hole pairs, left and right of the apical screw (*see* **Notes 10–12**).

7. Remove the two holding threads completely and make two small incisions (using a pair of Cottle-Masing scissors or a no. 15 scalpel blade), perforating the entire skinfold to let the two lower screws of the titanium frame come through (*see* **Notes 13** and **14**). The fixation of the latter two screws using the baby mosquitos helps to adapt the skinfold temporarily to the titanium frame, facilitating further surgical preparations. The bent tip of the Baby-Mosquito should then point towards the mouse and not the surgeon.

8. Use the cranial baby mosquito to turn the back titanium frame in a perpendicular position to the transilluminating light. The edge of the circular area being projected through the central window of the back titanium frame to the skin facing the surgeon now can be easily marked with a dotted line using a fine black permanent pen.

9. Place the mouse in right lateral position under the dissecting microscope (~sixfold magnification; *see* **Note 15**) with the skinfold pointing to the surgeon.

10. Using delicate dissecting forceps (Micro-Adson) and micro-scissors (FD 103), remove all layers of the skin completely (epidermis, dermis, subcutis, cutaneous muscle, parts of the subcutaneous fatty tissue) along the marked dotted line (*see* **Notes 16** and **17**).

11. Stop possible bleeding along the edge of the wound gently by using sterile Q-tips slightly moistened with isotonic saline. Now allow 3–5 min to elapse to ensure that no further bleed-

ing takes place. Meanwhile, to avoid drying out the operation field, perfuse this area with isotonic saline (*see* **Note 18**).

12. Absorb excess saline with dry Q-tips (*see* **Note 19**) and put the optical magnification on ~tenfold. *The next step is probably the most critical step of the chamber preparation*: Use microforceps (BD 331) and a pair of microscissors (Vannas) to carefully remove the last layer of subcutaneous fatty tissue that is connected to the underlying cutaneous muscle of the opposite skin (*see* **Notes 20–22**).

13. Close the chamber preparation like a sandwich with the second titanium frame: First, connect the lower two screws to the corresponding holes of the second frame and then the apical one (*see* **Note 23**). If no air bubbles are visible between the coverslip and the underlying cutaneous muscle, screw on the nuts to finally fix the two frames of the chamber together (*see* **Notes 24** and **25**).

14. Perform four holding sutures (polypropylene 4-0) to spread and fix the edges of the sandwiched skinfold to all four pairs of boreholes left and right from the apical screw in both chamber frames. After stitching but before closing the knots of the two central holding sutures, cut the old temporary holding sutures from **step 6** and remove them completely (*see* **Note 26**).

15. Place the operated mouse in its cage and leave it on a thermal pad outside the hood at least until the mouse has regained consciousness (*see* **Note 27**).

3.2 Tissue Implantation into the Chamber Preparation

The following section describes the explantation of a solid tumor from a donor mouse and the consecutive implantation of a chunk of this tumor into the dorsal skinfold chamber preparation of a recipient mouse (*see* **Notes 28** and **29**):

1. Allow at least 48 h to elapse for the animals to recover completely from surgery before implantation of any tissue into the dorsal skinfold chamber preparation. Exclude all animals from further treatment and sacrifice them when there are signs of bleeding, inflammation, or any other irritation at the implantation site (*see* **Note 30**). Only chambers meeting criteria of intact microcirculation [11] should be used as sites for implantation.

2. Sacrifice the donor mouse bearing a solid subcutaneous tumor according to official *Guidelines for Care and Use of Experimental Animals*. For 2–3 min, completely insert the dead animal into a 70 % alcohol solution for disinfection. Excise the desired tumor surgically under aseptic conditions in the hood and put it into a sterile petri dish previously filled with cold (~6 °C) Hank's balanced salt solution.

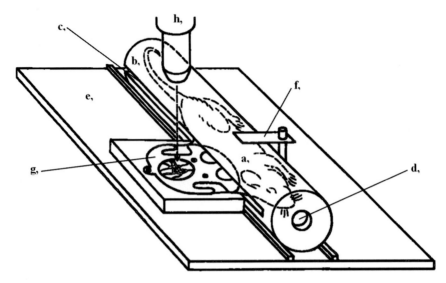

Fig. 3 A mouse fitted with a dorsal skinfold chamber inside a polycarbon tube fixed on the special mounting stage: *a* mouse, *b* slitted polycarbon tube, *c* slit, *d* breathing hole, *e* special mounting stage, *f* device to fix the polycarbon tube on the mounting stage, *g* dorsal skinfold chamber preparation, *h* objective of the intravital microscope

3. The dissecting microscope is only needed for this step of the protocol (magnification ~tenfold). Remove the capsule and all hemorrhagic or necrotic parts of the tumor with the help of microforceps and a pair of microscissors. Cut the remaining tumor into small chunks of a diameter no greater than about 0.5–1 mm (*see* **Note 31**).

4. Put the nonanesthetized recipient mouse in the slitted polycarbon tube (Fig. 3). An adhesive tape fixed across the slit right behind the chamber jutting out will prevent the animal escaping from the tube (*see* **Note 32**). Then, fix the chamber in a horizontal position in the special mounting stage, which will also serve as a stage to perform intravital microscopy of the implant at a later time (Fig. 3).

5. Remove the tension ring with the wrench. Use a 26-gauge needle as lever to lift the coverslip a few millimeters, finally grasping and removing it with microforceps.

6. Transfer one of the tumor chunks with another set of sterile microforceps onto the cutaneous muscle in the center of the open chamber (*see* **Note 33**).

7. With a new, sterile coverslip, reclose the chamber preparation. Before inserting the tension ring to fix the coverslip in position, make sure that there are no persisting air bubbles. These air bubbles can be removed with a Q-tip (*see* **Notes 24** and **34**).

8. Cover the central window of the back titanium frame, which has no coverslip inserted, with a piece of adhesive hypoallergenic tape (*see* **Notes 35** and **36**) and release the mouse back into its cage after removing the other adhesive tape from the tube.

9. Intravital microscopy of the implanted tumor chunk can be performed now repeatedly in the conscious or anesthetized animal by means of normal light and transillumination, or by means of epiillumination from a mercury lamp and a fluorescent filter set in combination with appropriate fluorescent dyes injected intravenously into the animal (*see* **Note 37**).

4 Notes

1. For optimal quality of the chamber preparation, one of the basic requirements is that the area of the dorsal skin associated with the chamber preparation lacks any injuries, scars, or other irritations. Therefore, prior to surgery, only mice from one brood should be held together in one cage, since mice from different broods tend to cause injury to one another. After chamber implantation, the animals must be housed separately in single cages; otherwise, they may destroy each other's chamber preparations by scratching and biting.

2. Work under a laminar flow hood and wear a surgical mask as well as rubber gloves to minimize the possibility of bacterial contamination of the chamber preparation. Between preparations of two different animals, all surgical instruments should be first cleaned mechanically with sterile nonwoven swabs soaked in alcohol and then sterilized with a dry sterilizer. The gloves should be washed with alcohol and changed from time to time.

3. To avoid cooling of the body temperature of the mouse, the anesthetized animal should be placed on a thermal pad (~37 °C) whenever possible.

4. While the depilatory cream is taking effect, clean and prepare the surgical instruments for the following chamber preparation.

5. To avoid irritation of the skin, wipe gently but unsparingly with fresh water-soaked swabs.

6. It is important to use depilatory cream as needed to remove all hairs. Otherwise, during later transillumination microscopy the remaining hair roots will show up as dark shadows, decreasing the optical quality of the region of interest.

7. Points of reference are the larger vessels of the skin, which run symmetrically to each other on the left and right side of the sagittal plane of the mouse.

8. The chamber should fit between the two holding sutures.

9. The weight of a baby mosquito fixed at the end of each holding thread is heavy enough to keep the skinfold in an upright position.

10. If possible, place the central window of this frame in a manner that it is lying centrally between the two main vascular trunks coming from caudal and cranial.

11. The skin between the two sutures should be unstressed and the apical screw of the chamber frame should just jut over the upper edge of the skinfold.

12. Because the two holding sutures have to be removed at a later time (*see* **step 14**) and to avoid local skin necrosis do not make these sutures too tight.

13. The location of larger skin vessels may be controlled with transilluminating light. These vessels should not be cut or damaged by the incisions.

14. There should not be any tension on the skin area between the two screws and holding sutures.

15. The two baby mosquitos may be used to adjust the skinfold and keep it level, parallel to the surface of the thermal pad.

16. Be sure to remove all macroscopic particles left inside or around the marked skin area before cutting. Loose hairs or small fibers of the nonwoven swab may be detected easily under the dissecting microscope and can be removed using microforceps.

17. Avoid hurting the underlying inside of the opposite skinfold. It may be advantageous to perform the initial incision in the center of the marked area. Then continue cutting towards and along the dotted line, respectively.

18. Allow enough time for bleeding to stop. The last layer of subcutaneous fatty tissue still protects the underlying cutaneous muscle of the opposite skin, which will later serve as the site of tissue implantation. After-bleeding at a later time onto the unprotected cutaneous muscle can easily destroy the chamber preparation.

19. Place the Q-tip close to the edge of the operation field. In doing so, touching of the vulnerable inside of the underlying skin is avoided.

20. Be sure to dissect the subcutaneous fatty tissue from the underlying cutaneous muscle by cutting, and not by pulling it away. Too much pulling may lead to disruption of small vessels of the muscle and thus to uncontrolled micro-bleeding.

21. **Steps 12** and **13** have to be performed as free of interruption as possible, to avoid drying and damage of the tissue layer

(cutaneous muscle), which will later be used as a bed for implanting other tissues.

22. To save time, it may be advantageous to remove all the remaining subcutaneous fatty tissue in toto, starting from caudal and toward cranial if you are cutting right-handed.

23. At the time of closure of the chamber preparation, the coverslip should already be inserted and fixed in the second titanium frame with the tension ring.

24. Usually, the cutaneous muscle should stick to the coverslip solely by adhesion forces, automatically expelling remaining air. After closure of the two chamber frames, small, persisting air bubbles may be carefully "pushed out" of the chamber from behind through the central window of the first titanium frame with a dry Q-tip. If you fail to remove all air bubbles, open the slit between both titanium frames for some millimeters, insert few drops of saline between the coverslip and the cutaneous muscle using a 26-gauge needle to drive out remaining air bubbles and close the chamber again.

25. The first nut should be screwed on the apical screw. Make sure not to tighten the nuts too much. This could result in local skin necrosis or a deficient blood flow to and from the skin being part of the chamber preparation.

26. Make at least 6–8 knots in each of the four holding sutures, since the mice sometimes try to chew through sutures.

27. It will take a maximum of 1–2 days for the mouse to get completely accustomed to its new "knapsack." To allow the mouse to eat easily during these first days after surgery, some food may be put directly on the floor of the cage. After this time period, the mouse should show normal behavior again (e.g., cleaning itself, eating, drinking, sleeping, playing, and climbing around in the cage).

28. To avoid immune reactions between the recipient animal and the tumor, use either isografted mouse carcinomas or immunodeficient mice as recipients (e.g., severe combined immunodeficient [SCID] mice).

29. Only fast-growing tumors are suitable for implantation since the mice must be sacrificed (on average) less than 30 days after the initial chamber implantation. Stimulated by the weight of the chamber, new skin will grow and lead to a lateral tipping over of the chamber preparation, causing reduced blood flow to and from the skinfold sandwiched between the two titanium frames. As a rule of thumb, solid tumors reaching visible size within 1–3 weeks after subcutaneous implantation may be suitable for implantation into the chamber preparation.

30. Daily weight monitoring may help to appraise the general state of health of the animal. After an initial loss of weight (<10 %), mice should stabilize again within the first 48 h after surgery. When bearing a tumor, a further loss of weight might be observed in these animals with increasing tumor volume over time.

31. To avoid warming of the tumor chunks before implantation in different animals, put the petri dish on ice from time to time.

32. Since it may be stressful for mice to be inserted into a tube during experiments, it is recommended to leave a tube in their cage 1 or 2 weeks prior to the chamber implantation so that they become accustomed to it.

33. To avoid drying of the cutaneous muscle and to facilitate air-bubble-free reclosure of the chamber, moisten it with few drops of saline.

34. Be sure that the tumor chunk implanted does not move away from its position in the center of the chamber.

35. The growing tumor may sometimes provoke an itching stimulus at the site of implantation. Using a tape may prevent injuries to the skin and implant of the mice caused by scratching.

36. Avoid direct contact between the tape and the skin of the mouse.

37. This may be an advantage in comparison to other non-transparent chamber models such as the cranial bone window model in mice [12] where intravital investigations may be performed exclusively on anesthetized animals using epiillumination.

References

1. Sandison JC (1924) A new method for the microscopic study of living growing tissues by the introduction of a transparent chamber in the rabbit's ear. Ana Rec 28:281–287

2. Menger MD, Lehr HA (1993) Scope and perspectives of intravital microscopy—bridge over from *in vitro* to *in vivo*. Immunol Today 14:519–522

3. Leunig M, Messmer K (1995) Intravital microscopy in tumor biology: current status and future perspectives [review]. Int J Oncol 6:413–417

4. Jain RK, Schlenger K, Höckel M, Yuan F (1997) Quantitative angiogenesis assays: progress and problems. Nat Med 3:1203–1208

5. Leunig M, Yuan F, Menger MD, Boucher Y, Goetz AE, Messmer K, Jain RK (1992) Angiogenesis, microvascular architecture, microhemodynamics, and interstitial fluid pressure during early growth of human adenocarcinoma LS174T in SCID mice. Cancer Res 52:6553–6560

6. Leunig M, Yuan F, Berk DA, Gerweck LE, Jain RK (1994) Angiogenesis and growth of isografted bone: quantitative *in vivo* assay in nude mice. Lab Invest 71:300–307

7. Endrich B, Asaishi K, Goetz AE, Messmer K (1980) Technical report: a new chamber technique for microvascular studies in unanesthetized hamsters. Res Exp Med 177:125–134

8. Sckell A, Leunig M, Fraitzl CR, Ganz R, Ballmer FT (1999) The connective-tissue envelope in revascularization of patellar tendon grafts. J Bone Joint Surg (Br) 81-B:915–920

9. Leunig M, Demhartner TJ, Sckell A, Fraitzl CR, Gries N, Schenk RK, Ganz R (1999) Quantitative assessment of angiogenesis and osteogenesis after transplantation of bone: comparison of isograft and allograft bone in mice. Acta Orthop Scand 70:374–380

10. Vajkoczy P, Menger MD, Simpson E, Messmer K (1995) Angiogenesis and vascularization of murine pancreatic islet isografts. Transplantation 60:123–127

11. Sewell IA (1966) Studies of the microcirculation using transparent tissue observation chambers inserted in the hamster cheek pouch. J Anat 100:839–856

12. Sckell A, Klenke FM (2009) The cranial bone window model: studying angiogenesis of primary and secondary bone tumors by intravital microscopy. Methods Mol Biol 467:343–355

Chapter 18

Quantitative Estimation of Tissue Blood Flow Rate

Gillian M. Tozer, Vivien E. Prise, and Vincent J. Cunningham

Abstract

The rate of blood flow through a tissue (F) is a critical parameter for assessing the functional efficiency of a blood vessel network following angiogenesis. This chapter aims to provide the principles behind the estimation of F, how F relates to other commonly used measures of tissue perfusion, and a practical approach for estimating F in laboratory animals, using small readily diffusible and metabolically inert radio-tracers. The methods described require relatively nonspecialized equipment. However, the analytical descriptions apply equally to complementary techniques involving more sophisticated noninvasive imaging.

Two techniques are described for the quantitative estimation of F based on measuring the rate of tissue uptake following intravenous administration of radioactive iodo-antipyrine (or other suitable tracer). The Tissue Equilibration Technique is the classical approach and the Indicator Fractionation Technique, which is simpler to perform, is a practical alternative in many cases. The experimental procedures and analytical methods for both techniques are given, as well as guidelines for choosing the most appropriate method.

Key words Backflux, Blood flow rate, Cannulation, Distribution volume, Indicator fractionation, Iodo-antipyrine, Partition coefficient, Radiotracer, Tissue equilibration

1 Introduction

The maturation phase of angiogenesis results in a functional blood vessel network. In established tumors, this network is abnormal, but nevertheless sufficient to support tumor growth and metastasis. Blood flow rate through the vascular network is a measure of its functional efficiency, knowledge of which is central to understanding the angiogenic process. Its quantitative estimation provides the basis for determining oxygen, nutrient, and drug delivery to tissue (see Note 1). In pathological angiogenesis, such as in tumors, blood flow rate is the most sensitive and relevant pharmacodynamic endpoint for determining the efficacy of drugs designed to disrupt blood vessel function. Therefore, estimation of blood flow rate is essential for both basic studies of the angiogenic process and applied studies of the effects of therapy. This chapter aims to provide the principles behind, and a practical approach to, the quantitative estimation of blood flow rate in experimental mice and rats.

Stewart G. Martin and Peter W. Hewett (eds.), *Angiogenesis Protocols*, Methods in Molecular Biology, vol. 1430, DOI 10.1007/978-1-4939-3628-1_18, © Springer Science+Business Media New York 2016

Blood flow rate is the rate of delivery of arterial blood to the capillary beds within a particular mass of tissue. It is typically measured in units of mls of blood per g of tissue per minute (ml blood/g tissue/min), or, alternatively, per unit volume of tissue (ml blood/ml tissue/min).

The average time taken for blood to pass through a particular capillary bed (capillary mean transit time (τ)) is the parameter that relates tissue blood flow rate (F in ml/g/min) to fractional blood volume of the tissue (V in ml/g). This classical relationship is known as the *central volume principle* [1]:

$$\tau = V / F \tag{1}$$

For different tissues, F can vary widely, for example it is approximately 0.1 ml/g/min in rat skin and 4.0 ml/g/min in rat kidney. From Eq. 1 and using a value for V of 0.03 ml/g for skin and 0.06 ml/g for kidney, τ is approximately 18 s and 0.9 s, respectively, for these tissues. From Eq. 1, τ is only indirectly proportional to F, if V is constant, and so neither τ nor V can be used to estimate F unless they are both measured simultaneously. The same considerations apply to parameters related to τ, such as *red blood cell velocity* (RBC velocity) and the *blood supply time* (BST); see below.

In order to estimate F, the most accurate approach is to measure the rate of delivery of an agent carried to the tissue by the blood. A contrast agent is injected into the blood-stream; its concentration time-course in arterial blood (input function) together with the kinetics of its uptake into tissue (tissue response function) are measured. F is then estimated from a mathematical model relating the tissue response function to the input function (see below). The contrast agent can be radio-active, whereby tissue concentrations can be measured by gamma or scintillation counting or by an external imaging system (e.g., a positron emitter for positron emission tomography). Alternatively, a contrast agent that is suitable for external magnetic resonance imaging, computed tomography, or ultrasound imaging can be used. Radio-active agents have the advantage that they can be administered at true tracer concentrations, therefore not interfering with physiological processes, and they do not necessarily need sophisticated imaging technology.

Some common methods for determining blood perfusion parameters are given below, most of which do not provide fully quantitative estimates of the blood flow rate, F:

1. Intravital microscopy is a specialized technique that enables direct visualization of tissue microcirculation, usually via surgically implanted transparent chambers or single-sided windows. This enables measurement of RBC velocity (μm/s) in individual capillaries, as well as the blood supply time (BST, defined below). RBC velocity is measured either by directly tracking individual fluorescently labelled red blood cells through vessel

segments [2] or matching the interference patterns of light reaching a camera through a slit system [3]. Modern computing techniques now enable comparison of optical signals at individual spatial locations with those in neighboring locations over time, so that two-dimensional maps of both speed and direction of blood flow can be constructed, based on similar principles to the classical slit system approach [4]. Measurements can be combined with measurements of red cell flux (number of red blood cells traversing a vessel segment per unit time) to calculate microvascular hematocrit [5] or with morphological measurements of individual blood vessel segments to obtain each segment's *volume flow rate* (F_{seg}), which assumes that RBCs are traveling with the bulk plasma flow [6]. Measurement of BST has been carried out in intravital studies of tumors, from images of the tumor vascular network over time, following the intravenous injection of a fluoresecent marker such as TRITC-dextran [7]. For each pixel of the vascular image, BST is defined as the time difference between the frame showing maximum fluorescence intensity and the frame showing maximum fluorescence intensity in a tumor-supplying artery, during a short timescale following intravenous injection. Both RBC and BST provide functional information on tumor perfusion, but are not directly related to F, as discussed above.

2. Laser Doppler flowmetry (LDF) provides a means of estimating *relative changes* in red cell velocity, e.g., following treatment, via surface or tissue-inserted probes. This measures a frequency shift in light reflected from moving red cells, which is a measure of average red cell velocity [8]. Again, it should be noted that changes in red cell velocity may not accurately reflect changes in F.

3. The fluorescent DNA-binding dye, Hoechst 33342, and certain carbo-cyanine dyes are examples of rapidly binding agents that have been used to determine a 'perfused vascular volume' (as a fraction of the total tissue volume) rather than blood flow rate per se. This method has been used especially in tumor studies [9, 10]. In this case, tissues are excised after several circulation times, following intra-venous injection of the dye, and functional vessels appear in tissue sections as fluorescent halos. Alternatively, a fluorescently tagged lectin that rapidly binds to endothelial cells in vivo can be used [11]. Conventional Chalkley point counting [12, 13] or image analysis provides the fractional tissue volume occupied by fluorescence. This is a useful measure of vascular function in many circumstances, but is insensitive because it cannot discriminate between perfused vessels with different flow rates.

4. For contrast agents that are confined to the blood-stream, methods based on Eq. 1 can be used to calculate F [14].

However, this is difficult in practice because τ is only a few seconds, requiring highly sensitive techniques for its measurement. Radioactive or colored 'microspheres' of diameters around 15–25 μm are a special class of contrast agents that are confined to the blood-stream because they should be trapped on first-pass through tissue. Therefore, following injection directly into a major artery, they distribute to tissues in direct proportion to the fraction of the cardiac output received by the tissues, enabling calculation of blood flow rate [15]. With this technique, care needs to be taken to ensure adequate mixing of the microspheres in the arterial blood (which is challenging in mice, for instance) and enough microspheres are lodged in the tissue regions of interest to obtain statistical validity. In the case of tumors, care needs to be taken to determine and correct for microspheres that are re-circulated due to lack of trapping in large diameter vessels [16, 17].

5. The principles used to calculate blood flow rate using microspheres can sometimes be used even when the indicator crosses the vascular wall into the tissue and re-circulates after the first pass through the tissue. If the tissue concentration of the indicator reaches a constant level that is maintained for the first minute or so after injection, this indicates that the extraction fraction by the tissue is equal to that of the whole body [18]. The fractional uptake of the indicator into the tissue must therefore equate to the blood flow fraction of the cardiac output received. In the original description of the technique, potassium and rubidium chloride behaved as 'pseudo-microspheres' in most normal tissues, with the notable exception of brain [18].

6. Small, lipid-soluble, metabolically inert molecules, which rapidly cross the vascular wall and diffuse through the extra-vascular space, are also useful as blood flow markers. In this case, the fraction of marker crossing the capillary vascular wall from the blood in a single pass through the tissue (extraction fraction, E) is close to 1.0, and for fully perfused tissue, the accessible volume fraction (α) of the tissue is also close to 1.0. The inert radioactive gas, [133]Xenon, or hydrogen can be administered by inhalation [19, 20]. However, safety issues with [133]Xenon and the necessity for tissue insertion of polarographic electrodes for hydrogen have limited their use. A practical approach, which has utility for accessing the spatial heterogeneity of tissue blood flow rate, is the intravenous administration of a small, lipid-soluble, inert molecule dissolved in saline. In this case, net uptake rate into tissue over a short time (seconds) after intra-venous injection is determined primarily by blood flow rate. Methods for quantitative estimation of tissue blood flow rate and related parameters using these agents are described below (*see* **Note 2**).

2 Materials

2.1 Radioactive Tracer Preparation

1. Any small, lipid-soluble molecule that can be suitably labelled and is not metabolized in tissue over the short time of the experiment can be used. Suitable radio-isotopes include ^{125}I and ^{14}C, where tissue and blood counts can be obtained using standard techniques and autoradiography/phosphor imaging can be applied if spatial variation in blood flow rate across a tissue of interest is required (*see* **Note 3**).

2. One that has been used commonly for both normal tissue studies (primarily brain [21]) and tumor studies [22] is iodo-antipyrine (IAP) (Fig. 1). ^{125}I-IAP is commercially available, for example from MP Biomedicals and ^{14}C-IAP (4-Iodo[*N-methyl*-14 C]antipyrine) from PerkinElmer. Alternatively, a technique for labelling IAP with ^{125}I is described by Trivedi [23] (*see* **Note 4**).

2.2 Animal Preparation

Large vessel cannulation is required for intravenous administration and arterial blood sampling. Materials required:

1. General anesthetic.

2. Heparinized saline for cannulae (add 0.3 ml of 1000 U/ml heparin to 10 ml saline).

3. General surgical equipment plus fine-angled forceps, small spring scissors, microvascular clip.

4. Polythene tubing cut to suitable lengths, size appropriate to that of the vessel being cannulated (e.g., for rat 0.58 mm internal diameter; 0.96 mm outside diameter).

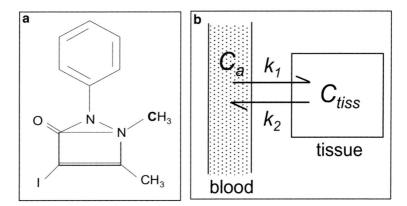

Fig. 1 Chemical structure of 4-Iodo[*N-methyl*-14C]antipyrine (**a**) and the compartmental model used for the quantitative estimation of *F* (**b**). When the extraction fraction, *E*, of a blood-borne tracer is 1.0, the rate constant k_1 represents *F*. k_2 represents the back-flux and C_a and C_{tiss} represent the arterial blood and tissue concentrations of the tracer, respectively. In this model, the tissue is a single well-mixed compartment

5. Dissecting microscope.

6. Cold light source.

7. Thermostatically controlled heating blanket, with rectal thermometer.

2.3 Blood Flow Assay

1. ^{125}I-labelled IAP (^{125}I-IAP) and suitable protective equipment.

2. Minimal dead-space glass syringe.

3. General dissecting instruments for excising tissue.

4. Stop-clock.

5. Lidded container containing saline-soaked gauze.

6. Gamma counter and suitable vials for tissue and blood samples.

7. Analytical balance.

8. Anesthetic (*see* **Note 5**).

9. Fraction collector set to collect at 1 s intervals.

10. Syringe pump for infusion and withdrawal.

11. 1000 U/ml heparin (use neat for rats and diluted 1 in 10 with saline for mice).

12. Injection saline.

13. High concentration solution of sodium pentobarbitone, e.g., Euthatal™.

---

3 Methods

3.1 Animal Preparation

In the rat, a tail artery and vein are most suitable for cannulation. In the mouse, either the carotid artery and jugular vein or femoral artery and vein can be used.

1. Prepare 30 cm lengths of cannulae. For rat, use 0.96 mm outside diameter (o.d.); 0.58 mm internal diameter (i.d.). The cannula wall may be shaved down at the tip and the end may be slightly bevelled to aid insertion. Use a microscope to ensure that there are no sharp edges. For mouse, use a short length of 0.61 mm o.d.; 0.28 mm i.d. cannula, stretched to a smaller diameter at the tip and connected to a longer length of 0.96 mm o.d.; 0.58 mm i.d. cannula to reduce resistance to flow. Attach each length to a 1 ml syringe filled with heparinized saline.

2. Anesthetize the animal, insert rectal thermocouple, and place on heating blanket. Also, an overhead lamp is a useful additional heat source.

3. Illuminate surgical area with a cold light source.

4. Expose the relevant vessel. For the rat tail, this involves making two 2 cm incisions through the skin, each side of the vessel, approximately 5 mm apart and approximately 2 cm from the base of the tail. Use artery forceps to clear the skin from the underlying connective tissue and cut the skin at the distal end to create a flap (*see* **Note 6**).

5. Keep exposed vessels moist at all times using warmed saline.

6. Free the vessel from surrounding connective tissue using fine blunt-end forceps.

7. Place two lengths of suture under the vessel, tying off the most distal to the heart, which can be used to apply slight tension to the vessel.

8. Occlude the vessels as far proximal as possible using a micro-vascular clip.

9. Using spring scissors, make a v-shaped cut in the vessel close to the distal knot and insert cannula. Advance cannula into the vessel approximately 2 cm or more, by removing clip (*see* **Note 7**).

10. Aspirate gently to ensure that blood is free flowing. It may not be possible to aspirate the vein, but a small volume of saline can be injected to check for patency.

11. Tie both sutures securely around the cannula. Use tape or tissue-compatible glue to secure the cannula to the skin distal to the distal suture. Close the wound.

3.2 Blood Flow Assay

Two alternatives are described; the classic tissue equilibration method for rats and the indicator fractionation method for rats or mice. Graham et al. [24] directly compared results obtained with these two techniques in a rat brain tumor model and the main advantages and disadvantages of each method are given in Table 1.

Table 1
Comparison of radiotracer methods for estimation of tissue blood flow rate (F)

Tissue equilibration technique	Indicator fractionation technique
Requires blood vessel cannulation	Requires blood vessel cannulation
Time-consuming	Relatively easy to perform
Calculations require curve fitting algorithms	Calculations require only a simple formula
Backflux is taken into account in the model	Very sensitive to errors associated with backflux
Relatively sensitive to timing errors and delay and dispersion effects	Reasonably tolerant of timing errors and delay and dispersion effects
Reasonably tolerant of imprecision in λ	Tolerant of imprecision in λ
Method of choice where λ is well-defined	Method of choice where λ is ill-defined, especially if F is low and α large

3.2.1 Tissue Equilibration Technique

Cannulation of two tail veins and one tail artery are required, as described above.

1. Remove ^{125}I-IAP from the freezer and bring slowly to room temperature. Using suitable containment and a low dead-space glass syringe, carefully remove required volume ^{125}I-IAP (0.2–0.3 MBq per rat) and dispense into a vial.

2. Evaporate the methanol using a very gentle stream of nitrogen and slowly add injection saline to the ^{125}I-IAP (0.8 ml per rat plus extra to account for syringe dead spaces, etc.). Gently mix.

3. Load a syringe of size suitable for infusion with the ^{125}I-IAP solution (needle must be suitable size for the venous cannula).

4. For anesthetized animals, keep warm, as described above. Check arterial blood pressure and heart rate by connecting the arterial cannula to a pressure transducer and recording device. Then clamp off the cannula and connect it to the fraction collector, loaded with pre-weighed glass tubes, for subsequent blood collection.

5. Inject and flush in 0.1 ml neat heparin (=100 I.U.) via one of the venous cannulae to ensure the blood flows freely from the arterial cannula.

6. Cut one of the venous cannulae to approx. 3 cm in length and connect a syringe containing approx. 0.5 ml Euthatal™. A small "T-connector" may also be used to allow drug administration via this cannula.

7. Set syringe pump speed to 1.6 ml/min (*see* **Note 8**). Carefully place the ^{125}I-IAP-containing syringe in the pump and connect it to the second venous cannula.

8. Start the stop-clock and unclamp the artery, checking that blood is free-flowing. At 5 s, start syringe pump and fraction collector (*see* **Note 9**). At 35 s, inject Euthatal™ and stop pump; rapidly excise tissues of interest and stop fraction collector (*see* **Note 10**). Place tissues in the lidded container to prevent drying. Weigh the blood tubes and cap them. Weigh the tissues and place them in gamma counting tubes.

9. Count the blood and tissue samples on the gamma counter.

3.2.2 Indicator Fractionation Technique

Cannulations of one artery and two veins are required. Alternatively, cannulae attached to shafts of hypodermic needles can be inserted into tail veins percutaneously instead of cannulating veins.

1. Follow points 1–3 above, preparing 0.07 MBq ^{125}I-IAP in 0.05 ml saline per mouse.

2. For anesthetized animals, keep warm, as described above. Check arterial blood pressure and heart rate by connecting the arterial cannula to a pressure transducer and recording device.

3. Set syringe pump speed to 150 μl/min (mouse) or 1.6 ml/min (rat).

4. Load a 250 μl syringe (for mouse) or 2 ml syringe (for rat) with approx 100 μl saline, attach a 23G needle and position in the pump.

5. Cut the venous cannula as short as possible and inject 0.05 ml of diluted heparin (≡5 I.U.) for mouse or 0.1 ml neat heparin (≡100 I.U.) for rat. Disconnect heparin syringe and attach ^{125}I-IAP syringe and a syringe containing injection saline via a T-piece. Connect syringe containing Euthatal™ to second venous cannula.

6. Clamp the artery cannula, disconnect it from the pressure transducer, and connect it to the syringe pump. Ensure that the pump is set to withdraw and allow it to withdraw very briefly to ensure that the cannula is patent and the syringe is positioned correctly.

7. Start the stop-clock and pump simultaneously. Check that the blood is flowing freely. At 3 s, inject 0.05 ml of ^{125}I-IAP for mouse or 0.2 ml for rat, as a rapid bolus via the venous cannula, followed immediately by 0.05 ml saline from the second syringe. At 13–18 s (*see* **Note 11**), inject Euthatal™ and immediately pull out full length of arterial cannula and excise the tissue of interest. All the blood should be retained in the cannula. Place the tissue in a pre-weighed gamma counting tube.

8. Attach a saline-filled syringe to the arterial cannula and eject all the blood into a gamma counting tube together with 1 ml of saline.

9. Count the blood and tissue samples on the gamma counter.

3.3 Blood Flow Analysis

3.3.1 Tissue Equilibration Technique

1. Analysis is based on a model which assumes a vascular compartment from which the input function derives a single (extravascular) well-mixed tissue compartment (Fig. 1). A small, highly soluble and inert tracer, such as IAP, is assumed to rapidly equilibrate between all blood components and the tissue compartment. In this case, the model based on Kety [25] describes the relationship between the tissue concentration of the tracer at time t, $C_t(t)$, and the arterial blood concentration of the tracer at time t, $C_a(t)$, by the Equation:

$$C_{\text{tiss}}(t) = k_1 C_a(t) \otimes \exp(-k_2 t) \tag{2}$$

where k_1 is tissue blood flow rate (F) and k_2 is $k_1/\alpha\lambda$; α is the effectively perfused fraction of tissue (i.e., the fraction of tissue that is immediately accessible to the tracer); and λ is the equilibrium partition coefficient of the tracer between tissue and blood; $\otimes$ denotes the convolution integral; in imaging studies, $\alpha\lambda$ is often referred to as the apparent volume of distribution

(VD$_{app}$) of the tracer in the tissue [26], $C_{tiss}(t)$ and $C_a(t)$ are expressed in radioactivity counts per g tissue and per ml blood, respectively, using 1.05 for the density of blood.

2. In this method, $C_{tiss}(t)$ is measured at only one time-point, i.e., after tissue excision. Hence, only one parameter, k_1 (F), can be estimated from the data (*see* **Note 12**). λ is approximated from literature values or estimated from separate experiment [27] and α is taken as 1.0 (*see* **Notes 12** and **13**). Studies have shown that the method is relatively insensitive to small changes in λ because of the short time-scale of the experiment [28]. Also *see* Table 1.

3. Solving Eq. 2: Data can be fitted to Eq. 2 using a simple 'Table Lookup Method'. In this method, since the input function is known, the expected tissue activity at the time of excision $C_{tiss}(T)$ can be calculated for each of a range of realistic values of F, using Eq. 2. Direct comparison of the observed $C_{tiss}(T)$ against the table then gives the required estimate of F (Fig. 2). Evaluation of the Integral in Eq. 2 requires a numerical integration routine, which are commonly available in statistical analysis packages, or can be programmed using computer applications such as MATLAB (The Mathworks, USA ©). A further issue, which needs to be taken into account when assessing the accuracy of this type of technique, is the possibility that the input function time course may not be accurately measured because of a time delay between the radioactivity reaching the tissue and reaching the blood collection tubes and because of smearing or dispersion effects occurring on the arterial cannula before blood collection. These delay and dispersion effects can be corrected for (*see* **Note 14**), but do add further complication to the analysis.

3.3.2 Indicator Fractionation Technique

1. This method was first used by Goldman and Sapirstein [29], with later modifications [30]. It simplifies the model used in Eq. 2 by assuming that the backflux of the tracer from tissue into the blood is negligible compared with its influx into the tissue, for a short period of time after injection of the tracer (*see* **Note 15**). Under these conditions, Eq. 2 reduces to:

$$k_1 = C_{tiss}(T) / \int_0^T C_a(t)\,dt \qquad (3)$$

where k_1, t, $C_a(t)$ are as defined above and $C_{tiss}(T)$ is concentration of tracer in the tissue at the end of the experiment (at time $t = T$).

2. T is typically set at 10–15 s, during which time collection of sufficient blood samples, as described for the tissue equilibration technique, is difficult. Instead, the constant withdrawal technique can be used. Here, blood is withdrawn from an artery at a constant rate using a pump from $t = -T$ to $t = T$,

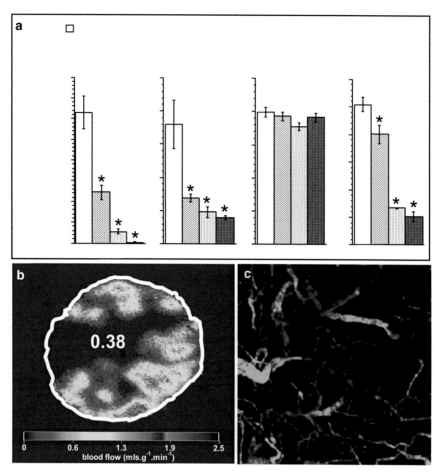

Fig. 2 Estimation of tissue blood flow rate (*F*) in the P22 rat sarcoma and several normal rat tissues using the tissue equilibration method with ^{125}I-IAP (**a**) and ^{14}C-IAP (**b**). Results in *panel* (**a**) were obtained by calculating C_{tiss} from gamma counts of large tissue samples. Data shows effects of the vascular disrupting agent combretastatin A4-P and the nitric oxide synthase inhibitor L-NNA plus the combination of the two. *Asterisk* represents a significant difference between treated and control untreated tumors. The image in *panel* (**b**) was obtained from an untreated P22 tumor by calculating multiple values for C_{tiss} from quantitative autoradiography of tumor sections. The mean *F* is 3.8 ml/g/min. The image in *panel* (**c**) illustrates the vascular networks in the P22 tumor obtained by multi-photon fluorescence microscopy

where $-T$ is the time at which the pump is started. A bolus injection of tracer is given at $t=0$. Under these conditions,

$$\int_0^T C_a(t)\,\mathrm{d}t = C_c\left(T+T^{'}\right) = C_c V_b / r = X / r \qquad (4)$$

where C_c is mean concentration of tracer in the blood sample; V_b is volume of blood collected; X is total radioactive counts in the collected blood sample, and r is rate of withdrawal of blood.

3. X/r can be substituted into Eq. 3 and blood flow rate, k_1, can be calculated from a knowledge of the counts, X, pump rate, r, and concentration of the tracer in the tissue at the end of the

experiment, $C_{tiss}(T)$. As for the tissue equilibration technique, there are inaccuracies in the measurement of the input function using this technique, associated with delay and dispersion along the plastic cannula. However, the constant withdrawal method means that definition of a concentration-time curve is not required and only the last part of the actual arterial time-course is lost by the blood sampling method. In addition, the experimental set-up of the indicator fractionation method means that the arterial cannula can be kept short, minimizing the delay involved.

3.3.3 Comparison of the Two Blood Flow Methods

The advantages and disadvantages of the classic tissue equilibration method and the indicator fractionation method are summarized in Table 1. Patlak et al. [28] carried out an evaluation of errors involved in the two techniques, which can be summarized as follows:

1. Errors in the tissue equilibration method are minimized if an optimal infusion schedule is used (a ramped schedule is best but a constant infusion is reasonable), timing is measured precisely, and corrections are made for delay and dispersion. Under these conditions, 10–15 % inaccuracy in the value used for λ is well-tolerated in the calculation of F. If precise measurement of λ can be made, this is the method of choice.

2. If λ cannot be measured reasonably accurately, the indicator fractionation technique may be the better option for estimating F. Errors associated with backflux are minimized by a short experimental time. Errors associated with imprecise timing are minimized by bolus administration of the tracer (so that arterial concentration is low at tissue excision). The constant withdrawal method is an added advantage for its simplicity and accuracy. However, backflux cannot be completely prevented (especially with bolus administration) and may introduce significant errors where F is high and/or α is low. Delay and dispersion effects are reasonably well-tolerated.

3. Both techniques require accurate measurement of tracer concentration in the tissue (C_{tiss}).

4 Notes

1. The methodology presented here is based on a single tissue compartment model. This can be derived from the classic Renkin-Crone unit capillary model [31, 32], which gives an explicit relationship between flow, the extraction of substances from blood into tissue, and the mean permeability surface area product of the capillary bed. Recent simulation studies show that the degree of local heterogeneity in capillary architecture and transit times of blood through the capillaries may also

affect the precise relationship between flow and extraction of substances from the blood into tissue [33, 34].

2. The basic experimental principles and analytical methods described here also apply to various external imaging techniques that are now available for use in small animals, e.g., positron emission tomography [35]. These techniques allow repeated evaluation of blood flow rate (and other pharmacodynamic end-points) in the same animal, as long as the biological and radioactive half-lives of the tracer are compatible with the time-scale of the experiment. In addition, they allow definition of more than one vascular parameter (*see* **Note 12** below).

3. Instead of obtaining a single value for the blood flow rate within a tissue (usually by calculating C_{tiss} from gamma counts of tissue activity), the variation of blood flow rate within a tissue can be obtained at high spatial resolution (~50 μm) by using a radiotracer that is suitable for autoradiography or phosphor imaging (Fig. 2). In these cases, C_{tiss} is obtained in raster fashion across tissue sections for calculation of corresponding k_1 (F) values [36].

4. Local radiation safety procedures need to be followed for all the techniques described to avoid contamination of personnel and equipment.

5. General anesthesia seriously affects mean arterial blood pressure in rodents, especially in mice, and its effects on tissue blood flow rates need to be considered in planning experiments. Animals can be allowed to recover from general anesthesia induced by inhalational anesthetics following cannulation, but, in this case, procedures for preventing cannula disturbance and minimising pain and distress to the animals need to be implemented [37].

6. Tail cannulations: the tail artery lies relatively deep within a cleft in the cartilage and requires an incision to be made through the overlying connective tissue for it to be accessible. Once freed, it is robust for cannulation; the vein is much more superficial and easily located, although more fragile than the artery and liable to constriction and tearing.

7. A topical vasodilator such as procaine can be used to aid cannula insertion.

8. Volumes of saline solutions of IAP for intravenous administration are chosen to compensate for rate of blood loss during the course of the experiments.

9. A constant infusion schedule for delivery of the radiotracer is described, as it is simple to achieve in practice. However, an infusion schedule that increases with time (ramped) could be employed because this reduces the influence of an incorrect value for λ, on the calculated value of F [28]. The movement

of the fraction collector and the infusion/withdrawal rates of the pump need to be carefully calibrated prior to experiments.

10. Timing errors can be significant if blood flow to tissues of interest is not stopped at the instant that the pump is stopped. Also *see* Table 1.

11. A short duration increases timing errors, but a longer one increases errors associated with backflux of the tracer from the tissue into blood. Also *see* Table 1.

12. A disadvantage of this particular technique is that the tissue concentration of the tracer is assayed at only one time point. Hence, as noted above, only one parameter (F) can be estimated, whereas values for α and λ have to be assumed. Other, more sophisticated techniques involving noninvasive imaging, such as PET (*see* **Note 2**), involve a full time course of the tissue to be assayed, allowing estimation of $\alpha\lambda$ (VD_{app}) for example, as well as F. This is of particular interest in the case of tumor blood flow, where the effectively perfused tissue fraction (α) (i.e., the fraction of tissue that is immediately accessible to the tracer) is often less than 1.0 because of large intercapillary distances or ischemic regions [26]. However, the spatial resolution of noninvasive imaging cannot compete with the high spatial resolution achievable with the invasive techniques described here (*see* **Note 3**). In the invasive technique, α is usually assumed to be 1.0. If it is actually less than 1.0, for a sampled tissue region of interest, the measured tissue concentration of the tracer, C_{tiss}, will be low because it is averaged over the whole region, including the inaccessible part. Thus, the calculated value of F will reflect average blood flow for the region and underestimate that of the perfused tissue fraction.

13. Beyond the short time-scale of these experiments, IAP redistributes in tissue in a space-dominated rather than a flow-dominated pattern. At equilibrium, IAP might be expected to distribute in proportion to the tissue water content, such that λ would be similar in all tissues and close to 1.0. However, experimental evidence indicates that, although λ for IAP is close to 1.0 in many tissues, it is somewhat variable [27]. This may relate to its high lipid solubility or it may be a reflection of problems in accessing true values of λ because of loss of the radioactive label from the molecule at long times (hours) after injection.

14. A simple model which can be used to describe dispersion effects is given by the equation:

$$C_m(t) = C_a(t) \otimes k_d \exp(-k_d t) \tag{5}$$

where $C_m(t)$ is the measured tracer concentration at the cannula outflow, $C_a(t)$ is the inflow concentration, and k_d (min^{-1})

is a dispersion constant, which is dependent on flow rate, length, and internal diameter of the cannula and the interaction with blood on its internal surface. k_d for a particular cannula and flow rate of blood can be calculated from Eq. 5, if experiments are undertaken in vitro whereby the blood pumped through a cannula at a particular rate is switched rapidly between labelled blood and unlabelled blood and the dispersion effect measured in the outflow. Results from such an experiment are presented in Table 2. Delay (t_d) can be estimated directly from the known volume of the cannula and the blood flow rate down the cannula. These values for k_d and t_d can be incorporated in Eq. 2 to give a working form of the Equation:

$$C_{\text{tiss}}(t) = (k_1 / k_d) C_m (t + t_d) + (1 - k_2 / k_d) k_1 C_m (t + t_d) \otimes \exp(-k_2 t) \quad (6)$$

If dispersion effects are small, i.e., large k_d, this Equation reduces to Eq. 2. If dispersion effects are marked, i.e., small k_d, then the Equation illustrates that failing to take dispersion effects into account has a marked effect on estimates of F. These issues were first quantitatively described by Meyer [38] (but note that a dispersion time constant equivalent to $1/k_d$ was used in this case).

15. In addition to using a short experimental time, T, errors associated with backflux in this technique are minimized if blood flow rate is low and there is a big accessible space in the tissue for the tracer (high α). Note that the short T gives the potential for large timing errors and great care must be taken to time the experiment accurately. As for the tissue equilibration technique, this means that blood flow to tissues needs to be stopped precisely at T. Interestingly, timing errors appear to be less of an issue with this technique than with the tissue equilibration technique for measurement of cerebral blood flow rate [28].

Table 2
Example of expressions used to calculate the dispersion constant k_d (min⁻¹) from the linear speed (ν in cm per min) of blood flowing down various types of cannulae

Type of cannula	200 mm length	300 mm length
0.50 mm internal diameter	$6.72 + 0.048 * \nu$	$3.84 + 0.036 * \nu$
0.58 mm internal diameter	$9.54 + 0.040 * \nu$	$-0.6 + 0.042 * \nu$

These expressions were obtained by pumping blood at several known flow rates through each type of cannula and rapidly switching between labelled and unlabelled blood. The time-activity curves for the blood flowing out of the cannulae were compared with the known inflow time-activity curves, using Eq. 5, to estimate k_d for each condition. The relationship between k_d and ν was described by the equation for a straight line, as shown. The cannulae used were prepared from Portex™ low-density polyethylene tubing

References

1. Stewart GN (1894) Researches on the circulation time in organs and on the influences which affect it: parts I-III. J Physiol (Lond) 15:1–89

2. Reyes-Aldasoro CC, Akerman S, Tozer GM (2008) Measuring the velocity of fluorescently labelled red blood cells with a keyhole tracking algorithm. J Microsc 229:162–173

3. Intaglietta M, Tompkins WR (1973) Microvascular measurements by video image shearing and splitting. Microvasc Res 5: 309–312

4. Fontanella AN, Schroeder T, Hochman DW, Chen RE, Hanna G, Haglund MM, Secomb TW, Palmer GM, Dewhirst MW (2013) Quantitative mapping of hemodynamics in the lung, brain, and dorsal window chamber-grown tumors using a novel, automated algorithm. Microcirculation 20:724–735

5. Brizel DM, Klitzman B, Cook JM, Edwards J, Rosner G, Dewhirst MW (1993) A comparison of tumor and normal tissue microvascular hematocrits and red cell fluxes in a rat window chamber model. Int J Radiat Oncol Biol Phys 25:269–276

6. Tozer GM, Prise VE, Wilson J, Cemazar M, Shan S, Dewhirst MW, Barber PR, Vojnovic B, Chaplin DJ (2001) Mechanisms associated with tumor vascular shut-down induced by combretastatin A-4 phosphate: intravital microscopy and measurement of vascular permeability. Cancer Res 61:6413–6422

7. Oye KS, Gulati G, Graff BA, Gaustad JV, Brurberg KG, Rofstad EK (2008) A novel method for mapping the heterogeneity in blood supply to normal and malignant tissues in the mouse dorsal window chamber. Microvasc Res 75:179–187

8. Stern MD (1975) In vivo evaluation of microcirculation by coherent light scattering. Nature 254:56–58

9. Smith KA, Hill SA, Begg AC, Denekamp J (1988) Validation of the fluorescent dye hoechst 33342 as a vascular space marker in tumours. Br J Cancer 57:247–253

10. Hill SA, Tozer GM, Chaplin DJ (2002) Preclinical evaluation of the antitumour activity of the novel vascular targeting agent Oxi 4503. Anticancer Res 22:1453–1458

11. Lunt SJ, Akerman S, Hill SA, Fisher M, Wright VJ, Reyes-Aldasoro CC, Tozer GM, Kanthou C (2011) Vascular effects dominate solid tumor response to treatment with combretastatin A-4-phosphate. Int J Cancer 129:1979–1989

12. Chalkley HW (1943) Method for quantitative morphologic analysis of tissues. J Natl Cancer Inst 4:47–53

13. Vermeulen PB, Gasparini G, Fox SB, Colpaert C, Marson LP, Gion M, Belien JA, de Waal RM, Van Marck E, Magnani E, Weidner N, Harris AL, Dirix LY (2002) Second international consensus on the methodology and criteria of evaluation of angiogenesis quantification in solid human tumours. Eur J Cancer 38: 1564–1579

14. Weiskoff RM (1993) Pitfalls in MR measurement of tissue blood flow with intravascular tracers: which mean transit time? Magn Reson Med 29:553–559

15. Messmer K (1979) Radioactive microspheres for regional blood flow measurements. Actual state and perspectives. Bibl Anat 18:194–197

16. Jirtle RL (1980) Blood flow to lymphatic metastases in conscious rats. Eur J Cancer 17:53–60

17. Jirtle RL, Hinshaw WM (1981) Estimation of malignant tissue blood flow with radioactively labelled microspheres. Eur J Cancer Clin Oncol 17:1353–1355

18. Sapirstein LA (1958) Regional blood flow by fractional distribution of indicators. Am J Physiol 193:161–168

19. Obrist WD, Thompson HK, King CH, Wang HS (1967) Determination of regional cerebral blood flow by inhalation of 133-xenon. Circ Res 20:124–135

20. Young W (1980) H2 clearance measurement of blood flow: a review of technique and polarographic principles. Stroke 11:552–564

21. Sakurada O, Kennedy C, Lehle J, Brown JD, Carbin JL, Sokoloff L (1978) Measurement of local cerebral blood flow with iodo [^{14}C] antipyrine. Am J Physiol 234:H59–H66

22. Tozer GM, Shaffi KM (1993) Modification of tumour blood flow using the hypertensive agent, angiotensin II. Br J Cancer 67: 981–988

23. Trivedi MA (1996) A rapid method for the synthesis of 4-iodoantipyrine. J Labelled Compd Radiopharm 38:489–496

24. Graham MM, Spence AM, Abbott GL, O'Gorman L, Muzi M (1987) Blood flow in an experimental rat brain tumor by tissue equilibration and indicator fractionation. J Neuro-Oncol 5:37–46

25. Kety SS (1960) Theory of blood tissue exchange and its application to measurements of blood flow. Methods Med Res 8:223–227

26. Tozer GM, Shaffi KM, Prise VE, Cunningham VJ (1994) Characterisation of tumour blood flow using a "tissue-isolated" preparation. Br J Cancer 70:1040–1046

27. Tozer GM, Morris C (1990) Blood flow and blood volume in a transplanted rat fibrosarcoma: comparison with various normal tissues. Radiother Oncol 17:153–166

28. Patlak CS, Blasberg RG, Fenstermacher JD (1984) An evaluation of errors in the determination of blood flow by the indicator fractionation and tissue equilibration (Kety) methods. J Cerebr Blood Flow Metab 4:47–60

29. Goldman H, Sapirstein LA (1973) Brain blood flow in the conscious and anaesthetized rat. Am J Physiol 224:122–126

30. Gjedde SB, Gjedde A (1980) Organ blood flow rates and cardiac output of the Balb/c mouse. Comp Biochem Physiol 67A:671–674

31. Renkin EM (1959) Transport of potassium-42 from blood to tissue in isolated mammalian skeletal muscles. Am J Physiol 197:1205–1210

32. Crone C (1963) The permeability of capillaries in various organs as determined by use of "indicator diffusion" method. Acta Physiol Scand 58:292–305

33. Jespersen SN, Ostergaard L (2012) The roles of cerebral blood flow, capillary transit time heterogeneity, and oxygen tension in brain oxygenation and metabolism. J Cerebr Blood flow Metab 32:264–277

34. Ostergaard L, Tietze A, Nielsen T, Drasbek KR, Mouridsen K, Jespersen SN, Horsman MR (2013) The relationship between tumor blood flow, angiogenesis, tumor hypoxia, and aerobic glycolysis. Cancer Res 73: 5618–5624

35. Herrero P, Kim J, Sharp TL, Engelbach JA, Lewis JS, Gropler RJ, Welch MJ (2006) Assessment of myocardial blood flow using 15O-water and 1-11C-acetate in rats with small-animal PET. J Nucl Med 47:477–485

36. Tozer GM, Prise VE, Wilson J, Locke RJ, Vojnovic B, Stratford MRL, Dennis MF, Chaplin DJ (1999) Combretastatin A-4 phosphate as a tumor vascular-targeting agent: early effects in tumors and normal tissues. Cancer Res 59:1626–1634

37. Richardson CA, Flecknell PA (2005) Anaesthesia and post-operative analgesia following experimental surgery in laboratory rodents: are we making progress? Altern Lab Anim 33:119–127

38. Meyer E (1989) Simultaneous correction for tracer arrival delay and dispersion in CBF measurements by the H215O autoradiographic method and dynamic PET. J Nucl Med 30:1069–1078

Chapter 19

Chorioallantoic Membrane Microtumor Model to Study the Mechanisms of Tumor Angiogenesis, Vascular Permeability, and Tumor Cell Intravasation

Elena I. Deryugina

Abstract

The mechanisms governing the development of angiogenic blood vessels, which not only deliver the nutrients to growing tumors but also provide the conduits for tumor cell dissemination, are still not fully resolved. The model systems based on the grafting of human tumor cells onto the chorioallantoic membrane (CAM) of the chick embryo offer several advantages to study complex processes underlying tumor angiogenesis and tumor cell dissemination. In particular, the CAM model described here allows for investigation of multiple microtumors as independent entities, thereby greatly facilitating quantification and statistical analyses of tumor neovascularization and cancer spreading. This CAM microtumor system was designed specifically to measure the level of tumor cell intravasation in combination with quantitative analyses of the microarchitecture and permeability of the intratumoral angiogenic blood vessels. By using this newly established microtumor model we have demonstrated the functional involvement of tumor matrix metalloproteinase-1 (MMP-1) and epidermal growth factor receptor (EGFR) in regulating the development of a distinct angiogenic vasculature capable of sustaining tumor cell intravasation and metastasis.

Key words Chick embryo, Chorioallantoic membrane, Animal models of cancer dissemination, Tumor angiogenesis, Vascular permeability, Tumor cell intravasation, Tumor invasion, Cancer metastasis, Matrix metalloproteinase 1, Epidermal growth factor receptor

1 Introduction

Tumor cell dissemination occurs through vascular routes and results in metastatic disease, a leading cause of cancer-related deaths. To establish secondary site metastases, aggressive tumor cells are believed to acquire certain genetic advantages, which allow them to successfully detach from the primary tumor, invade the local stroma, actively enter the vasculature (intravasation), survive in the circulation, exit from the capillary network of the secondary site (extravasation), and initiate proliferation (metastatic outgrowth) [1, 2]. Although studied extensively for more than a century, the specific mechanisms underlying different steps in cancer

Stewart G. Martin and Peter W. Hewett (eds.), *Angiogenesis Protocols*, Methods in Molecular Biology, vol. 1430,
DOI 10.1007/978-1-4939-3628-1_19, © Springer Science+Business Media New York 2016

progression are not fully understood and their necessity or sufficiency, active or passive mode of execution, and occurrence in permissive or instructive environments are still under debate.

A number of mammalian (mainly murine) models have been developed to study individual steps of the metastatic cascade [3], but most of these systems are costly, labor intensive, and require long-term commitment as well as sophisticated equipment. These limitations frequently lead to insufficiently powered studies and semiquantitative research. In this regard, chick embryo models offer practical alternatives allowing the rapid screening of the key molecules functionally involved in different processes of metastatic spread. First, the chick embryo is a naturally immunodeficient organism allowing human tumor cells to grow in the host without species-specific restrictions. Second, each step of the metastatic cascade (with the exception of cancer cell initiation) can be reproduced and quantified in the independent chick embryo models specifically tailored to study tumor growth, invasion, angiogenesis, tumor cell intravasation, and spontaneous metastasis, as well as survival of tumor cells in the circulation, their extravasation, and colonization of different internal organs [4]. All these models are based on the growth of human tumor cells in the specialized extraembryonic tissue, namely the chorioallantoic membrane (CAM), a highly vascularized organ serving as an embryonic analogue of adult lungs [5].

The mechanisms of tumor cell dissemination via vascular routes have been in the spotlight of cancer research for many decades, but are not fully resolved and present many challenges for researchers. The full nature of angiogenic vessels in tumors is still unclear with regard to whether they are formed *de novo* from activated endothelial cells of pre-existing vessels or co-opted from adjacent mature vasculature, or even created *bona fide* from cancer stem cells [6]. Furthermore, the tumor vasculature is considered to be immature since it is represented by torturous vessels with high levels of permeability and leakiness because of insufficient coverage with pericytes [7, 8]. On the other hand, a reduction in metastasis may be accompanied by a decrease in pericyte coverage and permeability of tumor-associated vessels [9]. Therefore, there is an intrinsic discrepancy in the notion that angiogenic vasculature of metastatic tumors is functionally underdeveloped, but should be attenuated further to prevent tumor cell spread. Correspondingly, not only do anti-angiogenic treatments commonly result in normalization of tumor vasculature [10], but they also may lead to increased spread of tumor cells [11]. Our research led us to a somewhat surprising conclusion that the angiogenic vasculature should actually be developed to a proper extent to accommodate efficient tumor cell intravasation and dissemination. These intravasation-sustaining vascular routes are represented by distinct angiogenic vessels with specific microarchitecture and a certain level of permeability.

1.1 The CAM Microtumor Model System

To address how the microarchitecture of tumor vessels and vascular permeability regulate the levels of tumor cell intravasation and metastasis, we developed a model system that allows the simultaneous quantification of all these parameters in individual microtumors. In this model system, the microtumors are initiated on the CAM of chick embryos following grafting of collagen droplets containing human tumor cells (Fig. 1a). If fluorescently labeled tumor cells are used, CAM microtumors can be visualized by epifluorescence microscopy along with chick embryo vessels highlighted with fluorescently conjugated lectins or dextrans of contrasting color (Fig. 1b–d, *left panels*). Within 5–7 days of grafting, rich vascular networks develop in the CAM microtumors allowing the aggressive tumor cells to intravasate and metastasize to the internal organs of the chick embryo (Fig 1b–d, *right panels*). When highly disseminating human HT-hi/diss fibrosarcoma and HEp3-hi/diss epidermoid carcinoma are used, the levels of intravasation and metastasis can be quantified as early as 5–6 days after cell grafting on the CAM (Fig. 1e). Disseminated human tumor cells are quantified in the portions of the CAM distal to the microtumors (intravasation) and in the internal organs such as liver (spontaneous metastasis) by human-specific real-time *Alu* qPCR conducted in comparison with chicken cell samples spiked with known numbers of human cells [12, 13].

The CAM also supports the growth of aggressive murine tumor cells. For example, the B16-F10 murine melanoma line gives rise to easily detectable dark brown microtumors (Fig. 2a, b). The B16-F10 microtumors rapidly develop well-integrated angiogenic vasculature that can be stained with fluorescently conjugated lectin and visualized as early as day 3–5 by epifluorescence microscopy (Fig. 2b, c).

By using the CAM microtumor model system, we have recently demonstrated that the intravasation and dissemination processes depend on the presence of angiogenic vasculature with certain microarchitecture characteristics, namely the vessels with a sizable lumen that can accommodate the volume of an intravasating tumor cell (Fig. 3a, *top*). Specifically, we have shown [14] that carcinoma-produced matrix metalloproteinase 1 (MMP-1) facilitates the formation of angiogenic blood vessels with a lumen of >15 μm in diameter (Fig. 3b), and that downregulation of MMP-1 production results in the development of collapsed, lumen-devoid intratumoral vessels (Fig. 3a, *bottom*, b), incapable of sustaining tumor cell intravasation and metastasis (Fig. 3c). Furthermore, to facilitate intravasation and metastasis, the intratumoral vasculature should have a certain level of permeability, which can be quantified by measuring the exudation of permeable and nonpermeable fluorescent dextrans injected into the chick embryo vasculature (Fig. 4). In HEp3-hi/diss microtumors, the intravasation-sustaining angiogenic vessels were permeable to low molecular weight

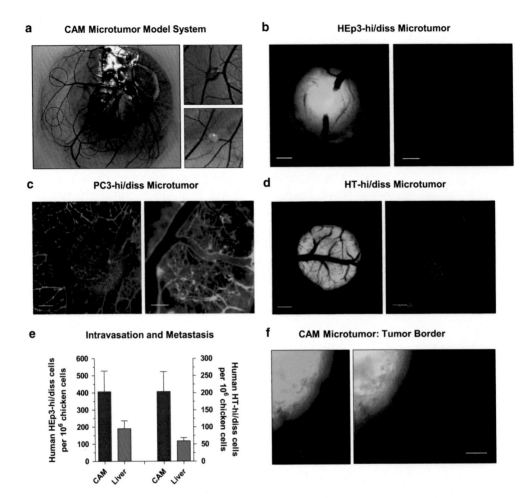

a CAM Microtumor Model System

b HEp3-hi/diss Microtumor

c PC3-hi/diss Microtumor

d HT-hi/diss Microtumor

e Intravasation and Metastasis

f CAM Microtumor: Tumor Border

Fig. 1 CAM microtumor model system. (**a**) Human microtumors developed on the CAM of the chick embryo incubated *ex ovo* (adapted from ref. 14). A 10-day-old chick embryo was grafted with five 10 μL droplets each containing 1×10^5 human HEp3-hi/diss epidermoid carcinoma cells. Within 6 days, CAM microtumors (*outlined by dotted circles in the left panel*) become visible to the naked eye and the vasculature becomes amenable for lectin or dextran injections. A couple of allantoic veins accessible for i.v. injections are indicated by the *arrows*. Note that allantoic veins, which bring the oxygenated blood towards the embryo, are bright red in color and should not be mistaken for allantoic arteries, which are dark red in color since they carry oxygen-depleted blood from the embryo for oxygenation in the CAM. Two enlarged microtumors are presented on the *right*. (**b-d**) CAM microtumors (*left panels*; both *red* and *green channels* are shown) and their respective intratumoral vasculature (*right panels*; only "vascular" channel is depicted) visualized in epifluorescent microscope are shown. (**b**) *Left*, CAM microtumor developed from HEp3-hi/diss epidermoid carcinoma cells labeled with CellTracker Green. CAM vasculature was highlighted with Rhodamine-conjugated *Lens culinaris* agglutinin (LCA). Bar, 250 μm. *Right*, Intratumoral vasculature at higher magnification. Bar, 100 μm. (**c**) *Left*, CAM microtumor developed from PC3-hi/diss prostate carcinoma cells stably transfected with tdTomato protein. CAM vasculature was highlighted with FITC-conjugated LCA. *Right panel*, Intratumoral vasculature at higher magnification. Bars, 250 μm. (**d**) *Left panel*, CAM microtumor developed from HT-hi/diss fibrosarcoma cells stably transfected with GFP. Bar, 500 μm. CAM vasculature was highlighted with Rhodamine-conjugated LCA. *Right panel*, Intratumoral vasculature at higher magnification. Bar, 100 μm. (**e**) Levels of tumor cell intravasation (CAM) and metastasis (liver) were determined by quantitative *Alu* PCR in the embryos bearing CAM microtumors developed from HEp3-hi/diss cells (bars on the *left*) or HT-hi/diss cells (bars on the *right*). (**f**) The lack of visible stromal invasion in the CAM microtumor model. Before grafting on the CAM, HEp3-hi/diss cells were labeled with CellTracker Green. After 6 days, the chick embryo vasculature was labeled with Rhodamine-conjugated LCA. *Left panel*: A portion of HEp3-hi/diss microtumor (*green*) presenting a well-defined smooth border is shown. *Right panel*: Fluorescence signals emitted by the tumor (*green*) and tumor-associated vasculature (*red*) are merged. The *yellow color* indicates overlap of intense green and red fluorescence signals. Bar, 100 μm

CAM Microtumor Model: Murine B16-F10 Melanoma

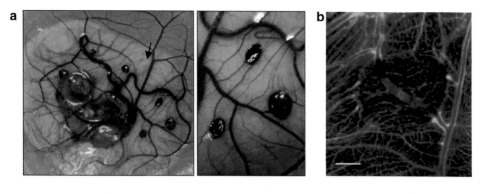

B16-F10 Microtumor: Day 3 after cell grafting

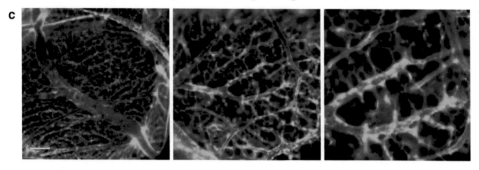

B16-F10 Microtumor: Day 5 after cell grafting

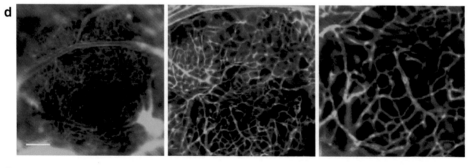

Fig. 2 CAM microtumor system adopted for murine tumor cells. (**a**) Murine microtumors developed on the CAM of the chick embryo incubated *ex ovo*. A 10-day-old chick embryo was grafted with six 10 μL droplets each containing from 1 to 2×10^5 murine B16-F10 melanoma cells. *Left panel:* Within 7 days, B16-F10 microtumors become visible as dark brown outgrowths. One of the allantoic veins accessible for i.v. injections is indicated by the *arrow*. *Right panel:* An enlarged portion of the CAM with three microtumors is shown. (**b**) Piece of the CAM containing B16-F10 microtumor was positioned on a glass slide and visualized in epifluorescent microscope at the low, 2× objective magnification. Bar, 500 μm. (**c** and **d**) B16-F10 microtumors 3 days (**c**) and 5 days (**d**) after cell grafting on the CAM. The intratumoral vasculature was labeled with FITC-conjugated LCA and immunofluorescent signal acquired monochromatically. *Note:* the intratumoral angiogenic vessels are sprouting on day 3, whereas on day 5, the vessel network appears to be more developed with fully connected capillaries. *Left*, *middle*, and *right* panels depict the same microtumor evaluated at different magnifications with 4×, 10×, and 20× objectives, respectively. Bars, 500 μm

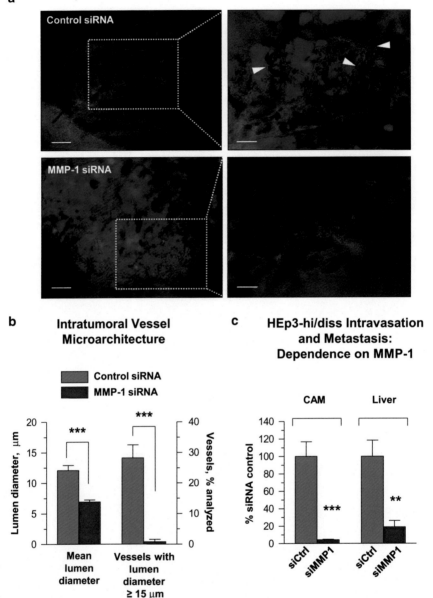

Fig. 3 Intravasation and metastasis of human tumor cells require the presence of intratumoral angiogenic vessels with specific microarchitecture (adapted from ref. 14). (**a**) Intratumoral vasculature within HEp3-hi/diss microtumors developing from control cells (treated with control siRNA; *upper panels*) or cells transfected with siRNA against MMP-1 (*bottom panels*). Before grafting, tumor cells were stained with CellTracker Green. The chick embryo vasculature was labeled with Rhodamine-conjugated LCA. *Left*, microtumors visualized with a 10× objective. Bars, 25 μm. *Right*, Enlarged portions of microtumors visualized with a 20× objective. Bars, 50 μm. *Arrowheads* in the *upper panel* depicting control tumor indicate dilated blood vessels containing tumor cells and erythrocytes. (**b**) Intratumoral vasculature in the CAM microtumors developed from control and MMP-1-silenced HEp3-hi/diss cells was analyzed for the mean lumen diameter (bars on the *left*) and, more specifically, for the percentage of vessels with a lumen of ≥15 μm in diameter (bars on the *right*). (**c**) Levels of intravasation (CAM) and metastasis (liver) of HEp3-hi/diss cells disseminating from CAM microtumors were measured with quantitative *Alu* PCR in the tissue samples harvested from embryos grafted with the MMP-1-competent cells (treated with control siRNA, siCtrl) or MMP-1 silenced (siMMP1) cells

dextran and this permeability was reduced significantly by either inhibition of MMP-1 production in tumor cells (Fig. 4a) or the use of a proteinase activated receptor-1 (PAR1) inhibitor (Fig. 4b).

The development of intratumoral vasculature with similar microarchitecture and permeability characteristics also underlies the mechanisms involved in dissemination of epidermal growth factor receptor (EGFR)-overexpressing cancer cells. Thus, functional inhibition of EGFR by RNA interference or small molecule inhibitors substantially diminished the production of vascular endothelial growth factor (VEGF) production and the development of VEGF-dependent permeable lumen-containing angiogenic vasculature in HT-hi/diss microtumors, concomitantly with a dramatic inhibition of tumor cell intravasation and metastasis.

Importantly, CAM microtumors develop as relatively compact structures without any visible invasion into the surrounding tissue (Fig. 1f). Because of this characteristic, our findings also indicate that spontaneous dissemination can occur exclusively via the intratumoral vasculature and independently of stromal invasion at the tumor border, a conventionally accepted prerequisite for the escape of aggressive cancer cells from the primary tumor before they reach the adjacent tumor-associated blood vessels for intravasation. In contrast to this view, the data from our CAM microtumor model are consistent with the notion that spontaneous metastasis may occur at those stages of cancer progression that precede a pronounced stromal invasion, i.e., much earlier than the development of the invasive front revealed by cancer pathologists during histological examination of resected primary tumors.

2 Materials

The use of chick embryo models as any animal model requires an Animal Protocol approved by the respective Institutional Animal Care and Use Committee (IACUC).

2.1 Embryonated Chicken Eggs

The embryonated ("hatching") eggs can be purchased from commercial sources such as Charles River Laboratories or from an accredited local poultry/chicken farm (*see* **Note 1**).

2.2 Refrigerator and Incubators

1. A small refrigerator adjusted to ~10 °C and ~80 % relative humidity, which are optimal for chick embryo development (*see* **Note 2**).

2. A rotating egg incubator set at 37.5 °C and 80 % relative humidity. Alternatively, if a specialized egg incubator is not available, a stationary thermostat or cell culture incubator (without a CO_2 supply) can be used and the eggs rotated manually 2-3 times a day.

a   Permeability of Intratumoral Vasculature: MMP-1 silencing

Control siRNA   **MMP-1 siRNA**

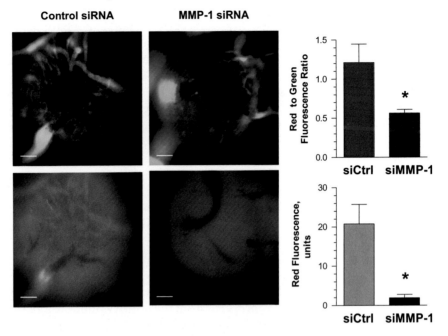

b   Permeability of Intratumoral Vasculature: PAR1 Antagonist

Vehicle Control   **PAR1 Antagonist**

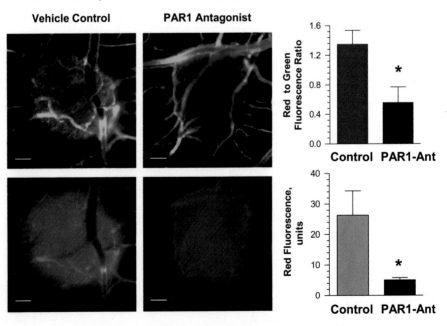

Fig. 4 Measurements of vascular permeability in the CAM microtumor model system (adapted from ref. 14). (**a**) CAM microtumors were generated from nonlabeled HEp3-hi/diss cells treated with control siRNA (*left panels*) or MMP-1 siRNA (*right panels*). Six days after cell grafting, tumor-bearing embryos were first inoculated with the permeable TRITC-conjugated dextran to determine the levels of vascular permeability (red fluorescence). After 1 h incubation, the embryos were inoculated with the nonpermeable FITC-conjugate dextran to determine the volume of perfusable vasculature (green fluorescence). Red and green fluorescence signals were acquired in

3. An incubator at 37.5 °C and 80 % relative humidity (without CO_2 supply) for the *ex ovo* development of chick embryos within weigh boats.

4. Tissue culture incubator adjusted to 37 °C, 5 % CO_2, and ~90 % relative humidity.

2.3 Ex Ovo Chick Embryo Culture

1. A shielded area or a "cloning" box with a glass front to accommodate cutting the eggshell.

2. A small drill (e.g., Dremel 300), with exchangeable circular blades (e.g., No. 409).

3. Weigh boats (3-1/2″), sprayed with 70 % ethanol, air dried, and sterilized under UV light in the laminar flow hood for at least 2 h.

4. Sterile square 100×100 mm Petri dishes. The lids and bottom halves of dishes are used to cover individual weigh boats.

2.4 Tissue Culture

1. Common tissue culture equipment and plastics including a laminar flow hood, pipettors and centrifuge, Petri dishes and cell culture flasks, serological pipettes.

2. Light microscope to evaluate cell cultures and count detached cells.

3. Hemocytometer (e.g., Neubauer) for counting cells.

4. A suitable tumor cell line (*see* **Note 3**).

5. Tumor cell growth medium: The recommended basal culture medium for the tumor line supplemented appropriately for the growth and maintenance of the tumor cells.

6. Basal medium with either 10 % fetal bovine serum (FBS), or 1 % bovine serum albumin (BSA), for washing cells prior to seeding on the CAM.

7. Trypsin-EDTA solution or nonenzymatic cell detachment solution to remove adherent tumor cells from flasks.

8. Dulbecco's calcium and magnesium-free phosphate-buffered saline (PBS). Autoclave to sterilize.

9. Native bovine skin type-I collagen solution: Acid-extracted collagen is neutralized on ice by mixing 8 parts (v/v) of collagen with 1 part 10× minimal essential medium (MEM) or PBS and

Fig. 4 (continued) individual microtumors visualized in epifluorescent microscope (*top panels*). To appreciate the difference in the levels of permeability, only red fluorescence signal is shown in *bottom panels*. Bars, 200 µm. *Graphs* on the *right* show quantification of red and green fluorescence in lysates of individual microtumors measured in a fluorometer. The ratio of red-to-green signal provides the measurement of vascular permeability independent of the volume of perfusable vasculature in individual microtumors. Data are means ± SEM. *$P < 0.05$. (**b**) Developing HEp3-hi/diss microtumors treated with either a PAR1 antagonist or vehicle control. Vascular permeability was measured as described above in (**a**). Bars, 500 µm. Data in the graphs are means ± SEM. *$P < 0.05$

1 part of 1 N NaOH. The use of 10× MEM allows the pH to be monitored while the 1 N NaOH is added. Add a volume of ultrapure water equivalent to that of the solution used to give a solution with a final collagen concentration of 2.0–2.2 mg/mL (*see* **Note 4**).

10. Fluorescent labeling of tumor cells: Transfect cells with a construct expressing a fluorescent protein (e.g., enhanced green fluorescent protein (EGFP) or red fluorescent protein, e.g., tdTomato) or transiently stain with fluorescent dyes such as CellTracker Green (CMFDA).

2.5 CAM Microtumor Analyses

1. One-mL "tuberculin" syringes and 30½ gauge needles to inject fluorescent lectins or dextrans into chick embryo vasculature.

2. Rhodamine- or fluorescein-conjugated (LCA) to highlight the chick embryo vasculature (*see* **Note 5**). Dilute the LCA with PBS to 0.1 mg/mL.

3. Low mol wt, permeable, fluorescent dextran: e.g., Tetramethylrhodamine (TRITC)-conjugated dextran of 155 kDa (Sigma; #T1287), or Rhodamine-conjugated dextran of 70 kDa (Sigma; #R9379).

4. High mol. wt., nonpermeable, fluorescent dextran, e.g., fluorescein isothiocyanate (FITC)-conjugated dextran 2000 kDa (*see* **Note 6**).

5. Small size surgical scissors and forceps to cut the CAM and manipulate portions containing microtumors.

6. Glass slides for microscopic evaluation and imaging of CAM microtumors (*see* **Note 7**).

7. Black polystyrene 96-well microplates for fluorescence measurements.

8. Microplate or single tube fluorometer.

9. Immunofluorescent microscope equipped with objectives to evaluate specimens under low (2–4×) and high (10–63×) magnification, and a video camera to record images for subsequent quantitative analyses.

10. Silver nitrate applicator (e.g., Grafco; #1590).

11. mRIPA buffer for lysing of microtumors: 50 mM Tris-HCl pH 7.4 containing 1 % (v/v) Triton X-100, 150 mM NaCl, and protease inhibitors.

3 Methods

3.1 Embryo Development and Preparation of CAM

The chick embryo has a relatively short development (~21 days), which is tightly regulated. Experiments should be planned carefully since the procedures need to be performed within a narrow time window allowing for only ~6 h variance.

1. Upon receipt, the embryonated eggs may be placed into a 37.5 °C/80 % humidity incubator in a horizontal position to initiate embryo development. During incubation, the eggs should be rotated 2-3 times per day. Alternatively, eggs can be stored in an upright position (rounded, air sack-containing, large end up) at 10 °C/80 % humidity for 1–2 weeks (*see* **Note 8**).

2. After a 3-day incubation at 37.5 °C remove the eggs from the incubator, spray with the 70 % alcohol, and dry with paper towels.

3. Leave eggs in the horizontal position for 30–60 min, either on a flat surface or in the incubator, to allow the embryo to situate itself on the top of the yolk (*see* **Note 9**).

4. Cut the eggshell at the bottom of the egg superficially with the drill blade. Carefully press against the bottom to deliver the contents of the egg into a weigh boat. Viable embryos are visible by naked eye and can be distinguished by their beating hearts and blood flow.

5. Cover the boat containing live embryo with a lid or bottom half of a square Petri dish and transfer the covered boats to the stationary 37.5 °C humidified incubator for 7 days (*see* **Note 10**).

6. The CAM begins to develop between day 5 and 6 of embryo development and appears as a blood vessel-containing sack above the *embryo proper*. After 7 days of *ex ovo* development (day 10 of embryo development), the embryos present a well-developed CAM that can spread out over almost all the exposed surface of the embryo-containing boat. At this stage, the CAM is fully developed and can be grafted with tumor cells. Day 9 embryos may also be used if the tumor cells are slow growing or large microtumors are required (*see* **Note 11**).

3.2 Establishing Microtumors on the CAM

To initiate microtumors, tumor cells are mixed with neutralized type I collagen solution and the mixture distributed in small droplets onto fully vascularized areas of the CAM (*see* **Note 12**) as described below.

1. Detach adherent tumor cells with trypsin-EDTA or enzyme-free detachment solution. Harvest and wash the cells in basal medium supplemented with 10 % FCS or 1 % BSA.

2. Centrifuge the cells and resuspend the cell pellet in serum-free medium and count using a hemocytometer.

3. Centrifuge the cells and discard the supernatant.

4. Resuspend the cell pellet in neutralized type I collagen solution to a final concentration of $1–2 \times 10^7$ cells/mL (2.0–2.2 mg/mL type-1 collagen).

5. Place four to six 10 μL droplets on the CAM of each embryo, thereby allowing the initiation of individual microtumors from

1 to 2×10^5 cells grafted (*see* Figs. 1a and 2a). Droplets should be placed between the large CAM vessels, but not too close to the edges of the CAM, or weigh boat walls (*see* **Note 13**). Do not disturb the embryos for 10–15 min to allow the collagen to solidify.

6. Return the embryos to the incubator for 5–7 days.

3.3 Investigation of Tumor Angiogenesis and Tumor Cell Intravasation and Metastasis

To study the mechanisms underlying the development of intratumoral angiogenic vasculature and tumor cell intravasation and metastasis, the developing microtumors can be treated topically with growth factors, cytokines, small inhibitor molecules, or blocking antibodies with or without additional rescuing components. To facilitate penetration into the tumor interior, the agents are applied in 1 % dimethyl sulphoxide (DMSO). The final concentrations of agent added to the CAM will depend on their specific characteristics, but volumes from 10 to 20 μL are enough to cover the area of growing microtumors. Treatment regimes can be daily, or every other day, during 4–6 days of tumor development and initiated at the time of tumor cell grafting or after an initial 24–48 h incubation. After 5–7 days of culturing *ex ovo*, the embryos with developing human microtumors are ready for quantification of the levels of intravasation and metastasis by *Alu* qPCR (Figs. 1e and 3c) along with quantitative characterization of the microarchitecture (Fig. 3b) and permeability (Fig. 4) of the intratumoral vasculature.

3.3.1 Tumor Cell Intravasation and Metastasis

1. The levels of human tumor cell intravasation can be determined using the DNA extracted from the CAM located most distal to the microtumors (see **Note 14**). Purified DNA can be used in human-specific *Alu* qPCR in parallel with the samples of human tumor cells spiked into the chicken tissue background.

2. Harvest samples from the embryo taking care to prevent contamination of the samples with tumor cells from undeveloped primary tumors. If highly disseminating tumor cells are used to initiate microtumors, the level of spontaneous metastasis can be measured by *Alu* qPCR in liver (the major internal organ where metastasizing tumor cells may be found) or lungs of the embryo.

3.3.2 Intratumoral Angiogenic Vasculature

To analyze the microarchitecture of the intratumoral vasculature, the embryos are grafted with fluorescently labeled tumor cells, e.g., stably expressing enhanced GFP. In this case, the vasculature of embryos should be stained with lectin conjugated with a contrasting red fluorescent tag, e.g., Rhodamine-conjugated LCA.

1. Inject 100–200 μL of 0.1 mg/mL LCA in PBS into one of the allantoic veins using a 30½-gauge needle and 1 mL syringe. The bleeding at the injection site can be stopped by pressing a silver nitrate applicator against the cut. After approximately 10 min, the vascular endothelium will be labeled, and CAM tissue can be harvested for microscopic evaluation.

2. Harvest ~ 2 x 2 cm samples of the CAM containing microtumors frand wash in PBS.

3. Place the samples on the glass slides for evaluation and imaging in fluorescent microscope (*see* **Note 6**).

4. Record digital images and analyze to quantify the different parameters of intratumoral angiogenic vasculature, including the lumen diameter, vessel length, and number of branching points.

3.4 Vascular Permeability and Tumor Cell Intravasation

To investigate the relationship between the level of tumor cell intravasation and the extent of vascular permeability, CAM microtumors are initiated with unlabeled cancer cells (*see* Subheading 3.2, **steps 1–6**).

1. After 5–7 days of tumor development, inject an allantoic vein with the low mol. wt. (permeable) fluorescent dextran, e.g., Rhodamine-conjugated dextran of 70 kDa. The bleeding at the injection site can be stopped by pressing a silver nitrate applicator against the cut.

2. Return injected embryos to the incubator, or leave in a warm room for 1 h.

3. Inject high mol. wt. (nonpermeable) dextran conjugated with the fluorochrome of the color contrasting the permeable dextran, e.g., 2000 kDa FITC-dextran into a different allantoic vein (*see* **Note 5**).

4. Immediately, harvest the CAM distal to the microtumors for *Alu* qPCR and then resect the microtumors from the CAM and place into mRIPA buffer (between 200 and 300 mL per tumor; 2–3 small tumors could be mixed together).

5. Mince the tumors with scissors and incubate at 4 °C for 45–60 min under constant rotation to extract the dextrans.

6. Clear the tumor lysates by centrifugation at 12,000–14,000 rpm for 15–20 min at 4 °C.

7. The lysates from individual microtumors are distributed into 96-well plate fluorescence microplates (100–150 µL per well) to measure fluorescence in a microplate fluorometer. Alternatively, a single tube fluorometer can be used for fluorescence measurement.

8. Levels of red fluorescence, reflecting amounts of tissue-retained low mol. wt. TRITC-conjugated dextran, are measured at 576 nm (excitation at 557 nm). Levels of green fluorescence, reflecting amounts of high mol. wt. FITC-conjugated dextran trapped in blood vessels, are measured at 516 nm (excitation at 492 nm). The relative "red" fluorescence indicates the level of vascular permeability, whereas the relative "green" fluorescence indicates the volume of perfusable vasculature (*see* **Note 15**).

4   Concluding Remarks

The newly established CAM microtumor model makes it possible to investigate the functional involvement of tumor-derived and host-delivered molecules in the development of a specific intratumoral neovasculature capable of sustaining tumor cell intravasation and metastatic dissemination. Our recent findings using this model indicate that tumor cell intravasation requires the presence of angiogenic vessels with a distinct size lumen and certain level of permeability. When this category of intratumoral vessels is lost due to changes in the tumor cells or tumor microenvironment, the levels of intravasation and metastasis are reduced dramatically. This requirement of functional angiogenic vessels with distinct microarchitecture was confirmed recently in our studies employing syngeneic murine tumors as well as an orthotopic model of human head and neck carcinoma in immunodeficient mice. These confirmatory findings highlight the advantages of the chick embryo models as a discovery-based model system. Finally, by using our CAM mictotumor model we have demonstrated the possibility of tumor cell dissemination via intratumoral angiogenic blood vessels and without involvement of tumor-adjacent blood vessels and stromal invasion of primary tumor cells towards these vessels. These important observations indicate that tumor cell metastasis can occur much earlier in the course of cancer progression than is currently appreciated based on histological examination of solid tumors in cancer patients.

5   Notes

1. We use the White Leghorn, but any breed of chicken can be used. It is important that the eggs are free of major pathogens that could affect embryo development and contaminate the laboratory environment and personnel. Specific pathogen-free (SPF) eggs are available from commercial sources.

2. The temperature can range from 7 to 13 °C; the relative humidity from 70 to 90 %. The required humidity can be achieved using a container filled with distilled water placed at the bottom of the fridge or thermostat. It is important to maintain conditions free of fungal growth.

3. Tumor cells should be capable of proliferation on the CAM and intravasation into angiogenic CAM vessels. However, if tumor cells generate microtumors, but are incapable of spontaneous dissemination, the CAM microtumor model will still allow study of angiogenesis.

4. Caution should be taken while handling the collagen solution to avoid trapping of air bubbles.

5. The vasculature should be highlighted with the lectin conjugated with a contrasting fluorophore to that of the tumor cells, e.g., Rhodamine-conjugated LCA (Vector; #W1012) could be used with EGFP labeled tumor cells.

6. The pair of permeable and nonpermeable fluorescent dextrans should have contrasting fluorophores to allow for the measurement of permeability versus total volume of perfusable vasculature. The measurement of both parameters allows the use of their ratio for comparison of permeability independent of differences in perfusable vasculature developed in the individual microtumors.

7. Glass slides should not be too thick to allow for epifluorescent examination of whole, nondissected, CAM microtumors.

8. Not all embryonated eggs give rise to viable embryos even if immediately placed into an incubator. Furthermore, one week of storage at 4 °C results in ~10 % eggs without viable embryos. Each day of storage prior to incubation usually causes an additional 1 % loss in embryo viability. This should be taken into consideration when calculating the number of embryos required for a particular experiment. To improve embryo viability, turn or tilt the stored eggs twice a day.

9. Chick embryos are rarely contaminated with bacteria, but are easily contaminated with fungi. Therefore, all procedures should be performed in a clean room with sterilized instruments and materials. In addition, for long-term handling of embryonated eggs or embryos developing *ex ovo*, the room should be maintained at 24–26 °C.

10. It is convenient to transport the embryo-containing boats on a shelf removed from the incubator.

11. Avoid shaking the live embryos, especially after several days of development *ex ovo*, to prevent the CAM sticking to the side of the weigh boat. This may cause excessive bleeding and premature death of the embryo.

12. In theory, any extracellular matrix protein(s) which will form gels (e.g., Matrigel) can be used for generation of microtumors; however, we do not know whether formed microtumors will be sufficiently transparent for epifluorescence microscopy.

13. The CAM constantly expands, enveloping the embryo, and therefore can bring microtumors placed too close to the edges to the bottom of the weigh boat.

14. Tumor cell intravasation occurs via angiogenic blood vessels that provide vascular conduits for escaped cancer cells from the primary tumor. Because of the nature of blood circulation in the chick embryo, the majority of tumor cells, which have entered the CAM vessels, become trapped in the capillary

network of the CAM, which serves as a repository of newly intravasated tumor cells.

15. The ratio of two parameters, namely the intensity of red fluorescence over the intensity of green fluorescence, gives the index that more accurately reflects the difference in vascular permeability since it becomes independent of total volume of perfusable vasculature in the individual microtumor.

Acknowledgements

This work was supported by NIH grants R01CA105412 and R01CA129484.

References

1. Valastyan S, Weinberg RA (2011) Tumor metastasis: molecular insights and evolving paradigms. Cell 147:275–292
2. Mina LA, Sledge GW Jr (2011) Rethinking the metastatic cascade as a therapeutic target. Nat Rev Clin Oncol 8:325–332
3. Francia G, Cruz-Munoz W, Man S, Xu P, Kerbel RS (2011) Mouse models of advanced spontaneous metastasis for experimental therapeutics. Nat Rev Cancer 11:135–141
4. Deryugina EI, Quigley JP (2008) Chick embryo chorioallantoic membrane model systems to study and visualize human tumor cell metastasis. Histochem Cell Biol 130:1119–1130
5. Romanoff AL (1960) The avian embryo. The Macmillan Co, New York
6. Weis SM, Cheresh DA (2011) Tumor angiogenesis: molecular pathways and therapeutic targets. Nat Med 17:1359–1370
7. Raza A, Franklin MJ, Dudek AZ (2010) Pericytes and vessel maturation during tumor angiogenesis and metastasis. Am J Hematol 85:593–598
8. Dudley AC (2012) Tumor endothelial cells. Cold Spring Harb Perspect Med 2:a006536
9. Chantrain CF, Henriet P, Jodele S et al (2006) Mechanisms of pericyte recruitment in tumour angiogenesis: a new role for metalloproteinases. Eur J Cancer 42:310–318
10. Carmeliet P, Jain RK (2011) Principles and mechanisms of vessel normalization for cancer and other angiogenic diseases. Nat Rev Drug Discov 10:417–427
11. Ebos JM, Lee CR, Cruz-Munoz W et al (2009) Accelerated metastasis after short-term treatment with a potent inhibitor of tumor angiogenesis. Cancer Cell 15:232–239
12. Conn EM, Botkjaer KA, Kupriyanova TA et al (2009) Comparative analysis of metastasis variants derived from human prostate carcinoma cells: roles in intravasation of VEGF-mediated angiogenesis and uPA-mediated invasion. Am J Pathol 175:1638–1652
13. Bekes EM, Schweighofer B, Kupriyanova TA et al (2011) Tumor-recruited neutrophils and neutrophil TIMP-free MMP-9 regulate coordinately the levels of tumor angiogenesis and efficiency of malignant cell intravasation. Am J Pathol 179:1455–1470
14. Juncker-Jensen A, Deryugina EI, Rimann I et al (2013) Tumor MMP-1 activates endothelial PAR1 to facilitate vascular intravasation and metastatic dissemination. Cancer Res 73:4196–4211

Chapter 20

The Rabbit Corneal Pocket Assay

Lucia Morbidelli and Marina Ziche

Abstract

The rabbit corneal micropocket angiogenesis assay uses the avascular cornea as a substrate canvas to study angiogenesis in vivo. Through the use of standardized slow-release pellets, a predictable angiogenic response is generated over the course of 1–2 weeks and then quantified. Uniform slow-release pellets are prepared by mixing purified angiogenic growth factors such as basic fibroblast growth factor or vascular endothelial growth factor and a synthetic polymer to allow slow release. A micropocket is surgically created in the rabbit cornea under anesthesia and a pellet implanted. On the days later, the angiogenic response is measured and qualified using a slit lamp, as well as the concomitant vascular phenotype or inflammatory features. The results of the assay are used to assess the ability of potential therapeutic molecules to modulate angiogenesis in vivo, both when released locally or given by ocular formulations or through systemic treatment. In this chapter, the experimental details of the rabbit cornea assay and technical implementations to the original protocol are described.

Key words Angiogenesis, Capillary, Endothelial cell, Angiogenic factors, Drug treatment, Slow release

1 Introduction

The cornea pocket assay consists in the placement of an angiogenesis inducer into a micropocket made in the cornea thickness and the evaluation of vascular outgrowth from the peripherally located limbal vessels toward the stimulus. The micropocket assay is a suitable model to characterize the pro- and antiangiogenic effect of modulators and drugs with potential application in other fields as wound healing or cancer or ocular disease. Specifically, a series of diseases are associated with corneal vascularization as traumatic injuries, corneal graft rejection, infections, and chronic inflammation.

Respect to the use of smaller animals (mouse, rat), the rabbit has a series of advantages. Due to the easy manipulation of rabbit cornea, not only purified growth factors/drugs are studied, but also tumor tissue samples and cell suspensions. Since the rabbits are amenable and do not require anesthesia for daily monitoring, the

Stewart G. Martin and Peter W. Hewett (eds.), *Angiogenesis Protocols*, Methods in Molecular Biology, vol. 1430, DOI 10.1007/978-1-4939-3628-1_20, © Springer Science+Business Media New York 2016

continuous observation of neovascular growth in the same animal allows the evaluation of drugs acting as suppressors or stimulators of angiogenesis with a reduction in the total number of animals needed for a statistical assessment.

The assay was chosen for the absence of a preexisting vascular pattern in New Zealand white rabbits and for the easy manipulation of the cornea and continuous monitoring of the neovascular growth. Our group settled a series of modifications of the original method [1], having set up protocols for the implant of multiple samples, including cell suspensions and tissue fragments. This technique, extensively used during the years, has been substantially modified to characterize angiogenesis inducers, to validate angiogenesis inhibitors, to study the interaction between different factors and the cellular, biochemical, and molecular mechanism of angiogenesis.

Refinement of drug formulation for local eye delivery and pharmacokinetic profile in eye components can be established.

In the following sections, the experimental details and protocols of the avascular cornea assay are presented.

2 The Rabbit Cornea Pocket Assay

First of all, the protocols and treatments must be approved by the local laboratory animal ethical board and the national agencies, according to the current laws and guidelines (European Directive 2010/63/EU or ARVO Statement for the Use of Animals in Ophthalmic and Vision Research, and 3R guidelines as in http://www.nc3rs.org.uk/), since the surgical procedure requires general anesthesia.

The micropocket assay is performed in albino rabbits (*see* **Note 1**) and requires the simultaneous presence of two qualified operators in all the steps (*see* **Note 2**).

2.1 Materials

2.1.1 Animals

New Zealand albino rabbits (Charles River, www.criver.com) of 1.5–2.5 kg (*see* **Note 3**).

2.1.2 Reagents and Drugs

1. Recombinant growth factors or drugs to be studied as slow release preparation must be dissolved in water or phosphate-buffered saline (PBS) or ethanol or methanol in highly concentrated solutions (0.1–1 mg/ml) (*see* **Note 4**).

2. Slow release polymer. Different polymers can be used (*see* **Note 5** for a comparison with other polymers). Ethylene-vinyl-acetate copolymer (Elvax-40) (DuPont de Nemours, Wilmington, DE, www.dupont.com) should be previously prepared and tested for biocompatibility [2] (*see* **Note 6**).

3. Xilazine solution (20 mg/ml) (Xilor).

4. Zoletil-20, a combination of a dissociative anesthetic agent, tiletamine hypochloride, and a tranquilizer, zolazepam hypochloride (each at 10 mg/ml).

5. The local anesthetic benoxinate or oxybuprocain chlorohydrate (0.4 % solution).

6. Tanax (T-61), an euthanasic mixture containing embutramide (200 mg/ml), mebenzonium iodide (50 mg/ml), and tetracaine hydrochloride (5 mg/ml).

7. Fixative: 4 % paraformaldehyde in PBS, pH 7.4.

8. Liquid nitrogen, isopentane, and OCT tissue-teck medium or similar.

2.1.3 Facilities, Equipments, and Materials

1. Cell culture facility equipped with vertical laminal flow hood and autoclave.

2. Animal facility equipped with a sterile surgical room.

3. Disposable scalpel for ocular microsurgery (n° 10–11, Aesculap).

4. Sterile forceps, silver spatula, microsurgery scissors, microspatula (*see* Fig. 1 for details).

5. Teflon plates (10 × 10 cm) and 6 cm glass Petri dishes.

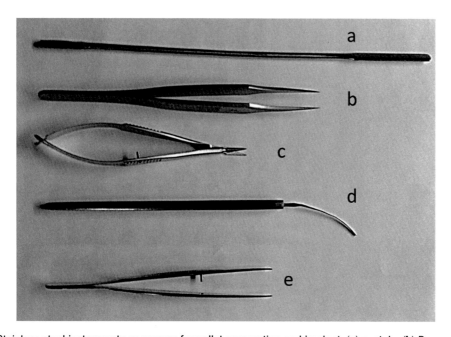

Fig. 1 Stainless steel instruments necessary for pellet preparation and implant: (**a**) spatula, (**b**) Dumont tweezers, and (**c**) Vannas scissors for pellet preparation and manipulation, (**d**) pliable iris spatula for ocular microsurgery and micropocket creation, (**e**) microforceps to keep open the edge of the corneal micropocket during pellet implantation

6. Vacuum.

7. Latex dental dam for endodontic procedures (DentalTrey, www.dentaltrey.com).

8. Insulin syringes.

9. Slit lamp stereomicroscope equipped with a digital camera.

10. Histology equipment and materials.

2.2 Protocols

2.2.1 Sample Preparation

The material under test can be in the form of slow-release pellets incorporating recombinant growth factors, cell suspensions, or tissue samples.

– Preparation of slow release pellets: In order to be implanted in the cornea, angiogenic factors (i.e. VEGF, FGF-2, cytokines, or other molecules) have to be prepared in a semisolid state, enabling surgical implantation and gradual release of the factor from the polymer. Pellets (implants)-bearing molecules to be tested are prepared under a laminar flow hood according to the following steps. A given amount of the compound to be tested is previously dried on a flat teflon surface. Then a predetermined volume of polymer casting solution (10 μl/pellet) is mixed with the dried compound on the teflon plate by the use of a stainless steel spatula. After drying, the film sequestering the compound is cut into $1 \times 1 \times 0.5$ mm homogeneous pieces under a stereomicroscope by the use of Vannas scissors and Dumont n. 5 tweezers. The pellets (in open glass Petri dishes) are left under vacuum at 4 °C overnight to remove residual solvent. Empty pellets of polymer are used as negative controls, while, depending on the experimental design, VEGF or FGF-2-containing pellets are used as positive controls (*see* **Note 7**).

– When testing the co-release of different molecules from the same pellet, the two substances are let to dry closely in the teflon plate and then incorporated in the same polymer preparation (*see* **Note 8**).

– Preparation of cell suspension: The intrinsic angiogenic potential due to different stages of tumor progression or to the expression of genes or gene products have been documented by our group as well as by others [3–6]. Prepare a cell suspension by trypsinization of confluent cell monolayers to a final dilution of 2–5×10^5 cells in 5 μl. When using cells, angiogenic response can be graded based also on the number of cells implanted into the corneal stroma.

– Preparation of tissue samples: Tissue samples of animal and human origin have been successfully implanted into the rabbit cornea to produce angiogenesis [7–10]. When tissues are tested, fragments are removed within 2 h from patients or animals and kept at 4 °C in complete medium. Fragments of 2–3 mg are obtained by cutting the fresh tissue samples under sterile conditions by the use of microdissection instruments under a stereomicroscope.

2.2.2 Surgery

1. Anesthetise animals with Xilor (0.5 ml, i.m.) followed by Zoletil (5 mg/kg i.m.) or alternatively sodium pentothal (10 mg/kg, i.v.). The deepness of anesthesia is checked as reflex to pressure (*see* **Note 9**).

2. Each eye is enucleated by the use of a dental dam and a local anesthetic (0.4 % benoxinate) is instilled just before surgery.

3. The pellet implantation procedure starts with a linear intrastromal incision, parallel to the corneoscleral limbus (linear keratotomy), using a surgical blade (disposable scalpel n. 10). The corneal micropocket for the pellet implant is produced with a 1.5 mm pliable silver spatula with smooth edge blade in the lower half of the cornea (*see* **Note 10**).

4. Pellet implant: The implant is introduced through the keratotomy line, parallel to the corneal epithelium and under it, in the external third of the stroma, up to 2 mm from the limbus. One single pellet is selected from the Petri dish using Dumont n. 5 tweezers and then introduced in the corneal pocket. Microforceps are used to keep open the edge of the cut. Locate the implant at 2 mm from the limbus to avoid false positives due to mechanical stress and to favor the gradient diffusion of test substances in the tissue, toward the endothelial cells at the limbal plexus.

5. When two factors are tested simultaneously, make two independent and parallel micropockets (*see* **Note 11** for different protocols).

6. Cell or tissue implant: The pocket is produced with an enlarged base (4 mm) to allocate cell samples. To reduce corneal tension before cell or tissue implant, a small amount (20–50 µl) of the aqueous humor can be drained from the anterior chamber with an insulin syringe.

7. By using a micropipette, introduce 5 µl containing $2–5 \times 10^5$ cells in medium supplemented with 10 % serum in the corneal micropocket. When the overexpression of growth factors/inhibitors by stable transfection of specific cDNA is studied, one eye is implanted with transfected cells and the other with the wild-type or vector-transduced cell line. Suitable cell lines for these experiments are mammary carcinoma cells (MCF-7) [11], lymphoma Burkitt's cells (DG75) [4], and Chinese hamster ovary cells (CHO) [6]. It might be necessary to evaluate the angiogenic potential of drug-treated cells. In these experiments, cell monolayers are pharmacologically treated before

the implant (18–24 h). One eye is implanted with treated cells and the contralateral with control cells [11, 12].

8. Tissue fragments are inserted in the corneal pocket with the aid of Dumont n. 5 tweezer. The angiogenic activity of tumor samples is compared with healthy tissue [9].

2.2.3 Quantification of Neovascular Growth

1. Subsequent daily observation of the implants is made with a slit lamp stereomicroscope without anesthesia. The clinical evolution of the implants and of the ocular lesions is recorded and the presence of corneal reactions, such as redness, corneal edema, the intensity of the corneal cellular infiltrate, and the total area of neovascularization, are scored. The use of slit lamp stereomicroscope and of awake animals allows the observation of newly formed vessels during time with prolonged monitoring, up to 1 month. When studying tumor-induced angiogenesis, neovascularization accompanies tumor cell growth (*see* Fig. 2).

2. An angiogenic response is scored positive when budding of vessels from the limbal plexus occurs after 3–4 days and capillaries progress to reach the implanted pellet in 7–10 days. Implants that fail to produce a neovascular growth within 10 days are considered negative, while implants showing an inflammatory reaction are discarded.

3. During each observation, the number of positive implants over the total implants performed is scored.

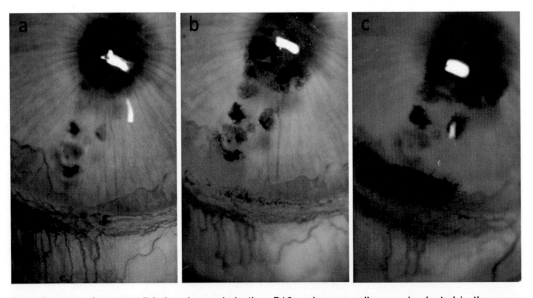

Fig. 2 Example of tumor cell-induced vascularization. B16 melanoma cells were implanted in the cornea stroma and neovascularization and cell growth were monitored during time. Images were taken at day 3 (**a**), 6 (**b**), and 10 (**c**). Note in (**c**) the increased intensity of melanin bearing cells (*brown-black spots*)

4. The potency of angiogenic activity is evaluated on the basis of the number and growth rate of newly formed capillaries, and an angiogenic score is calculated by the formula [vessel density × distance from limbus] [11, 13]. A density value of 1 corresponds to 0 to 25 vessels per cornea, 2 from 25 to 50, 3 from 50 to 75, 4 from 75 to 100, and 5 for more than 100 vessels. The distance from the limbus is graded (in mm) with the aid of an ocular grid.

5. To understand the mechanism of progression and/or regression by drug treatment, the two parameters (density and length) are considered separately, thus documenting the activity of treatment on endothelial cell proliferation (density) respect to elongation and organization (length).

6. The anterior ocular pole images are computer-analyzed at fixed times on animals under anesthesia. An advanced video camera connected to a color video monitor and a computer with video-bluster and special capture software is used to record corneal responses. In order to extract the vascular tree from every image, the following graphic processing is required:

   - adjustment of contrast and brightness, in order to highlight the vascular tree (image conversion in a gray scale format can be helpful in this stage);

   - image extraction of the vascular tree (Fig. 3).

   Commercially available software (i.e. Corel Photo Paint and Corel Draw; Adobe Photoshop and National Institute of Health Image J1.38X) can be used for these purposes [14].

2.2.4 Histological Examination and Immunohistochemical Analysis

Depending on the experimental design, histological or immunohistochemical analysis of corneal sections can be performed at fixed times during angiogenesis progression or at the end of the observations [11].

1. Animals are sacrificed with intravenous injection of 0.5 ml of Tanax or sodium pentothal (bolus 30 mg/kg).

2. The corneas are removed, oriented, and marked (*see* **Note 12**), immediately frozen in isopentane cooled in liquid nitrogen for 10 s, and stored at −80 °C in OCT tissue-teck medium. If required, the cornea can be fixed in paraformaldehyde.

3. Seven-μm-thick cryostat sections are stained with hematoxylin and eosin and adjacent sections can be used for immunohistochemical staining (*see* **Note 13**).

2.2.5 Gene and Protein Expression

At fixed times or at the end of the experiment, corneas can be removed and snap frozen in liquid nitrogen. By using standard extraction reagent and buffer, mRNAs and proteins can be extracted to assess gene transcription and protein expression during vascular and tissue responses or following drug treatment.

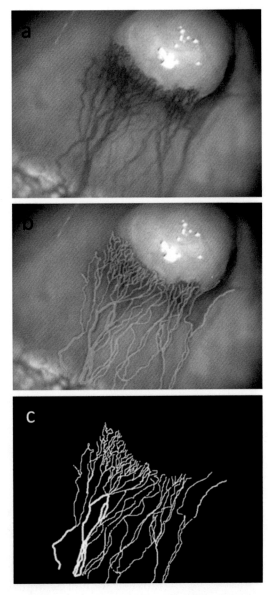

Fig. 3 Picture of neovascular growth induced by VEGF (100 ng/pellet) taken at day 10 (**a**). Panels (**b**) and (**c**) represent the same image processed by computerized image analysis for vessel extraction from the background

2.3 Drug Treatments and Pharmacokinetic Studies

When performing drug treatments for ocular pathologies and to validate stimuli or signalling pathway, different approaches can be followed.

1. Eye drops: isotonic buffers (i.e. PBS without calcium and magnesium) at physiological pH can be used to dissolve drugs to be studied for their ability to modulate corneal angiogenesis. Depending on drug nature and half-life, eye drops treatments can be performed twice-fivefold a day, soon after an angiogenic

stimulus has been implanted in the cornea stroma. Awake animals are immobilized in appropriate contention boxes. By the use of a sterile pipet, 50–100 µl of the drug solution is put in the subconjunctival space by pulling the lower lid. The eye is then kept close for at least 30 s to avoid liquid dispersion and drop out.

2. Ointment and gels: Simple eye ointment contains liquid paraffin (mineral oil) and wool fat (lanolin) in a yellow soft paraffin base (see **Note 14**). These ingredients produce a transparent, lubricating, and moistening film on the surface of the eyeball. Drug mixing is performed under hood and insulin syringes are prepared. 100 µl of ointment are poured in the subconjunctival space once or twice a day. Eye lids are closed and gently frictioned to form a film of the ointment or gel on the eye surface [15].

3. Intravitreal injections (30–50 µl/eye) can be also performed under general anesthesia to study drug stability in the vitreous and diffusion to retina or to the anterior chamber, to obtain data closely related to human ocular pharmacology.

4. When drugs or genes transduced by viral vectors have to be locally tested, microinjection of concentrated solutions is performed by the use of insulin syringes equipped with 30 G needles. After the removal of aqueous humor, a volume of 10 µl is injected within the corneal stroma in the space between the limbus and the pellet implant [16].

5. At fixed times after treatment started, following animal sacrifice, all the eye tissues (cornea, aqueous humor, lens, vitreous humor, retina) can be isolated and frozen in liquid nitrogen, and tissue homogenates assessed for drug distribution and metabolism.

3 Notes

1. Cornea has been found avascular in all strains of rabbits examined so far. In albino rabbits, the newly formed vessels are clearly visible on the background of the iris.

2. Operator skill for pellet manipulation, surgery, and monitoring of angiogenesis is required. All procedures and observations are conducted in a double-masked manner and the code identities are revealed only after the end of the experiment and data elaboration.

3. Body weight: in the range 1.8–2.5 kg for an easy handling and prompt recovery from anesthesia. Sex: except when hormone dependency of cells or tumors is a prerequisite of the experimental setting, males are used. Check with your animal facility

and veterinary doctor whether only specific pathogen-free (SPF) animals are admitted.

4. Sterility of materials and procedures is crucial to avoid nonspecific responses. DMSO should be avoided since incompatible with Elvax-40 polymerization and handling.

5. Ethylene vinyl acetate (Elvax-40) is the copolymer of ethylene and vinyl acetate (40 %). Elvax-40 is a polymer that approaches elastomeric materials in softness and flexibility, used in biomedical engineering applications as a drug slow delivery device. While the polymer is not biodegradable within the body, it is quite inert and causes little or no reaction following implantation.

   Polyvinylalcohol and polyhydroxyethyl-methacrylate (Hydron) can be used instead of Elvax-40 [17]. Hydron is hydrophobic; however, when the polymer is subjected to water, it will swell due to the molecule's hydrophilic pendant group. Depending on the physical and chemical structure of the polymer, it is capable of absorbing from 10 % to 600 % water relative to the dry weight. Because of this property, it was one of the first materials to be successfully used in the manufacture of flexible contact lenses. When comparing the release kinetics of proteins from Elvax-40 and Hydron, the release from Hydron was the most rapid, while it was the most slow from Elvax, which continued to release the incorporated protein till 100 days [2]. Polyvinylalcohol had an intermediate behavior [2].

   In our experience, the polymer of hydroxyethyl-methacrylate gave less satisfactory results than Elvax-40. The release of molecules (i.e. gangliosides) from Elvax-40 pellets implanted in the cornea is in the order of 30–40 % in the first 48 h and then remains constant [18]. Initially, the most superficial molecules are released, then water moves from the tissue inside the pellet matrix, leading the inner molecules on pellet surface [2].

6. Elvax-40 preparation and testing:

   – Weight 1 g of Elvax-40 (purchased as dry beads), extensively wash it in absolute alcohol for 100-fold at 37 °C, and dissolve in 10 ml of methylene chloride to prepare a 10 % casting stock solution.

   – Test the Elvax-40 preparation for its biocompatibility [2]. The casting solution is eligible for routine use if no implant performed with this preparation induces the slightest or histological reaction in the rabbit cornea during 14-day examination period.

7. Variability among growth factors in inducing angiogenesis has been found considering different angiogenic factors, different providers, and batch of preparation. Usually, the dose of VEGF or FGF-2 able to give a positive angiogenic response varies in the range 200–400 ng/pellet.

8. When two factors are co-released from the same pellet, the advice is to check before pellet implant if drug release in vitro is modified respect to the single molecule, as described in [19].

9. Immobilization (in appropriate contention box) during anesthetic procedure and observation is important to avoid self-induced injury.

10. Make the cut in the cornea in correspondence of the pupil and orient the micropocket toward the lower eyelid for an easy daily observation.

11. The modulation of the angiogenic responses by different stimuli can be assessed in the rabbit cornea assay (a) by implanting single pellets releasing both the angiogenic stimulus and the inhibitor [20–22], (b) by implanting in the same cornea two pellets placed in parallel micropockets and releasing different molecules [18, 23], and (c) through the addition or removal of single pellets in multiple implants [18].

12. Before embedding in OCT-tissue teck medium, pellets should be removed and corneas sampled and marked (i.e. with a cotton thread) for subsequent orientation at the cryostat once embedded in OCT medium.

13. Validated antibodies in our experience are anti-CD31 Ab (Dako, 200 µg/ml) (marker of neovascularization), anti-RAM11 Ab (Dako, 1.2 µg/ml) (marker of inflammation), anti $\alpha5\beta1$ integrin Ab (Chemicon, 1:50) (adhesion molecule expressed in epithelial and endothelial cells).

14. When using eye ointment, take into account that paraffin-based preparations have to be used for short time (1 week) to avoid toxicity by excipients.

Acknowledgments

The work was supported by the Italian Ministry of University (MIUR), the Italian Association for Cancer Research (AIRC), and Istituto Toscano Tumori (ITT).

References

1. Gimbrone M Jr, Cotran R, Leapman SB, Folkman J (1974) Tumor growth and neovascularization: an experimental model using the rabbit cornea. J Natl Cancer Inst 52:413–427

2. Langer R, Folkman J (1976) Polymers for the sustained release of proteins and other macromolecules. Nature 363:797–800

3. Brem SS, Gullino PM, Medina D (1977) Angiogenesis: a marker for neoplastic transfor-mation of mammary papillary hyperplasia. Science 195(4281):880–882

4. Cervenak L, Morbidelli L, Donati D, Donnini S, Kambayashi T, Wilson J, Axelson H, Castanos-Velez E, Ljunggren HG, De Waal Malefyt R, Granger HJ, Ziche M, Bejarano MT (2000) Abolished angiogenicity and tumorigenicity of Burkitt lymphoma by Interleukin-10. Blood 96:2568–2573

5. Woolard J, Wang WY, Bevan HS, Qiu Y, Morbidelli L, Pritchard-Jones RO, Cui TG, Suggono M, Waine E, Perrin R, Foster R, Digby-Bell J, Shields JD, Whittles CE, Mushens RE, Gillatt DA, Ziche M, Harper SJ, Bates DO (2004) VEGF165b, an inhibitory vascular endothelial growth factor splice variant: mechanism of action, in vivo effect on angiogenesis and endogenous protein expression. Cancer Res 64(21):7822–7835

6. Marconcini L, Marchio S, Morbidelli L, Cartocci E, Albini A, Ziche M, Bussolino F, Oliviero S (1999) c-fos-induced growth factor/vascular endothelial growth factor D induces angiogenesis in vivo and in vitro. Proc Natl Acad Sci U S A 96(17):9671–9676

7. Brem H, Folkman J (1975) Inhibition of tumor angiogenesis mediated by cartilage. J Exp Med 141(2):427–439

8. Bard RH, Mydlo JH, Freed SZ (1986) Detection of tumor angiogenesis factor in adenocarcinoma of kidney. Urology 27(5):447–450

9. Gallo O, Masini E, Morbidelli L, Franchi A, Fini-Storchi I, Vergari WA, Ziche M (1998) Role of nitric oxide in angiogenesis and tumor progression in head and neck cancer. J Natl Cancer Inst 90:587–596

10. da Silva BB, da Silva Júnior RG, Borges US, da Silveira Filho MA, Pimentel IC, Gebrim LH, Simões Mde J, Baracat EC (2005) Quantification of angiogenesis induced in rabbit cornea by breast carcinoma of women treated with tamoxifen. J Surg Oncol 90(2):77–80

11. Ziche M, Morbidelli L, Choudhuri R, Zhang H-T, Donnini S, Granger HJ, Bicknell R (1997) Nitric oxide-synthase lies downstream of vascular endothelial growth factor but not basic fibroblast growth factor induced angiogenesis. J Clin Invest 99:2625–2634

12. Lasagna N, Fantappiè O, Solazzo M, Morbidelli L, Marchetti S, Cipriani G, Ziche M, Mozzanti R (2006) Hepatocyte growth factor and inducible nitric oxide synthase are involved in multidrug resistance-induced angiogenesis in hepatocellular carcinoma cell lines. Cancer Res 66(5):2673–2682

13. Ziche M, Morbidelli L, Masini E, Amerini S, Granger HJ, Maggi CA, Geppetti P, Ledda F (1994) Nitric oxide mediates angiogenesis in vivo and endothelial cell growth and migration in vitro promoted by substance P. J Clin Invest 94:2036–2044

14. Monti M, Donnini S, Morbidelli L, Giachetti A, Mochly-Rosen D, Mignatti P, Ziche M (2013) PKCε activation promotes FGF-2 exocytosis and induces endothelial cell proliferation and sprouting. J Mol Cell Cardiol 63:107–117

15. Presta M, Rusnati M, Belleri M, Morbidelli L, Ziche M, Ribatti D (1999) Purine analog 6-methylmercaptopurine ribose inhibits early and late phases of the angiogenesis process. Cancer Res 59(10):2417–2424

16. Ziche M, Morbidelli L (2009) Molecular regulation of tumour angiogenesis by nitric oxide. Eur Cytokine Netw 20(4):164–170

17. Rogers MS, Birsner AE, D'Amato RJ (2007) The mouse cornea micropocket angiogenesis assay. Nat Protoc 2(10):2545–2550

18. Ziche M, Alessandri G, Gullino PM (1989) Gangliosides promote the angiogenic response. Lab Invest 61:629–634

19. Taraboletti G, Morbidelli L, Donnini S, Parenti A, Granger HJ, Giavazzi R, Ziche M (2000) The heparin binding 25 kDa fragment of thrombospondin-1 promotes angiogenesis and modulates gelatinases and TIMP-2 in endothelial cells. FASEB J 14:1674–1676

20. Morbidelli L, Donnini S, Chillemi F, Giachetti A, Ziche M (2003) Angiosuppressive and angiostimulatory effects exerted by synthetic partial sequences of endostatin. Clin Cancer Res 9(14):5358–5369

21. Bagli E, Stefaniotou M, Morbidelli L, Ziche M, Psillas K, Murphy C, Fotsis T (2004) Luteolin inhibits vascular endothelial growth factor-induced angiogenesis; inhibition of endothelial cell survival and proliferation by targeting phosphatidylinositol 3′-kinase activity. Cancer Res 64(21):7936–7946

22. Donnini S, Finetti F, Lusini L, Morbidelli L, Cheynier V, Barron D, Williamson G, Waltenberger J, Ziche M (2006) Divergent effects of quercetin conjugates on angiogenesis. Br J Nutr 95(5):1016–1023

23. Cantara S, Donnini S, Morbidelli L, Giachetti A, Schulz R, Memo M, Ziche M (2004) Physiological levels of amyloid peptides stimulate the angiogenic response through FGF-2. FASEB J 18(15):1943–1945

Chapter 21

The Corneal Micropocket Assay: A Model of Angiogenesis and Lymphangiogenesis

Shintaro Nakao and Ali Hafezi-Moghadam

Abstract

The cornea is a transparent tissue that lacks blood and lymphatic vessels. In addition, the cornea is readily accessible, which makes it convenient for direct visualization of angiogenesis and lymphangiogenesis. The corneal micropocket assay is a commonly used quantitative technique, in which a growth factor containing pellet is micro-surgically implanted into the cornea of a rodent. Subsequently, the growth of the preexisting limbal vessels toward the growth factor is visualized by live microscopy or immunohistochemistry. Recently, there has been significant interest in the process of lymphangiogenesis and the factors that regulate it. To facilitate these studies, we introduce a novel technique for visualization of the immune response during growth factor induced angiogenesis and lymphangiogenesis in the cornea.

Key words Lymphangiogenesis, Imaging, Leukocyte recruitment, VEGF, LYVE-1, Cornea, Conjunctiva, Limbal vessel, Growth factor

1 Introduction

The cornea, a transparent and avascular tissue, is made of extracellular matrix, keratocytes, and leukocytes. Indeed, avascularity of the cornea is a prerequisite for its transparency. In contrast to the normal cornea, the conjunctiva contains blood and lymphatic vessels. The growth of these vessels can accompany various corneal disorders and cause significant loss of visual acuity. In corneal diseases, a number of cytokines and growth factors are released, which lead to the accumulation of neutrophils, macrophages, and lymphocytes. The presence of inflammatory cells in the cornea goes often hand in hand with growth of new vessels. The underlying mechanisms are beginning to be understood.

Recently, we investigated the lymphangiogenic potential of inbred mouse strains after implantation of various growth factors in the cornea [1]. Our results revealed a significant heterogeneity in preexisting lymphatics and the growth of new lymphatic vessels, depending on the genetic background [1]. With the use of

Stewart G. Martin and Peter W. Hewett (eds.), *Angiogenesis Protocols*, Methods in Molecular Biology, vol. 1430, DOI 10.1007/978-1-4939-3628-1_21, © Springer Science+Business Media New York 2016

the corneal micropocket assay we were able to establish a new link between angiogenesis and lymphangiogenesis that finely regulates lymphatic growth in spatial and temporal relation to angiogenic vessels [2]. Our studies further revealed a unique organization of the inflammatory cells during angiogenesis and lymphangiogenesis [3, 4]. Although various cytokines, e.g., vascular endothelial growth factor (VEGF)-A and interleukin (IL)-1β, are known to induce angiogenesis and lymphangiogenesis in vivo, their role in leukocyte infiltration during angiogenic response has not been studied. In our recent studies we showed how to investigate the cytokine-dependent leukocyte recruitment during angiogenesis and lymphangiogenesis [3, 4]. Our results in the corneal micropocket assay revealed that interleukin-1β (IL-1β) induces infiltration of both neutrophils and macrophages during the angiogenesis, whereas macrophages but not neutrophils infiltrated in VEGF-A-induced angiogenesis. In the same assay we were able to show that infiltration of a subset of leukocytes that was key to lymphatic growth was regulated by the vascular adhesion protein-1 (VAP-1) [5].

In summary, we have been able to gain these insights by using the corneal micropocket assay, a powerful and reliable technique, the details of which are described in the following protocol.

2 Materials

2.1 Growth Factor/Cytokine Pellets

1. Poly(2-hydroxyethyl methacrylate) (PHM) solution: Add the PHM powder (see **Note 1**, P3932, Sigma) to ethanol to give a 12 % (w/v) solution. Vortex until the powder is completely dissolved.

2. Dulbecco's phosphate buffered saline (PBS): Sterilize by autoclaving or filtration through a sterile 0.22 μm filter.

3. Recombinant growth factors and cytokines: Growth factors, e.g., murine $VEGF_{164}$ and cytokines, such as monocyte chemoattractant protein-1 (MCP-1), or IL-1β can be implanted, depending on the experimental design.

4. PBS containing 0.1 % bovine serum albumin (BSA): Dissolve 0.1 % (w/v) BSA (high grade, IgG-free, low endotoxin suitable for in vivo/in vitro use, e.g., A2058, Sigma) in PBS. Sterilize by filtration through a 0.22 μm sterile filter.

5. Parafilm M.

6. White paper.

2.2 Pellet Implantation

1. Forceps (e.g., 11252-23; Fine Science Tools).

2. Blade holder (e.g., 10052-11; Fine Science Tools).

3. Blade (e.g., 10050-00; Fine Science Tools).

4. Pin holder (e.g., 26018-17; Fine Science Tools).

5. Insert Pin (e.g., 26007-02; Fine Science Tools).

2.3 Immuno-histochemistry

1. 4 % (w/v) paraformaldehyde (PFA).

2. Methanol.

3. PBS containing 10 % (v/v) goat serum and 1 % (v/v) Triton X-100.

4. Antibodies for immunostaining of the corneal flat mounts.

   (a) α-Mouse CD31 antibody (e.g., 550274; BD Pharmingen).

   (b) α-Mouse LYVE-1 Ab (e.g., 103-PA50AG; RELIATech GmbH).

   (c) Alexa Fluor 488 goat α-rat IgG (e.g., 52955A; Invitrogen).

   (d) Alexa Fluor 647 goat α-rabbit IgG (e.g., A21244; Invitrogen).

5. Mounting medium (e.g., TA-030-FM, Mountant Permafluor; Lab Vision Corporation).

2.4 Ex Vivo Leukocyte Assay

1. Acridine orange (AO) solution: Make up a stock solution according to manufacturer's instructions and dilute to 0.2 mg/mL in PBS.

2. Rhodamine-labeled Concanavalin A (Con A) lectin: Dilute rhodamine-labeled ConA to 10 μg/mL in PBS (pH 7.4).

3 Methods

3.1 Growth Factor/Cytokine Pellets

Perform all procedures at room temperature (*see* **Note 2**). To create a mold for the pellet, cut squares of 1 mm × 1 mm from Parafilm M. Use 1 mm × 1 mm graph paper and place the Parafilm on the paper. First, cut out the 1 mm × 1 mm frame as shown in Fig. 1. In this space the cytokine pellet will be prepared and sectioned into smaller pieces (*dotted lines*).

1. Mix the desired cytokine or growth factor with the 12 % PHM solution at a 1:1 ratio. For instance, to implant 200 ng of VEGF, prepare 1.8 μg of VEGF in 3 μL of sterile PBS containing 0.1 % BSA added to 3 μL of 12 % PHM solution. Add 1 μL of the prepared cytokine–PHM mixture into each square mold that was cut out of Parafilm and allow the mixture to dry at room temperature. Repeat this step several times with the remainder of the cytokine–hydron mixture.

2. Cut the pellet into nine equal pieces (*see* Fig. 1). However, since it is technically challenging to cut the square into exactly nine equal pieces, the four corner pieces should be discarded and the remaining other five pieces (*gray*) used as pellets for implantation.

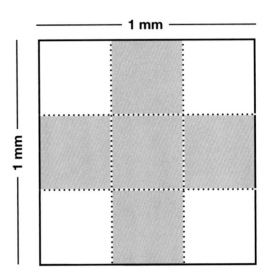

Fig. 1 Scheme for preparing the growth factor/cytokine pellets for implantation into the mouse cornea. The 1 mm should be divided into 9 equally. This is hard to achieve and we use only the pellets represented by the *gray squares*

3.2 Pellet Implantation

1. Anesthetize a mouse by intraperitoneal injection ketamine (100 mg/kg) and xylazine (10 mg/kg) (*see* **Note 3**).

2. Cut the top of cornea until the stroma with a surgical blade. Take great care not to penetrate into the anterior chamber of the eye (*see* **Note 4**).

3. Create a pocket in the corneal stroma. This can be accomplished using the pointed tips of a pair of fine forceps. Advance the closed forceps through the cut into the stroma parallel to the surface of the cornea and gently allow the tips to open, then close again for retraction.

4. Implant one of the prepared pellets into the mouse cornea. To prevent the pellet from spontaneously coming back out of the pocket, push it to the blind end of the pocket (*see* **Note 2**).

5. The angiogenic response becomes visible within the first 6 days after pellet implantation. Recently, we characterized the time course of the angiogenic growth (*see* Ref. 6).

3.3 Immuno-histochemistry

1. Enucleate eyes and fix with 4 % PFA for 30 min at 4 °C.

2. For whole-mount preparation expose the corneas by removing the rest of the eye.

3. Wash the tissues with PBS three times for 5 min and place them in methanol for 20 min.

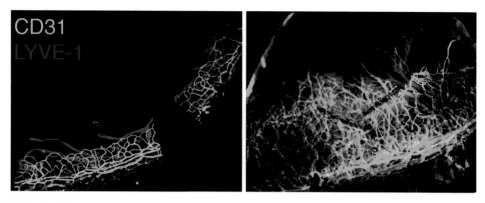

Fig. 2 Blood and lymphatic vessels in the mouse cornea. Corneal flatmount of a C57B6/J mouse illustrates CD31 positive blood and LYVE-1 positive lymphatic vessels. *Left image*, preexisting blood and lymphatic vessels. *Right image*, newly grown blood and lymphatic vessels in response to cytokine implantation

4. Incubate overnight at 4 °C with α-mouse CD31 mAb (5 µg/mL) and α-mouse LYVE-1 Ab (4 µg/mL) diluted in PBS containing 10 % goat serum and 1 % Triton X-100.

5. Wash tissues four times for 20 min in PBS followed by incubation with Alexa Fluor 488 goat α-rat IgG (~20 µg/mL) and Alexa Fluor 647 goat α-rabbit IgG (20 µg/mL) overnight at 4 °C.

6. Place radial cuts in the peripheral cornea to allow flat mounting on a glass slide using a mounting medium. Examine the flat-mounted tissues by fluorescence microscopy and obtain photomicrographs for quantification purposes. A representative micrograph is shown below (Fig. 2).

3.4 Ex Vivo Leukocyte Transmigration Assay

1. Anesthetize the pellet implanted mice (*see* Subheading 3.1, step 1).

2. Inject 500 µL of AO solution (0.2 mg/mL) intravenously (*see* Ref. 7).

3. Two hours after AO injection, perfuse the animals with 10 µg/mL rhodamine-labeled ConA. Open the chest cavity and place a 24-gauge perfusion needle into the aorta. Allow drainage by opening the right atrium.

4. Perfuse with 10 mL PBS to wash out blood cells in the vessels.

5. After PBS perfusion, perfuse with 5 mL rhodamine-labeled ConA and then again with 1 mL PBS to remove residual unbound Con A.

6. Immediately after perfusion, remove the corneas and prepare flat mounts using a mounting medium.

7. Quantify the number of transmigrated leukocytes into the corneal stroma.

4 Notes

1. Various studies have used Hydron (IFN Sciences) for this assay, however, since Hydron is no longer commercially available, we recommend the use of poly(2-hydroxyethyl methacrylate) (PHM) for this assay.

2. It is important that the pellets are freshly prepared for each assay in order to achieve optimal results.

3. Surgical procedures should be performed by trained and licensed individuals according to the relevant local ethical guidelines for animal care and experimentation.

4. The corneal micropocket assay is a reliable and reproducible in vivo technique. However, it requires a high level of precision and microsurgical skills. Although it will eventually generate robust results, it takes considerable effort to master this in vivo assay. It is critical that all procedural sources of variability, such as those related to the microsurgical skills are minimized. In our experience, to become proficient in this technique, tens of hours of practice are required.

   Once proficient, a number of creative variations in the design will become possible. For instance, recently we introduced a variation of this technique, which allows quantitation of leukocyte extravasation from angiogenic vessels and their reentry into the lymphatic vessels [6, 7].

References

1. Nakao S, Maruyama K, Zandi S, Melhorn MI, Taher M, Noda K, Nusayr E, Doetschman T, Hafezi-Moghadam A (2010) Lymphangiogenesis and angiogenesis: concurrence and/or dependence? Studies in inbred mouse strains. FASEB J 24:504–513

2. Nakao S, Zandi S, Hata Y, Kawahara S, Arita R, Schering A, Sun D, Melhorn MI, Ito Y, Lara-Castillo N, Ishibashi T, Hafezi-Moghadam A (2011) Blood vessel endothelial VEGFR-2 delays lymphangiogenesis: an endogenous trapping mechanism links lymph- and angiogenesis. Blood 117:1081–1090

3. Nakao S, Kuwano T, Tsutsumi-Miyahara C, Ueda S, Kimura YN, Hamano S, Sonoda KH, Saijo Y, Nukiwa T, Strieter RM, Ishibashi T, Kuwano M, Ono M (2005) Infiltration of COX-2-expressing macrophages is a prerequisite for IL-1 beta-induced neovascularization and tumor growth. J Clin Invest 115:2979–2991

4. Nakao S, Hata Y, Miura M, Noda K, Kimura YN, Kawahara S, Kita T, Hisatomi T, Nakazawa T, Jin Y, Dana MR, Kuwano M, Ono M, Ishibashi T, Hafezi-Moghadam A (2007) Dexamethasone inhibits interleukin-1beta-induced corneal neovascularization: role of nuclear factor-kappaB-activated stromal cells in inflammatory angiogenesis. Am J Pathol 171:1058–1065

5. Nakao S, Noda K, Zandi S, Sun D, Taher M, Schering A, Xie F, Mashima Y, Hafezi-Moghadam A (2011) VAP-1-mediated M2 macrophage infiltration underlies IL-1beta- but not VEGF-A-induced lymph- and angiogenesis. Am J Pathol 178:1913–1921

6. Nakao S, Zandi S, Lara-Castillo N, Taher M, Ishibashi T, Hafezi-Moghadam A (2012) Larger therapeutic window for steroid versus VEGF-A inhibitor in inflammatory angiogenesis: surprisingly similar impact on leukocyte infiltration. Invest Ophthalmol Vis Sci 53:3296–3302

7. Nakao S, Zandi S, Faez S, Kohno R, Hafezi-Moghadam A (2012) Discontinuous LYVE-1 expression in corneal limbal lymphatics: dual function as microvalves and immunological hot spots. FASEB J 26:808–817

Chapter 22

Models of Oxygen Induced Retinopathy in Rodents

Melissa V. Gammons and David O. Bates

Abstract

Much of the knowledge we have gained into the development of pathological ocular angiogenesis has come from the development of in vivo models that enable functional assessment of key components of signaling pathways in disease progression. Indeed, rodent models have facilitated identification of several therapeutics that target pathological angiogenesis. Two of the most widely used rodent models of oxygen induced retinopathy (OIR), Smith's mouse model and Penn's rat model reproducibly induce neovascularization reminiscent of the disease retinopathy of prematurity (ROP). In this chapter we discuss development of ROP in humans and compare features with that of the rat and mouse models, focusing both on the benefits and caveats of using such models. Furthermore, we discuss in detail the methodology of both procedures and discuss the importance of various features of the model.

Key words Retinopathy of prematurity, Oxygen-induced retinopathy, Neovascularization, Angiogenic cascade

1 Introduction

Retinopathy of Prematurity (ROP) is a potentially blinding disease affecting premature infants. Like other retinopathies including age-related macular degeneration (AMD) and proliferative diabetic retinopathy (PDR), ROP is characterized by pathological ocular angiogenesis and retinal neovascularization (NV) [1, 2]. Premature birth alters the development of the retinal vasculature resulting in the delayed maturation of vessels, retinal ischemia at the peripheral retina causes a decrease in available oxygen that is detected in the retina by Müller cells which, in turn secrete large amounts of growth factors, namely vascular endothelial growth factor (VEGF). VEGF is a potent endothelial cell-selective mitogen detected by receptors on retinal microvascular endothelial cells which, together with other signals leads to the induction of an angiogenic cascade [3]. The angiogenic signaling cascade makes for an attractive therapeutic target in ocular diseases that exhibit pathological neovascularization.

Stewart G. Martin and Peter W. Hewett (eds.), *Angiogenesis Protocols*, Methods in Molecular Biology, vol. 1430,
DOI 10.1007/978-1-4939-3628-1_22, © Springer Science+Business Media New York 2016

Knowledge gained from in vivo models of retinal diseases have yielded much of what we know about physiological and pathological blood vessel growth in the retina [4], knowledge that has also impacted our understanding of non-ocular angiogenic conditions. In this chapter we discuss two oxygen-induced retinopathy (OIR) models of ROP that have been developed over the years and have both been used to identify and therapeutically target molecular mechanisms that control the angiogenic cascade [5, 6]. We outline in detail both the advantages and disadvantages of each model, and the methodological procedures necessary to complete each model.

1.1 Human ROP Characteristics

ROP alters the normal development of retinal blood vessels in premature infants. Premature infants are susceptible to ROP due to the incomplete development of the retinal vasculature on birth. During normal retinal development vascularization begins at approximately 16-weeks gestational age initially provided by the hyaloid vessels traversing from the optic nerve to the anterior segment. Hyaloid vessels must regress for clear vision which usually occurs at 34-weeks gestational age, and by this time the intraretinal vasculature is well advanced; however, vessels do not reach the ora serrata (serrated junction between the retina and the ciliary body) until week-40. At birth the hyperoxic postnatal environment (partial pressure of dissolved arterial oxygen (PaO_2) is 55–80 mmHg compared with the uterine environment PaO_2 of 30 mmHg) is believed to stimulate reduction of growth factor production [7]. In premature infants this occurs prior to the completion of the vascular development and retards developmental angiogenesis.

ROP is a biphasic disease; initially characterized by the hyperoxic postnatal environment inducing vasoattenuation or the cessation of retinal vascular development [8, 9], as the vasculature is incomplete the retina becomes hypoxic inducing vasoproliferation and preretinal NV [10]. This preretinal NV predisposes the infant to plasma leakage, intravitreal hemorrhages, retinal detachment, and in some cases subsequent visual loss [11].

Improvements to the survival rates of premature babies have resulted in an increase in the prevalence of ROP in recent years. The Supplemental Therapeutic Oxygen for Premature Retinopathy of Prematurity (STOP-ROP) multicenter trail in 2004 showed 65 % of premature babies routinely placed in oxygen therapy on birth went on to develop ROP (Dale Phelps, personal communication, information derived from the STOP trial [12]). However, the development and severity of ROP is a multifactorial process where there is not one sole determinant, factors including temporal development and gene expression combine with clinical care to impact ROP pathogenesis.

1.2 Experimental Models of ROP

In 1954, Patz first demonstrated preretinal NV in rats that had been exposed to a constant level of extreme hyperoxia [13]. Similar results were seen in the same year in mouse models of retinal NV [14, 15]. However, such results were difficult to replicate in the mouse for many years. Ashton failed to produce a proliferative retinopathy using oxygen concentration of 80–90 % [16].

The inconsistencies and lack of quantification associated with this model led Smith and colleagues [6] to redefine exposure parameters to ensure consistent reproducible neovascularization in the mouse retina [6]. Smith's oxygen induced retinopathy (OIR) model was used on C57BL6 mice after hyaloid regression; in brief 1-week-old (P7) mice are exposed to 75 % oxygen for 5 days (P7–P12), during the first 48 h immature capillaries in the central retina rapidly regress leading to central vaso-obliteration, in the latter part of the hyperoxic exposure revascularization of this central retina begins. In addition, during this exposure vessels constrict to regulate PO_2. At P12 mice are returned to room air, leading to hypoxia in the vaso-obliterated central retina and an induction of Hif-1α-dependent angiogenic signaling. Hyperoxia exposure produced a far more profound and reproducible retinal NV in P7–P12 pups compared to previous experiments on P0 pups [6]. Strict regulation of this protocol successfully models the defining disease characteristics observed during the development of human ROP in an efficient experimental setting. This model his since been adopted as the benchmark model of mouse OIR, contributing to over 30,000 publications since its description in 1994 (Google Scholar search on keywords "oxygen induced retinopathy mice," June 19, 2014).

The development of the rat model of ROP experienced similar issues, Ashton and Blach failed to reproduce Patz's initial data [17], concluding previous investigators claims were "inadequately substantiated on the evidence provided," due to the inconsistent nature of the vasoproliferative response. Again like the mouse model the rat model has experienced numerous changes to make it more reliable, including alterations to staining employing histochemical methods as described by Flower et al., using ADPase staining [18]. The constant level of extreme hyperoxia in rat models, like in the mouse, led Ricci and colleagues to observe "*marked peripheral retinal neovascularization*" after 5 days of continuous exposure to 80 % oxygen in newborn rat pups [19]. Both rat and mouse models demonstrate retinal NV and boast different advantages and disadvantages making both therapeutically applicable. However, both also lack pathophysiological features observed clinically in newborn babies, a factor that then sparked Penn's enthusiasm into making the rat model more clinically robust.

Premature infants who develop the ROP experience rapid fluctuations in PaO_2 resulting in alternating periods of hyperoxemia and hypoxemia. This exposure paradigm is not explicitly defined in previous models where rats are maintained in severe hyperoxia (80 %) for 5 days, although they may experience relative hypoxia during the experiment if oxygen levels are not meticulously monitored. During periods where gas cylinders were changed or bedding and water replaced the oxygen levels within the chamber will rapidly fluctuate, and this may explain why various investigators saw inconsistencies in neovascular growth.

This led Penn and colleagues to pose the theory that it was in fact the transient fluctuations in oxygen levels during the experiment that stimulated neovascular growth and so he set up an experiment where newborn rats where exposed to a cycle of 80 % oxygen for 12 h followed by 40 % oxygen for 12 h for a number of days followed by a short period of time in normoxia. Indeed 66 % of rats developed preretinal NV during this insult compared to none exposed to a continuous 80 % followed by room air [5]. Penn subsequently went on to show that the range of PaO_2 variation determines the severity of OIR in newborn rats, more specifically showing fluctuations between 80 % and 40 % in healthy animals with normal lung functioning do not reflect the PaO_2 experienced by neonates in the neonatal intensive-care unit (NICU). Consequently he set the oxygen levels to fluctuate between 50 % and 10 % on a 24 h basis; this not only better reflected the arterial blood gases experienced by sick premature infants in the NICU but also resulted in a greater retardation of retinal blood vessel development and increase NV severity comparatively. After 4 days postexposure in room air, the incidence of preretinal neovascularization was 97 % in the 50/10 % rats and 72 % in the 80/40 % group [4, 5, 20].

Exposure to the 50/10 regime retards both the superficial and deep retinal vessels, thus causing the avascular periphery. Removal to normal air causes neovascular growth at the vascular boundary, the joining of neovascular tufts results in a ridge of preretinal vessel growth that is highly reminiscent of the mesenchymal ridge seen in human ROP [21].

1.3 Benefits and Caveats of the Rat and Mouse Models

Both the mouse OIR developed by Smith and the rat 50/10 OIR model developed by Penn have proved useful scientific models which have aided in furthering knowledge of pathological oxygen-induced ocular angiogenesis, and in testing novel treatments for therapeutic application. However, like most scientific models, each boasts both advantages and disadvantages, and requires careful control of a number of variables.

Unlike the development of the human disease the pathology induced in both models are induced by oxygen fluctuations postpartum. Eyes open once initial development of retinal vasculature is complete, in humans this coincides with birth whereas in rats and mice this starts to occur at around P12 in normoxic conditions. The high ex-uterine oxygen environment premature babies experience prior to complete retinal development can thus be easily modeled in pups after birth by manipulation of oxygen levels, making rodents   excellent   for   studying   pathological   as   well   as developmental angiogenesis. Both models, when correctly monitored, provide a robust induction of pathological neovascularization in a time-efficient manner.

In both models the severity induced can vary in different wild-type strains [22, 23] and because of vendor-related sub-strain differences [24, 25]. Smith et al. observed higher NV induction using C57Bl/6 wild-type mice from Jackson Laboratories (Bar Harbor, ME) compared to Taconic Farms (Germantown, NY) [25], and Penn et al. noted greater induction of NV in Sprague-Dawley rats from Charles River (Charles River Laboratories, Wilmington, MA) compared with several other vendors, observations that were confirmed in 2002 by a study comparing Charles River to Harlan [24]. For both the rat and mouse model the susceptibility to, and severity of, NV is very much dependent on genetic background and environment.

The mouse model is both highly reproducible and less complex than the rat model in terms of protocol. A strong advantage in using the mouse model is the ability to study gene knockouts relatively easily. Knockout mice allow investigation into the importance of a gene throughout various stages of pathological development (providing the knockout is not embryonic lethal), furthermore rescue experiments can be achieved in inducible knockouts where genes can be turned on and off at different times during development.

With current advances in technology the rat genome is becoming increasingly easier to manipulate and transgenic rats are available; however, at present for gene-knockout studies the mouse model is easily the preferred choice.

In the mouse model, pups do not enter the protocol until P7, so the hyaloid vessels have already significantly regressed and do not need to be removed from the retina following enucleation. In the rat model, however, pups are exposed to alternating oxygen levels immediately after birth, which substantially delays the regression of the hyaloid vessels thus requiring their careful removal during dissection. Despite the development of the improved 50/10 OIR model, getting a reproducible system requires detailed precise following of the methodology. For example as the procedure begins at birth it is important litters are monitored carefully, litter size is important for pathological induction thus it is often necessary to coordinate dams to give birth at a similar time so that litters can be pooled if necessary (observations from experimental data).

An adequate model should replicate key characteristics observed in the human pathology. ROP is a biphasic disease consisting of vaso-obliteration followed by neovascularization; two phases reflected in both the rat [5] and mouse [6] models of OIR. The rat model is widely accepted as the most clinically relevant model of ROP, OIR rat pups present of a ridge of preretinal vessel growth at P20 that is highly reminiscent of the mesenchymal ridge seen in human ROP [21].

However, a clear disadvantage of both models from a clinical perspective is that the pathology observed recovers after time. In the rat model NV regresses after P25 [26], and in the mouse model regression begins at P17 with little or no vaso-obliteration or NV present by P25 [27]. If one requires a model that tests drug dependence and/or the effect of drug withdrawal over a prolonged period of time, then this model is not going to be appropriate.

1.4 Assessment of Induced Pathology

Assessment of the pathology must be carried out blinded to treatment for both rat and mouse models. The retinae need to be coded so that the measurer is unaware of the treatment of the retinae.

1.4.1 Rat Model

One of the factors that make this model more robust compared to other OIR models is the ability to quantify retinal NV with clinical correlation. One method, although only semiquantitative and poses the possibility of becoming subjective, gives a quick read out of NV severity that correlates well with more quantitative and reliable methods. In this method a clock face is superimposed onto a flat mounted retina with each retinal quadrant containing 3 clock hours, investigators sum the number of clock hours containing the pathology. The standard method for quantifying NV in rodent OIR is counting cell nuclei above the internal limiting membrane in histologic sections; however, the much quicker clock-hour analysis demonstrated high correlation ($r_s = 0.95$, $P = 0.0001$) with nuclei counts [28]. Initial PRNV is very difficult to quantify by looking at images, as you are looking for growth through the retina into the vitreous it is essential that you are able to focus the microscope up and down to properly visualize the abnormal growth. Although severe NV can be observed at low magnification you really need to analyze each retina at least at 40× magnification to be sure. More recent publications have more quantitatively outlined NV tufts and expressed as a percentage of total retinal area [29].

1.4.2 Mouse Model

In the mouse model two main features are assessed; during the first phase the area of vaso-obliteration can be determined (P8–P12) and during the second phase the area of NV can be assessed (P14 onwards). These features can both be assessed in retinae at P17 during maximal proliferation [6]. Vaso-obliteration and NV can be measured in whole retinal flat mounts stained for endothelial cell marker, isolectin-B4, where the vaso-obliterated and NV areas can be separately outlined and expressed as a percentage of the total retinal area using image-processing software (for example ImageJ or Photoshop) [25].

This chapter describes in detail the methodological procedures necessary to complete both the mouse and the rat model, both which have significantly advanced knowledge surrounding the factors involved in the progression of retinopathy.

2 Materials

For oxygen exposure protocol

- Oxycycler (Biosperix) or ProOx110 (Biospherix)—if using a ProOx110 a Perspex chamber to house the cages will need to be made; we recommend this be sized to comfortably house two cages.

- Carbon dioxide sensor—calibrated (this may be included with your oxygen controller).

- Rat/mouse housing consumables—bedding, water, feed, replacement cages, etc.

- Animals: Rat strain—Sprague-Dawley (Recommended vendor Charles River), Mouse strain—C57/B6 (Recommended vendor: Jackson Laboratories).

- Nitrogen and oxygen cylinders (BOC).

For intravitreal injection (if necessary)

- 10 μl Hamilton syringe (WPI).

- 35 gauge needle (for murine eyes).

- 33 gauge needle (for rat eyes).

- Injectable drug made up in vehicle.

- Vehicle control (whatever the treatment is made up in minus the active ingredient).

- Isoflourane rig connected to mask that will comfortably fit the pup.

- Heat mat.

For dissection and staining

- Phosphate buffered saline (PBS).

- Angled Vannas scissors (WPI).

- 15° ophthalmic knife (for initial incision into eye, if not available one can use a needle).

- Forceps x2—fine pointed, select to suit individual preferences.

- Petri dishes 35 mm.

- 96-well round bottom well plate (or similar—for storage of retinae during staining protocol).

- 4 % paraformaldehyde in PBS.

- PBS–Triton X-100: 1× PBS pH 7.4, 1 % bovine serum albumin, 1 % Triton X-100.

- PBlec: 1× PBS pH 6.8, 10 mM calcium chloride, 10 mM magnesium chloride, 0.5 % Triton-X100.

- Isolectin IB4.
- Streptavidin conjugated Alexa Fluor 488 or 546/555/594.
- Microscope slides
- Coverslips
- Vectashield mounting medium.

3   Methodology

3.1   Oxygen Exposure Protocol for Rat 50/10 OIR

1. Order several Sprague-Dawley (*see* **Note 1**) prepartum, multiparous female rats, and receive them at approximately gestational Day 13 or 14 upon arrival. Stress of shipping can severely affected pregnant dams so changes in environment any closer to birth should be avoided. Alternatively rats can be bred in-house following standard animal husbandry procedures.

2. Check cages three times daily for newborn pups, dams usually give birth in the early hours of the morning; however, this is not always the case.

3. After parturition (ideally within 4 h), place mother and pups into an oxygen chamber (*see* **Note 2**), and begin specific oxygen-exposure profile (50–10 % O_2 alternate every 24 h). The mother and pups will remain in the oxygen chamber for maximum 14 days. If multiple mothers whelp within 24 h of each other, all pups will be pooled and each mother will be given 17 randomized pups. The minimum number of pups per litter at the beginning of each study is targeted at 14 pups per mother. This will help to ensure optimal litter sizes; litter size is a dependent variable on preretinal NV in this model [30]. Normal lighting cycles will be maintained, as well as standard room temperature (22 °C). On Day 7/0, animals are weighted (pups are expected to weigh less than aged matched conventionally reared controls), and given fresh bedding and water (*see* **Note 3**) (Fig. 1a).

4. Fourteen days postpartum (Day 14/0) or occasionally less than 14 days for some specific studies, the mother and pups will be removed from the oxygen chamber and placed in room air conditions for up to 6 days (Days 14/0–14/6). Again, animals are weighted and cages are changed. If the pup numbers drop below 13, this litter is usually not used for a formal study, since the NV score will be low for many pups.

5. At Day 14/0 or occasionally at another time point that is designed for the individual study, begin the delivery of test compounds via most appropriate route of administration and at predetermined frequency (e.g., topical drops, intravitreal injection(s)—*see* **Note 4**, subcutaneous injection(s) and pump

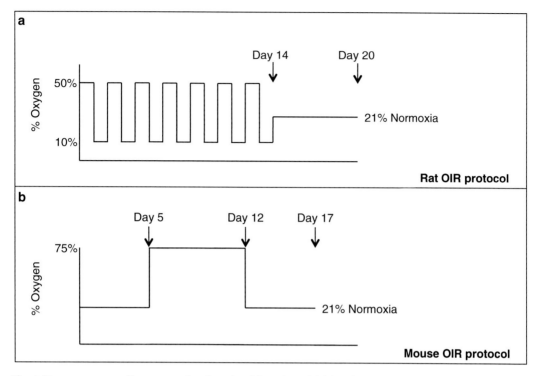

Fig. 1 Diagram representing oxygen insult protocol for rat model (**a**) and mouse model (**b**)

implantation, intraperitoneal injection(s), or oral gavage). If the procedure requires anesthesia, isoflourane inhalation is used.

6. After predetermined period of room air exposure, most frequently 6 days (*see* **Note 5**) (Day 14/6), euthanize mother and pups and enucleate eyes (pups only) for histologic and biochemical evaluations.

3.2 Oxygen Exposure Protocol for Mouse OIR

1. Order male and female mice of desired breed from desired vendor (*see* **Note 1**) for in house breeding or order prepartum, multiparous female mice, and receive them at approximately gestational Day 13 or 14 upon arrival.

2. Mice should be checked daily for pups and the day of parturition recorded, including litter size. Seven day old pups are placed into an oxygen chamber with their nursing dams and subjected to 5 days of 75 % oxygen—oxygen sensors should be calibrated prior to each experiment according to manufacturer's specification. During these 5 days the chamber should not be opened unless to replace food, water or bedding (*see* **Note 5**) (Fig. 1b).

3. At P12 mice are removed from the chamber to room air and retinal regrowth will occur. Pups should be weighed and

documented. As with the rat model treatments can be administered on or after exposure to normoxia—the same guidelines listed above should be adhered to (*see* Subheading 3.1, **step 5**).

4. By P17 neovascularization begins to spontaneously regress and is resolved by P25 so experiments should be concluded at P17 for maximal NV. Pups should, as with the rat model, be culled by lethal injection or isoflourane inhalation (*see* **Note 6**).

3.3 Intraocular Injections

1. It is important to collect and sterilize all equipment prior to placing the pups under anesthesia. Sterilizing injection equipment is essential to minimize risk of endophthalmitis.

2. Ensure pups are under anesthesia (*see* **Note 6**).

3. Using a 33-gauge needle for rat pups or a 35-gauge needle for mouse pups (NanoFil; World Precision Instruments, Sarasota, FL), connected to the Hamilton 10 μl syringe (WPI) insert at 90° posterior to the limbus to avoid lens damage, angle the needle at 45° so the tip sits beneath the lens in the vitreous (*see* **Note 7**).

4. Between 2–5 μl injections for rats and 1 μl injections for mice should be performed on both eyes using a Hamilton syringe (WPI) (*see* **Note 8**), one eye acting as a control (*see* **Note 9**).

5. The injection should be done slowly and is often easier if another investigator is present to push the syringe plunger. After the injection the needle should be held steady in the eye for approximately 30 s.

6. Animals should be monitored until recovery (~2 h) and then returned to normoxia. Any issues with the injection for example large backflow upon removal of the needle, external or internal vessel hemorrhaging of vessels should be noted and if necessary eyes excluded from the study.

3.4 Retinal Dissection and Flat-Mounting

1. Rats or mice should be culled with an overdose of inhalation isoflourane anesthetic (*see* **Note 10**), weighed and eyes removed.

2. Eyes should be fixed in 4 % PFA dissolved in 1× PBS (pH 7.4) for 60 min and enucleation performed using fine curved forceps and curved tip scissors (World Precision Instruments, WPI). Care needs to be taken to avoid excessive pressure on the globe. Retinal flat mounts are prepared with a modification of the methods of Chan-Ling [31].

3. Dissection is performed in a petri dish adapted to hold an eye, in 1–2 ml of chilled 1× PBS (4 °C) under a dissection microscope. Peribulbar fat and connective tissue are dissected away to expose the sclera. A short stump of optic nerve is left protruding from the globe. An incision is made immediately anterior to the corneoscleral limbus with a 15° ophthalmic scalpel.

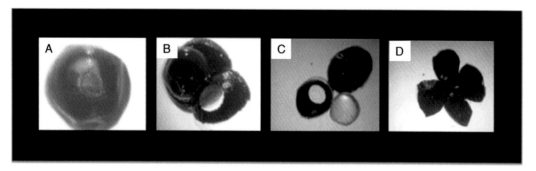

Fig. 2 Dissection of a mouse eye. (**a**) Whole eye removed. (**b**) Incision made at the level of the ora serrata, and cut to expose lens. (**c**) Lens and anterior chamber removed. (**d**) Four incisions made in the posterior cup prior to separating retina from sclero-choroidal complex

The cornea is grasped with fine-toothed forceps and an incision is continued circumferentially with curved-tip microdissection scissors. The cornea, iris, and crystalline lens are discarded (Fig. 2).

4. For rat preps only: forceps are used to avulse the hyaloid vessels and vitreous gel from their retinal attachments (*see* **Note 11**): particular attention should be paid to the regions posterior to the ora serrata and around the optic nerve head, where the attachments are most marked.

5. Four equally spaced radial relaxing incisions, extending two thirds of the way from the retinal periphery to the optic nerve head, are made with scissors to flatten the eyecup. The sclero-choroid is removed leaving just the retina, and two pairs of fine forceps used simultaneously to tease away residual vitreous and hyaloid vessels.

6. The fully dissected retina is returned to 1× PBS for staining.

3.5 Isolectin-B4 Histochemistry

1. Fixed retinas in 1× PBS are permeabilized for 2 h in PBS–Triton X-100 and retinae then washed twice in PBlec.

2. Fluorophore-conjugated isolectin-B4 with specificity for alpha-galactosylated glycoprotein residues on vascular endothelial cells and macrophages (*Griffonia simplicifolia* type I isolectin-IB4), should be diluted (4 μg/ml) in PBlec, 50 μl added to each retina and retinae left overnight in the dark at 4 °C (*see* **Note 12**).

3. The following day retinas are washed five times in 1× PBS and secondary antibody Alexa Fluor 488 conjugate (Molecular Probes, OR, USA) added 2 μg/ml in 1× PBS for at least 2 h.

4. Stained retinas should be washed and carefully flat-mounted on microscope slides, drop Vectashield (Vector Labs) onto the retinas and coverslip. Coverslips require sealing with nail varnish to prevent retinas from drying out.

3.6 Analysis and Quantification of Retinal Flat Mounts

3.6.1 Rat 50/10 OIR

1. Early stages of PRNV will be observed as small round clumps of cells protruding through the retinal surface, these have adopted the term "popcorn" swellings based on their phenotype.

2. PRNV always occurs close to the edge of vascular development, just prior to the avascular area in a brush border type manner. It is important to note that the budding of the most peripheral vessels is not PRNV, PRNV occurs slightly more centrally.

3. PRNV characteristically occurs at the ends of veins, not at the end of arteries. Arteries are identifiable as following oxygen insult the periarteriolar space is increased and you see an absence of vessels surrounding the arteries.

4. Initial "popcorn" swelling merge to form PRNV brush borders, characterized by dark staining.

5. In a pup who has undergone the 50/10 OIR insult one should expect to see PRNV in at least 6 clock hours (Fig. 3a).

6. For more quantitative analysis retinal flat mounts are imaged and if necessary separate images merged to form a full retinal image, the area of retina possessing PRNV is calculated and expressed over total retinal area (Fig. 3b and c).

7. Other features of pathology can be further detailed, for example vascular area, vessel tortuosity and vessel diameter [29].

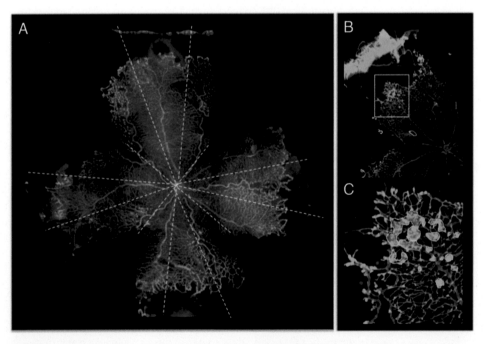

Fig. 3 Rat OIR analysis. (**a**) Stained flat-mounted retina split into clock hours for analysis. (**b**) Identification of pre-retinal neovascularization at the periphery. (**c**) Pre-retinal neovascularization outlined for area quantification

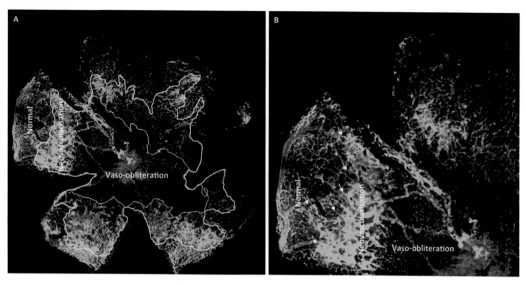

Fig. 4 Mouse OIR analysis. (**a**) Low power visualization (4× objective) of regions of the retina stained with IB4. The VOA is delineated from the NVA by an *orange line*. The NVA is separated from the Normal region by a *white line*. (**b**) Higher power (10× objective) with *arrows* showing where the neovascular front is

3.6.2   Mouse OIR

1. Two main features can be clearly observed in retinal flat mounts (usually from P17 pups) stained for isolectin-B4, areas of vaso-obliteration and neovascularization (Fig. 4).

2. Retinal flat mounts are imaged and images captured as formats compatible with ImageJ [32], and if necessary separate images merged to form a full retinal image. The central vaso-obliterated area (VOA, observed by a lack of vessels present) is outlined using the freehand selection tool in NIH image, and the area determined using the measure command in ImageJ.

3. The same is done for the neovascular area (NVA) where each area of angiogenesis, determined by the overgrowth and lack of definition of individual microvessels (*see* Fig. 4b) is outlined using the freehand selection tool and this area calculated using ImageJ. The area of normal vasculature (normal), as defined by the clear two layers of vessel network stained without areas of vascular growth is also measured.

4. Each of the calculated areas can be expressed as a percentage of the total retinal area (i.e., $100 \times VOA/(VOA + NVA + normal)$). Some more automated methods have been developed and are discussed in [25].

4   Notes

1. Studies have identified that the same strain of rat or mouse ordered from a different supplier can have profound differences in pathology reproducibility. Smith et al. observed higher

NV induction using C57Bl/6 wild-type mice from Jackson Laboratories compared to Taconic Farms [25], and Penn et al., noted greater induction of NV in Sprague-Dawley rats from Charles River. We were able to successfully induce NV using C57/B6 and Sprague-Dawley from Charles River, though we did not compare strains in house [29, 33].

2. Should a Biospherix oxycycler not be available a similar system can be created using one or two ProOx controllers connected to a gas input (nitrogen for 10 % oxygen for 50 %) and an in-house built perspex chamber. The caveat being oxygen levels require daily manual switching and close monitoring to ensure correct oxygen levels are maintained. Should a manual operation procedure be used take care to ensure chamber is ventilated to prevent humidity and carbon dioxide build up (Carbon dioxide can be scrubbed using soda lime).

3. Unnecessary opening of the oxygen chamber during the procedure should be avoided to prevent pups acclimatizing to normoxia. Bedding and water will require changing at least once during the procedure, this should be done quickly and whilst changing the oxygen levels from 10 % to 50 % for the rat model.

4. Conventionally reared rat pups open their eyes following complete development of the retinal vascularization at P12 (postnatal day 12); however, pups exposed to alternating oxygen levels during development experience retarded growth of the retinal vasculature and thus delayed opening of the eyes. It may be necessary to surgically open the eyes in order to administer intravitreal injections; it is of paramount importance to perform control injections in the contralateral eye of the same pup as intralitter variation is often high in this model.

5. The 50/10 OIR model develops vascular tortuosity at P12, peripheral avascular retina at p14 and intravitreous neovascularization at P18. Note that in this model pre-retinal neovascularization recovers by P25 [26], thus to observe differences in pathology we recommend ending the experiment at P20.

6. For isoflourane anesthesia we recommend 2–4 % isoflourane with an oxygen flow of 1–2 l/min.

7. If available use an operating/surgical microscope to visualize the correct position of the needle in the eye during the injection. You should be able to move the tip of the needle without altering the position of the lens. If you can't do this it is likely you have injected into the lens and thus the injection is void. It will be clear if the injection has penetrated the lens as it will cloud the eye. You must also take care to avoid hemorrhaging of large vessels as you penetrate the eye.

8. For rat pups we recommend a maximum injection volume of 5 µl but a preferred volume of 1–3 µl and for mice pups a recommended injection volume of 1 µl, larger volumes will significantly increase intraocular pressure (IOP) and disrupt retinal vessels. The larger the volume the greater the chance of losing drug via backflow during removal of the needle post-injection. Injection volume will be determined by the characteristics of the injected drugs.

9. Control will depend of the vehicle used to dissolve or suspend the drug, for example DMSO. If injecting antibodies we recommend using the equivalent concentration of IgG from the animal the antibody was raised in. If your treatment requires more than one injection take care not to inject into the same location as the previous injection.

10. Avoid methods such as cervical dislocation as the pressure in the eye will increase and lead to the rupture of small blood vessels.

11. Removal of the hyaloid vessels is important for staining of the retinal vessels. Usually the vessels regress during development; however, this is retarded under the variable oxygen conditions experienced during this procedure. Take care to carefully remove all the vessels for a clean prep.

12. If necessary, time can be reduced to 2 h at room temperature.

References

1. Aiello LP (2005) Angiogenic pathways in diabetic retinopathy. N Engl J Med 353(8): 839–841

2. Gariano RF, Gardner TW (2005) Retinal angiogenesis in development and disease. Nature 438(7070):960–966

3. Qazi Y, Maddula S, Ambati BK (2009) Mediators of ocular angiogenesis. J Genet 88(4):495–515

4. Barnett JM, Yanni SE, Penn JS (2010) The development of the rat model of retinopathy of prematurity. Doc Ophthalmol 120(1):3–12

5. Penn JS, Henry MM, Tolman BL (1994) Exposure to alternating hypoxia and hyperoxia causes severe proliferative retinopathy in the newborn rat. Pediatr Res 36(6):724–731

6. Smith LEH, Wesolowski E, McLellan A, Kostyk SK, Damato R, Sullivan R, Damore PA (1994) Oxygen-induced retinopathy in the mouse. Invest Ophthalmol Vis Sci 35(1):101–111

7. Payne JW, Patz A (1977) Fluorescein angiography in retrolental fibroplasia. Int Ophthalmol Clin 17(2):121–135

8. Chanling T, Stone J (1993) Retinopathy of prematurity—origins in the architecture of the retina. Prog Ret Res 12:155–178

9. Hardy P, Dumont I, Bhattacharya M, Hou X, Lachapelle P, Varma DR, Chemtob S (2000) Oxidants, nitric oxide and prostanoids in the developing ocular vasculature: a basis for ischemic retinopathy. Cardiovasc Res 47(3):489–509

10. Penn JS, Thum LA (1989) The rat as an animal model for retinopathy of prematurity. Prog Clin Biol Res 314:623–642

11. Foos RY (1985) Chronic retinopathy of prematurity. Ophthalmology 92(4):563–574

12. Phelps DL, Lindblad A, Bradford JD, Wood NE, Oden NL, Cole C, MacKinnon B, Yaffe A, Everett DF, Wright L, Krulewitch C, Brozanski BS, Young T, Scott M, Hawkins BS, Begg CB, Bell EF, Buckley EG, Hay WW, Kushner BJ, Snouck-Hurgronje L, Taylor CR, Lindblad AS, Bachy C, Berlin SH, Brandt D, Guzzey M, Henson L, Jolles B, Stine E, Thomas-Sharp C, Van Lare J, Stewart J, Alexander T, Anderson C, Ashrafzadeh M, Baumal C, Bhatt A, Blocker R, Brown E, Cordova M, Dacey MP, Duker J, Eagle J, Eichenwald E, Faherty C, Fujii A, Gray J, Hartnett ME, Harvey-Wilkes K, Hetrick J, Hughes M, Ip M, Izatt S, Lacy R, Levy J, Margolis T, McAlmon KR, McCabe O, Moore M, Niffenegger J, Nikou S, Petersen R,

Petit K, Pierce EA, Powers B, Pursley M, Reichel E, Remis L, Rivellese M, Shephard BA, Sorkin J, Stark A, Struzik S, Tran T, VanderVeen DK, Vreeland P, Wilker R, Wright J, Kim V, Desai V, Rutledge B, McClead RE, Fellows R, Biel M, Bremer DL, Maddox R, Mann B, McGregor ML, Nye C, Peterman P, Rogers GL, Rosenberg EM, Seguin J, Stephen S, Anderson CW, Cordero L, Spitzer AR, Corcoran L, Cullen J et al (2000) Supplemental therapeutic oxygen for prethreshold retinopathy of prematurity (STOP-ROP), a randomized, controlled trial. I. Primary outcomes. Pediatrics 105(2):295–310

13. Patz A (1954) Oxygen studies in retrolental fibroplasia. IV. Clinical and experimental observations—the 1st Edward L. Holmes memorial lecture. Am J Ophthalmol 38(3):291–308

14. Gyllensten LJ, Hellstrom BE (1954) Experimental approach to the pathogenesis of retrolental fibroplasia. I. Changes of the eye induced by exposure of newborn mice to concentrated oxygen. Acta Paediatr Suppl 43(100):131–148

15. Gerschman R, Nadig PW, Snell AC, Nye SW (1954) Effect of high oxygen concentrations on eyes of newborn mice. Am J Physiol 179(1):115–118

16. Ashton N, Coomes EN, Garner A, Oliver DO (1968) Retinopathy due to progressive systemic sclerosis'. J Pathol Bacteriol 96(2):259–268

17. Ashton N, Blach R (1961) Studies on developing retinal vessels: VIII. Effect of oxygen on the retinal vessels of the ratling. Br J Ophthalmol 45(5):321–340

18. Flower RW, McLeod DS, Lutty GA, Goldberg B, Wajer SD (1985) Postnatal retinal vascular development of the puppy. Invest Ophthalmol Vis Sci 26(7):957–968

19. Ricci B, Calogero G (1988) Oxygen-induced retinopathy in newborn rats—effects of prolonged normobaric and hyperbaric-oxygen supplementation. Pediatrics 82(2):193–198

20. Penn JS, Henry MM, Wall PT, Tolman BL (1995) The range of pao2 variation determines the severity of oxygen-induced retinopathy in newborn rats. Invest Ophthalmol Vis Sci 36(10):2063–2070

21. Cunningham S, Fleck BW, Elton RA, McIntosh N (1995) Transcutaneous oxygen levels in retinopathy of prematurity. Lancet 346(8988): 1464–1465

22. Chan CK, Pham LN, Zhou J, Spee C, Ryan SJ, Hinton DR (2005) Differential expression of pro- and antiangiogenic factors in mouse strain-dependent hypoxia-induced retinal neovascularization. Lab Invest 85:721–733

23. Gao G, Li Y, Gee S, Dudley A, Fant J, Crosson C, Ma JX (2002) Down-regulation of vascular endothelial growth factor and up-regulation of pigment epithelium-derived factor: a possible mechanism for the anti-angiogenic activity of plasminogen kringle 5. J Biol Chem 277:9492–9497

24. Kitzmann A, Leske D, Chen Y, Kendall A, Lanier W, Holmes J (2002) Incidence and severity of neovascularization in oxygen- and metabolic acidosis-induced retinopathy depend on rat source. Curr Eye Res 25:215–220

25. Stahl A, Connor KM, Sapieha P, Chen J, Dennison RJ, Krah NM, Seaward MR, Willett KL, Aderman CM, Guerin KI, Hua J, Löfqvist C, Hellström A, Smith LEH (2010) The mouse retina as an angiogenesis model. Invest Ophthalmol Vis Sci 51(6):2813–2826

26. Geisen P, Peterson LJ, Martiniuk D, Uppal A, Saito Y, Hartnett ME (2008) Neutralizing antibody to VEGF reduces intravitreous neovascularization and may not interfere with ongoing intraretinal vascularization in a rat model of retinopathy of prematurity. Mol Vis 14(41–43):345–357

27. Davies MH, Stempel AJ, Powers MR (2008) MCP-1 deficiency delays regression of pathologic retinal neovascularization in a model of ischemic retinopathy. Invest Ophthalmol Vis Sci 49:4195–4202

28. Zhang SC, Leske DA, Holmes JM (2000) Neovascularization grading methods in a rat model of retinopathy of prematurity. Invest Ophthalmol Vis Sci 41(3):887–891

29. Gammons MV, Dick AD, Harper SJ, Bates DO (2013) SRPK1 inhibition modulates VEGF splicing to reduce pathological neovascularization in a rat model of retinopathy of prematurity. Invest Ophthalmol Vis Sci 54(8):5797–5806

30. Holmes JM, Duffner LA (1996) The effect of postnatal growth retardation on abnormal neovascularization in the oxygen exposed neonatal rat. Curr Eye Res 15:403–409

31. ChanLing T (1997) Glial, vascular, and neuronal cytogenesis in whole-mounted cat retina. Microsc Res Tech 36(1):1–16

32. Schneider CA, Rasband WS, Eliceiri KW (2012) NIH Image to ImageJ: 25 years of image analysis. Nat Methods 9:671–675

33. Rennel ES, Waine E, Guan H, Schuler Y, Leenders W, Woolard J, Sugiono M, Gillatt D, Kleinerman ES, Bates DO, Harper SJ (2008) The endogenous anti-angiogenic VEGF isoform, VEGF(165)b inhibits human tumour growth in mice. Br J Cancer 98(7):1250–1257

Chapter 23

The Sponge Implant Model of Angiogenesis

Silvia Passos Andrade and Mônica Alves Neves Diniz Ferreira

Abstract

The host response observed after the application of an appropriate stimulus, such as mechanical injury or injection of neoplastic or normal tissue implants, has allowed the cataloging of a number of molecules and cells involved in the vascularization of normal repair or neoplastic tissue. Implantation of sponge matrices has been adopted as a model for the accurate quantification of angiogenic and fibrogenic responses, as they may occur during wound healing, in vivo. Such implants are particularly useful because they offer scope for modulating the environment within which angiogenesis occurs. Sponge implantation model has been optimized and adapted to characterize essential components and their roles in blood vessels formation in a variety of physiological and pathological conditions. As a direct consequence of advances in genetic manipulation, mouse models (i.e., knockouts, SCID, nude) have provided resources to delineate the mechanisms regulating the healing associated with implants. Here we outline the usefulness of the sponge implant model of angiogenesis and detailed description of the methodology.

Key words Angiogenesis, Sponge model, Blood flow, Myeloperoxidase, *N*-Acetylglucosaminidase, Cytokine, Chemokine, Collagen, Intraperitoneal adhesion

1 Introduction

The realization that advances in angiogenesis research depend on making assays quantitative, and reproducible, in vitro and vivo has led to the development of new techniques and improvement of old or current models to comply with such requirements. Implantation technique for assessment of inflammatory process is a well established surgical procedure. Grindlay and Waugh [1], Woessner and Boucek [2], and Edwards [3] were the first to use polyvinyl sponge implants, in dogs, rats, and rabbits, as a framework for assessing the ingrowth of vascularized connective tissue and measurement of enzyme activities in the newly formed fibrovascular tissue.

This technique has further been developed to determine other biochemical variables in fibrovascular tissue including collagen metabolism [4], fibronectin deposition [5], and proteoglycan turnover [6]. Additionally, the technique has been employed to characterize the sequence of histological changes in granulation

Stewart G. Martin and Peter W. Hewett (eds.), *Angiogenesis Protocols*, Methods in Molecular Biology, vol. 1430, DOI 10.1007/978-1-4939-3628-1_23, © Springer Science+Business Media New York 2016

tissue formation [7] and to monitor the kinetics of cellular proliferation [8]. The extent of neutrophil and macrophage accumulation in the sponge compartment has also been made possible by assaying the inflammatory enzymes, myeloperoxidase (MPO), and N-acetyl-β-d-glucosaminidase (NAG) [9–11]. The effects of various anti-inflammatory agents on leukocyte migration and production of prostaglandin-like activity has also been evaluated in sponge implants [12]. The model of acute inflammation has been particularly useful and allows the collection and examination of both cellular and fluid phases of the exudate formed within the sponge [12]. It was also modified to study more chronic inflammatory responses and the evolution of the granulation tissue [12]. Intraperitoneal implantation of synthetic matrix in mice has been shown to induce adhesion-like tissue. The various components (inflammatory, angiogenic, and fibrogenic) of the peritoneal fibrovascular tissue, as well as the effects of various compounds on these processes have been studied [13–15]. Sponge implantation has also been used as a framework to host rodent cell lines [16–18]. The advantage of the implantation technique for the purpose of investigating tumor-induced angiogenesis is that the assessment of the relative contributions of the tumor cells to early changes in the implant blood-flow can be detected even before visible growth of the tumor mass is evident. Using the ^{133}Xe clearance technique [19] or fluorescein diffusion method, the development of Colon 26, melanoma B16, or Ehrlich tumor has been evaluated in terms of characterizing solid tumor hemodynamic features.

2 Materials

2.1 Sponge Matrix

A number of different sponge matrices have been used for inducing fibrovascular growth and to host tumor cells (*see* **Note 1**). The synthetic materials are mainly polyvinyl alcohol, cellulose acetate, polyester, polyether, and polyurethane alone or in combination. In our laboratory we use sponge disks made of polyether polyurethane. This type of material possesses the following characteristics: uniform pore size and intercommunicating pore structure, ability to resist chemical treatment, biocompatibility. The attachment of a polythene cannula in the center of the sponge disk may provide free access to the interior of the implants (*see* Subheading 2.2 and Fig. 1a).

2.2 Cannula (Optional)

1. Polythene tubing for cannula—1.4 mm internal diameter; 1.2 cm long.

2. Polythene tubing for plug—1.2 mm internal diameter; 0.6 cm long.

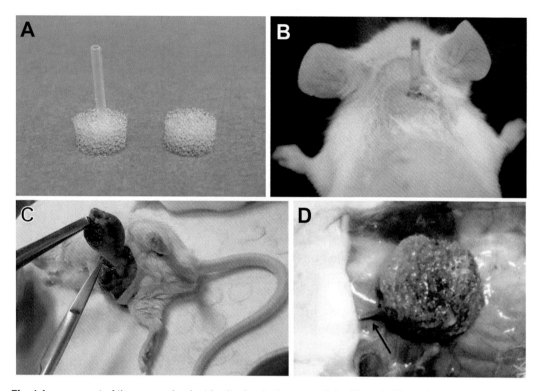

Fig. 1 Arrangement of the sponge implant in situ. Implant sponge disk with and without the cannula (**a**) and the subcutaneous arrangement of the implant in a mouse (**b**). Intraperitoneal implantation of synthetic matrix in mice showing an adhesion-like tissue (**c**). A vascularized sponge 14 days post implantation is shown in (**d**). Blood vessels of various sizes can be seen infiltrating the sponge

3. 5-0 silk sutures for attachment of the cannula to the center of the sponge disks, for holding the sponge disk in place following implantation and for closing the surgical incision (*see* **Note 2**). Please note: If a cannula is attached, the outside end must be sealed (Fig. 1a and b).

2.3 Anesthesia

1. Rats are anesthetized using intramuscular injection of 0.5 mL/kg of 0.315 mg/kg fentanyl citrate and 10 mg/mL fluanisone.

2. Mice are anesthetized using the combination of fentanyl citrate and fluanisone acetate plus 5 mg/mL of midazolam hydrochloride each at a dose of 0.5 mL/kg or with subcutaneous injection of xylazine (10 mg/kg) and ketamine (60 mg/kg) solution.

3 Methods

3.1 Preparation of Sponge Implants

Circular sponge disks are cut from a sheet of sponge using a cork-borer. Usually the diameter and thickness of the disk depend on the animal used. For mice and rats the recommended dimensions range from 8 mm × 4 mm and 12 mm × 6 mm, respectively.

The sponge disks are sterilized by boiling in distilled water for 10 min, then placed in sterile glass petri dishes and irradiated overnight under ultraviolet light in a laminar flow hood. For the cannulated implant, use a segment of polythene tubing (1.4 mm internal diameter). This is secured to the interior of each sponge disk (i.e., midway through its thickness) by three 5-0 silk sutures, in such a way that the tube is perpendicular to the disk face and its open end is sealed with a removable plastic plug made of a smaller polythene tubing (1.2 mm internal diameter). The cannula allows accurate injection of tracers, tests substances and withdrawn of fluid into and from the interior of the implant (Fig. 1a).

3.2 Surgical Procedure for Sponge Disk Implantation

1. Implantation of sponge disks is performed with aseptic techniques following induction of anesthesia.

2. The hair on the dorsal (for subcutaneous implants) or abdominal (for intraperitoneal implants) side is shaved and the skin wiped with 70 % (v/v) ethanol in distilled water.

3. A through 1 cm long dorsal incision or in the peritoneal cavity by means of a 1 cm long mid-line incision in the linea alba of the abdomen is made and through it one subcutaneous pocket is prepared by blunt dissection using a pair of curved scissors.

4. A sterilized sponge implant is then inserted into the pocket.

5. For the cannulated implant, a small incision previously made on the cervical side of the pocket allows the cannula to be exteriorized.

6. The base of the cannula is sutured to the animal skin to immobilize the sponge implant. Finally, the cannula is plugged with a smaller sealed polythene tubing, so as to prevent infection.

7. The mid-line incision is closed by three interrupted 5-0 silk stitches and the animals kept singly with free access to food and water after recovery from anesthesia.

8. The sponge disks (cannulated or not) implanted subcutaneously or intraperitoneally can be left in situ for periods ranging from days to weeks. Images of the sponge disk in situ and its arrangement after implantation are shown (Fig. 1a–d).

3.3 Estimating Blood Flow Development by Efflux of Sodium Fluorescein Applied Intraimplant

Low molecular fluorochrome-complexed tracers or fluorogenic dyes have provided additional methods for detecting new blood vessels. Compared with radioactive isotope compounds, the advantages of fluorescent dyes are evident. Fluorescence is relatively atoxic, nonradioactive, and inexpensive [20]. The measurement of fluorochrome-generated emission in the bloodstream following its application in the sponge implant compartment at various intervals postimplantation reflects the degree of local blood flow development and the interaction of the angiogenic site with the systemic circulation [21]. This approach can be used to study sponge-induced angiogenesis quantitatively and to investigate the pharmacological reactivity of the neovasculature (*see* **Note 3**).

Measurements of the extent of vascularization of sponge implants are made by estimating $t_{1/2}$ (min) of the fluorescence peak in the systemic circulation following intraimplant injection of sodium fluorescein (50 µL/kg of a sterile solution of sodium fluorescein 10 %) at fixed time intervals (for example, days 1, 4, 7, 10, and 14) postimplantation.

1. Anesthetize the animal.

2. Determine blood background fluorescence by piercing the extremity of the tail and collecting 5 µL of blood with a heparinized yellow tip. Transfer the blood sample to a centrifuge tube contained 1 mL of isotonic saline (0.9 %).

3. At time 0, administer sodium fluorescein (50 µL of a sterile solution of 10 % sodium fluorescein per kg weight) to anesthetized animals.

4. After 1 min collect the first blood sample following dye injection as in **step 2**.

5. At 3 min collect a second blood sample. Repeat this procedure every 2–3 min for 25–30 min.

6. Centrifuge blood samples for 10 min at $1400 \times g$ (2000 rpm). Keep the supernatant for fluorescence determination (excitation 485 nm/emission 519 nm).

7. From the fluorescence values estimate the time for the fluorescence to peak in the bloodstream (absorption) and the time required for the elimination of the dye from the systemic circulation (elimination). These parameters are expressed in terms of half-time ($t_{1/2}$; time taken for the fluorescence to reach or to decay 50 % of the peak value in the systemic circulation).

3.4 Estimating Biochemical Parameters in the Implants

Quantitation of various biochemical parameters further supports the functional characterization of the fibrovascular tissue that infiltrates the implants and has been used to corroborate assessment of angiogenesis [12, 22–26].

1. Remove the implants at any time postimplantation as required (*see* **Note 4**).

2. Immediately upon removal, weigh, homogenize in 2 mL of Drabkin reagent, and centrifuge the tissue in isotonic physiological solution (saline 0.9 % or phosphate-buffered saline).

3. Store the supernatant of the homogenate at −20 °C for later analysis of:

   • hemoglobin (vascular index);

   • cytokines;

   • chemokines (angiogenic factors).

4. Store the pellet of the homogenate at −20 °C for determination of pro-inflammatory enzymes myeloperoxidase (MPO—neutrophils influx); N-acetyl-β-d-glucosaminidase (NAG—macrophages recruitment).

3.4.1 Hemoglobin Determination

The vascularization of the implant can be assessed by measuring the amount of hemoglobin contained in the tissue, using the Drabkin method.

1. After centrifuging the homogenate (implant + 2 mL Drabkin reagent) at $12,000 \times g$ for 20 min, filter it in a 0.22-μm Millipore filter.

2. Determine hemoglobin concentration by measuring absorbance at 540 nm using an ELISA plate reader and compare against a standard curve of hemoglobin.

3.4.2 Cytokine Determination

1. Take 50 μL of the supernatant previously homogenized in Drabkin reagent (to remove hemoglobin) and centrifuged ($12,000 \times g$, 20 min at 4 °C) and add 500 μL of PBS pH 7.4 containing 0.05 % Tween 20, centrifuged at $12,000 \times g$ at 4 °C for 30 min.

2. The amount of the cytokines and the chemokines in each sample is determined using Immunoassay Kits and following the manufacturer's protocol.

3. Dilutions of cell-free supernatants are added in duplicates to ELISA plates coated with a specific murine polyclonal antibody against the cyto/chemokine, followed by the addition of a second polyclonal antibody against the cyto/chemokine. After washing to remove any unbound antibody-enzyme reagent, a substrate solution (a 1:1 solution of hydrogen peroxide and tetramethylbenzidine) is added to the wells. The reaction is terminated with 50 μL/well of 1 M H_2SO_4. Plates are read at 492 nm in a spectrophotometer. Standards are 0.5-log10 dilutions of recombinant murine chemokines from 7.5 pg/mL to 1000 pg/mL (100 μL). Express the results as pg/mg wet tissue.

3.4.3 Measurement of MPO and NAG Activities

The extent of neutrophil accumulation in the implants is measured by assaying myeloperoxidase (MPO) and N-acetyl-β-d-glucosaminidase (NAG) activities in whole tissue.

MPO Activity

1. After processing the supernatant of the tissue for hemoglobin determination (*see* above), a part of the corresponding pellet is weighed, homogenized in pH 4.7 buffer (0.1 M NaCl; 0.02 M Na_3PO_4; 0.015 M Na_2EDTA) and centrifuged at $12,000 \times g$ for 20 min at 4 °C.

2. The pellets are then resuspended in 0.05 M sodium phosphate buffer (pH 5.4) containing 0.5 % hexa-1,6-bis-decyltrimethyl-ammonium bromide (HTAB, Sigma). The suspensions are freeze-thawed three times using liquid nitrogen and finally centrifuged at $10,000 \times g$ for 20 min at 4 °C.

3. MPO activity in the resulting supernatant is assayed by mixing 25 μL of 3,3′-5,5′-tetramethylbenzidine (TMB), prepared in dimethylsulfoxide (DMSO) in a final concentration of 1.6 mM; with addition of 100 μL H_2O_2 in a final concentration of 0.003 % v/v, dissolved in sodium phosphate buffer (pH 5.4) and 25 μL of the supernatant from the tissue sample.

4. The assay is carried out in a 96-well microplate and is started by adding the supernatant sample to the H_2O_2 and TMB solution and incubated for 5 min at 37 °C.

5. The reaction is terminated by adding 100 μL 4 M H_2SO_4 at 4 °C and is quantified colorimetrically at 450 nm in a spectrophotometer. Results can be expressed as change in OD per gram of wet tissue.

NAG Activity

Numbers of monocytes/macrophages can be quantitated by measuring the levels of the lysosomal enzyme NAG, present at high levels in activated macrophages [9, 27].

1. Part of the pellet, remaining after the hemoglobin measurement, is kept for this assay. These pellets are weighed, homogenized in NaCl solution (0.9 % w/v) containing 0.1 % v/v Triton X-100 (Promega) and centrifuged ($3000 \times g$; 10 min, 4 °C).

2. Samples of the resulting supernatant (100 μL) are incubated for 10 min with 100 μL *p*-nitrophenyl-*N*-acetyl-d-glucosaminide (Sigma), prepared in 0.1 M of citrate/sodium phosphate buffer (pH 4.5) in a final concentration of 2.24 mM.

3. The reaction is terminated by the addition of 100 μL 0.2 M glycine buffer pH 10.6. Hydrolysis of the substrate is determined by measuring the color absorption at 405 nm. NAG activity is finally expressed as OD per gram wet tissue.

3.4.4 Soluble Collagen/ Collagen Measurements

Total soluble collagen can be measured in whole tissue homogenates by the Sirius Red reagent based-assay [13, 28, 29].

1. Anesthetize the animal.

2. Remove the implants at any time postimplantation as required.

3. Immediately upon removal, weigh, homogenize in 1 mL of PBS.

4. 50 μL of sample is mixed with 50 μL of Sirius Red reagent by gentle inversion.

5. The collagen–dye complex precipitates by centrifugation at $5000 \times g$ for 10 min. The supernatants are drained off and discarded.

6. The pellet is washed with 500 μL of ethanol (99 % pure and methanol free).

7. 1 mL of a 0.5 M NaOH solution is added to the remaining pellet of collagen-bound dye.

8. After solubilization, samples are transferred to a 96-well plate and read at 540 nm.

9. The calibration curve is set up on the basis of a gelatin standard.

10. The results are expressed as μg collagen/mg wet tissue.

3.5 Histological Analysis of the Implants

To establish the sequential development of granulation tissue and blood vessels in the implants, several histologic techniques have been employed.

1. Kill the animals bearing the implants.

2. Dissect the implants free of adherent tissue, fix in formalin (10 % w/v in isotonic saline), and embed them in paraffin.

3. Cut the sections (5–8 μm) from halfway through the sponge's thickness.

4. Stain and process for light microscopy studies.

   Routine histological stains such as Hematoxylin and eosin (HE), Masson's and Gomori's trichrome, Dominici blue, and picrosirius red staining can be used for determining the implant fibrovascular tissue main features (area, vessels, inflammatory cells, and collagen content) (Fig. 2a–d).

3.5.1 Detection of Endothelial Cells/Blood Vessels

Detection of endothelial cells/blood vessels can be performed using a monoclonal antibody against CD 31.

1. The sponge implant is fixed in 10 % buffered formalin (pH 7.4) and processed for paraffin embedding.

2. Tissue sections (5 μm) are dewaxed and antigen retrieval is performed in citrate buffer (pH 6).

3. The slides are boiled in citrate buffer for 25 min at 95 °C and then, cooled for 1 h in the same buffer.

4. Sections are incubated for 5 min in 3 % hydrogen peroxide to quench endogenous tissue peroxidase.

5. Nonspecific binding is blocked by using normal goat serum for 10 min (1:10 in phosphate-buffered saline) with 1 % bovine serum albumin (in phosphate-buffered saline).

6. The sections are then immunostained with monoclonal antibody to CD31 (1:40 dilution, Dako) for 60 min at room temperature.

7. After washing in Tris–HCl buffer, sections are incubated for 30 min at room temperature with biotinylated Link Universal Streptavidin-HRP (Dako).

8. The reactions are revealed by applying 3,3′-diaminobenzidine (DAB) in chromogen solution (Dako).

9. Sections are counterstained with hematoxylin and mounted in Permount.

10. Immunostaining is performed manually, and the spleen is used as a positive control. Negative controls are carried out with omission of the primary antibody, resulting in no detectable staining. The expression of these proteins is evaluated on the basis of the extent of cytoplasmic immunolabeling in endothelial cells forming lumen in six high-power fields, regardless of staining intensity (400×). Vascular structures labeled with CD31 are marked in brown (Fig. 2b).

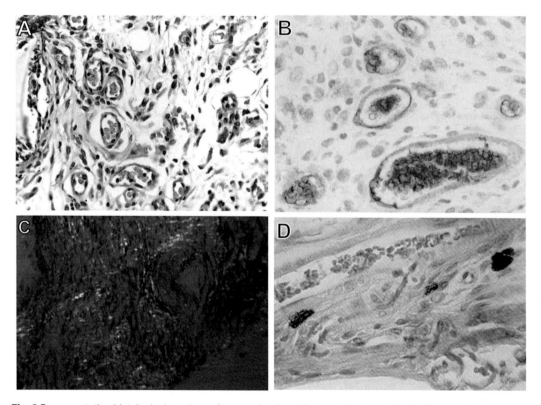

Fig. 2 Representative histological sections of sponge implant. Hematoxylin and eosin for fibrovascular tissue (**a**), blood vessels stained with CD31 (**b**), picrosirius red staining for collagen (**c**), and Domici blue for mast cells (**d**)

4   Notes

1. Depending on the material, the inflammatory response can cause excessive matrix deposition and unwanted fibrosis. Because of the variety of the materials used (size, structure, composition, porosity) the pattern of the response varies widely.

2. The attachment of the cannula to the center of the sponge disk is facilitated by making in one end of the polythene tubing "teeth-like" (usually 4) structures, in such a way that the thread is secured in one of them and then sutured to the sponge.

3. To eliminate acute effects of the vasoactive substances tested (vasodilation or vasoconstriction) they should be given 6–8 h prior blood flow measurement.

4. To avoid possible "contamination" of blood spilled during in after the surgical procedure and/or with surrounding preexisting vessels, removal of the implants must be done 30 min after the animals' death.

Acknowledgments

This work was supported by grant from CNPq and FAPEMIG—Brazil.

References

1. Grindlay JH, Waugh JM (1951) Plastic sponge which acts as a framework for living tissue; experimental studies and preliminary report of use to reinforce abdominal aneurysms experimental studies and preliminary report of use to reinforce abdominal aneurysms. AMA Arch Surg 63:288–297

2. Woessner JF Jr, Boucek RJ (1959) Enzyme activities of rat connective tissue obtained from subcutaneously implanted polyvinyl sponge. J Biol Chem 234:3296–3300

3. Edwards RH, Sarmenta SS, Hass GM (1960) Stimulation of granulation tissue growth by tissue extracts; study by intramuscular wounds in rabbits. Arch Pathol 69:286–302

4. Paulini K, Korner B, Beneke G et al (1974) A quantitative study of the growth of connective tissue: Investigations on polyester-polyurethane sponges. Connect Tissue Res 2:257–264

5. Holund B, Clemmensen I, Junker P et al (1982) Fibronectin in experimental granulation tissue. Acta Pathol Microbiol Immunol Scand A 90:159–165

6. Bollet AJ, Goodwin JF, Simpson WF et al (1958) Mucopolysaccharide, protein and DNA concentration of granulation tissue induced by polyvinyl sponges. Proc Soc Exp Biol Med 99:418–421

7. Holund B, Junker P, Garbarsch C et al (1979) Formation of granulation tissue in subcutaneously implanted sponges in rats. Acta Pathol Microbiol Scand A 87:367–374

8. Davidson JM, Klagsbrun M, Hill KE et al (1985) Accelerated wound repair, cell proliferation and collagen accumulation are produced by a cartilage-derived growth factor. J Cell Biol 100:1219–1227

9. Bailey PJ (1988) Sponge implants as models. Methods Enzymol 162:327–334

10. Belo AV, Barcelos LS, Ferreira MAND et al (2004) Inhibition of inflammatory angiogenesis by distant subcutaneous tumor in mice. Life Sci 74:2827–2837

11. Ferreira MAND, Barcelo LS, Campos PP et al (2004) Sponge-induced angiogenesis and inflammation in PAF receptor-deficient mice (PAFR-KO). Br J Pharmacol 141:1185–1192

12. Ford-Hutchinson AW, Walker JA, Smith JA (1977) Assessment of anti-inflammatory activity by sponge implantation techniques. J Pharmacol Methods 1:3–7

13. Mendes JB, Rocha MA, Araújo FA et al (2009) Differential effects of rolipram on chronic subcutaneous inflammatory angiogenesis and on peritoneal adhesion in mice. Microvasc Res 78:265–271

14. Araújo FA, Rocha MA, Ferreira MA et al (2011) Implant-induced intraperitoneal inflammatory angiogenesis is attenuated by fluvastatin. Clin Exp Pharmacol Physiol 38:262–268

15. Oviedo-Socarrás T, Vasconcelos AC, Barbosa IX et al (2014) Diabetes alters inflammation, angiogenesis, and fibrogenesis in intraperitoneal implants in rats. Microvasc Res 93:23–29. doi:10.1016/j.mvr.2014.02.011

16. Mahadevan V, Hart IR, Lewis GP (1989) Factors influencing blood supply in wound granuloma quantitated by a new in vivo technique. Cancer Res 49:415–419

17. Andrade SP, Bakhle YS, Hart I et al (1992) Effects of tumour cells and vasoconstrictor responses in sponge implants in mice. Br J Cancer 66:821–826

18. Lage AP, Andrade SP (2000) Assessment of angiogenesis and tumor growth in conscious mice by a fluorimetric method. Microvasc Res 59:278–285

19. Kety SS (1949) Measurement of regional circulation by local clearance of radioactive sodium. Am Heart J 38:321–331

20. McGrath JC, Arribas S, Daly CJ (1996) Fluorescent ligands for the study of receptors. TIPS 17:393–399

21. Andrade SP, Machado RD, Teixeira AS et al (1997) Sponge-induced angiogenesis in mice and the pharmacological reactivity of the neovasculature quantitated by a fluorimetric method. Microvasc Res 54:253–261

22. Andrade SP, Vieira LBGB, Bakhle Y et al (1992) Effects of platelet activating factor (PAF) and other vasoconstrictors on a model of angiogenesis in the mouse. Int J Exp Pathol 73:503–513

23. Andrade SP, Cardoso CC, Machado RDP et al (1996) Angitensin-II-induced angiogenesis in sponge implants in mice. Int J Microcirc Clin Exp 16:302–307

24. Hu D-E, Hiley CR, Smither RL et al (1995) Correlation of 133Xe clearance, blood flow and histology in rat sponge model for angiogenesis. Lab Invest 72:601–610

25. Buckley A, Davidison JM, Kamerath CD et al (1985) Sustained release of epidermal growth factor accelerates wound repair. Proc Natl Acad Sci U S A 82:7340–7344

26. Plunkett ML, Halley JA (1990) Na in vivo quantitative angiogenesis model using tumor cells entrapped in alginate. Lab Invest 62:510–517

27. Belo AV, Barcelos LS, Teixeira MM et al (2004) Differential effects of antiangiogenic compounds in neovascularization, leukocyte recruitment, VEGF production, and tumor growth in mice. Cancer Invest 22:723–729

28. Phillips RJ, Burdick MD, Hong K et al (2004) Circulating fibrocytes traffic to the lungs in response to CXCL12 and mediate fibrosis. J Clin Invest 114:438–446

29. Campos PP, Bakhle YS, Andrade SP (2008) Mechanisms of wound healing responses in lupus-prone New Zealand White mouse strain. Wound Repair Regen 16:416–424

Chapter 24

Measurement of Angiogenesis, Arteriolargenesis, and Lymphangiogenesis Phenotypes by Use of Two-Dimensional Mesenteric Angiogenesis Assay

Andrew V. Benest and David O. Bates

Abstract

Successful therapeutic angiogenesis requires an understanding of how the myriad interactions of growth factors released during angiogenesis combine to form a mature vascular bed. This requires a model in which multiple physiological and cell biological parameters can be identified. The adenoviral-mediated mesenteric angiogenesis assay as described here is ideal for that purpose. The clear, thin, and relatively avascular mesenteric panel can be used to measure increased vessel perfusion by intravital microscopy. In addition, high-powered microvessel analysis is carried out by immunostaining of features essential for the study of angiogenesis or lymphangiogenesis (including endothelium, pericyte, smooth muscle cell area, and proliferation), allowing functional data to be obtained in conjunction with high-power microvessel ultrastructural analysis. Therefore, the mesenteric angiogenesis model offers a robust system to analyze the morphological changes associated with angiogenesis, induced by different agents.

Key words Angiogenesis, Lymphangiogenesis, Arteriolargenesis, Pericyte, Mural cell

1 Introduction

The proangiogenic effects of endothelial growth factor overexpression have been characterized in a wide range of models. For instance, vascular endothelial growth factor (VEGF) has been overexpressed in the rat mesentery [1], skeletal muscle [2, 3], cardiac muscle [4], trachea [5], skin [6–8], and corneal eye pocket [9, 10]. With the exception of the trachea and mesenteric models, a reductionist approach has relied on quantifying the increase in vessel density as the angiogenic response. Although the primary end point of angiogenesis should be increased vessel density, this does not reveal any details regarding the manner in which angiogenesis proceeds, such as whether growth is by a sprouting or nonsprouting mechanism. Consequently it represents a single statement of the angioarchitecture (vessel density within tissue), rather than the

Stewart G. Martin and Peter W. Hewett (eds.), *Angiogenesis Protocols*, Methods in Molecular Biology, vol. 1430, DOI 10.1007/978-1-4939-3628-1_24, © Springer Science+Business Media New York 2016

phenotype of the vessels themselves. Moreover, in many of these systems it has proven difficult, if not impossible, to quantify the increase in parameters such as vessel branch point density, proliferating endothelial cells (PECs), and so on. Furthermore, the physiological characterization of vessels undergoing angiogenesis within the microvessel network is normally not possible. The mesentery is currently used for physiological recordings of microvessel permeability [11], compliance [12], vasoreactivity [13], and conducted vasodilation [14], so an angiogenesis assay in the same tissue allows the determination of functionality of the neovessels formed, measuring parameters such as vascular reactivity, hydraulic conductivity, vessel compliance, and more.

To illustrate this, we describe an angiogenesis assay that compares different growth factors: angiopoietin 1 (Ang-1) [15], VEGF [1, 15], neurotrophins [16], VEGF-C [17], members of the kallikrein family [18], or wide ranging vascular modulating agents (endothelial nitric oxide synthase, eNOS) [19] . Details of the work here have been published [1, 15, 17, 19], but the methodology is described here in detail. For a full description of the angiogenic phenotype to be considered (e.g., type of vessel growth, time course, mechanisms), the use of a two-dimensional microvascular network such as the mesentery offers significant advantages. Although models using the tracheal microvessel network are able to rely on a well-studied system, the method of perfusion, fixation, and the three-dimensional networks mean that (1) only perfused vessels are analyzed, (2) measurements of perfusion are indirect, and (3) the tracheal vasculature cannot be visualized in vivo.

For therapeutic purposes, the most successful end result of angiogenesis would be a direct increase in tissue perfusion. The rat mesentery surpasses the limitations of the tracheal system and offers the ability to quantify the vessel phenotype and how this might be manifested anatomically. We can image the perfused mesenteric microvasculature on day 1 and day 7 in vivo, and, if required, perform physiological measurements such as those for permeability [11], perfusion [1], and vessel reactivity [13, 20], then stain and image the mesentery by confocal fluorescence or electron microscopy. This enables quantitative assessments of vessel density, branching, sprouting, length and diameter, and the degree and constituents of mural support, which can be linked to the degree of tissue perfusion. Furthermore, its noteworthy that the rat mesentery has a functional lymphatic system which has enabled the interaction between the blood and lymphatic circulations to be studied in vivo [17] and in vitro [21], and demonstrates a novel lymphangiogenesis modification to the standard angiogenesis assay.

2 Materials

Prepare all solutions using ultrapure water. For solutions to be used in vivo autoclaved water is essential. Prepare and store all reagents at room temperature, unless otherwise indicated. Diligently follow all waste disposal regulations when disposing of waste materials.

2.1 Solutions

1. Mammalian Ringer solution: (millimolar concentrations: 132.0 NaCl, 4.6 KCl, 1.27 $MgSO_4$, 2.0 $CaCl_2$, 25.0 $NaHCO_3$, 5.5 D-glucose, 3.07 HEPES acid, and 2.37 HEPES sodium salt, pH corrected to 7.45 ± 0.02 with 0.115 M NaOH) (*see* **Note 1**).

2. 4 % paraformaldehyde (in 1× PBS, pH 7.4) (*see* **Note 2**).

3. PBX: 0.5 % Triton X-100 in PBS.

4. Block Solution: PBX + 1 % BSA or 10 % normal goat serum.

2.2 Antibodies and Conjugates

1. *Griffonia simplicifolia* isolectin IB4 (GSI-IB4, Molecular Probes, Cambridge, UK) mouse monoclonal Ki-67 (Novocastra Lab, Newcastle upon Tyne, UK, NCL-L-Ki67-MM1), NG2 (Chemicon, Temecula, CA, USA, MAB5384, 5 μg/mL) and α-smooth muscle actin (DAKO, Glostrup, Denmark, M 0851, 1.4 μg/mL). Rabbit anti-mouse lymphatic vessel endothelial hyaluronic acid receptor (LYVE-1) (10 μg/μL, a kind gift from Dr David Jackson, University of Oxford).

2. Alexa Fluor 405, 488 or tetramethylrhodamine isothiocyanate (TRITC)-labeled streptavidin (1 μg/mL, S-870, Molecular Probes) and Alexa Fluor 488, 350 goat anti-mouse immunoglobulin G (IgG; 2 μg/mL, Molecular Probes) are used as secondary detection antibodies to lectin, Ki67, and VEGF and smooth muscle actin, respectively.

3 Methods

3.1 Angiogenesis, Arteriolargenesis, and Lymphangiogenesis Preparation

All surgical procedures need to be performed using sterile equipment, and an aseptic technique adhered to. It is recommended that all solutions used are prewarmed and stable at room temperature before commencing surgery.

1. Anesthetise male Wistar rats (300–350 g), although larger rats (350–450 g) are more suitable for lymphangiogenic assays [17] using 5 % isoflurane inhalation, anesthesia maintain with 2 % isoflurane.

2. Insert a rectal temperature probe connected to a thermostatically controlled heated blanket (Harvard CCM79 animal blan-

ket control unit, Kent, UK) to maintain core body temperature at 37 °C.

3. Shave (electric clippers) and sterilize (chlorhexidine, pevidine, or ethanol).

4. Perform laparotomy, creating an incision approximately 1.5–2 cm long.

5. Tease out a small region of the small intestine and gently drape over a quartz pillar (*see* Fig. 1a). Within each mesenteric panel is a flowing microvessel bed (Fig. 1a and b) Image the panel using a digital camera attached to the microscope. Damp, cotton buds (Q-tips) are ideal for gut manipulation.

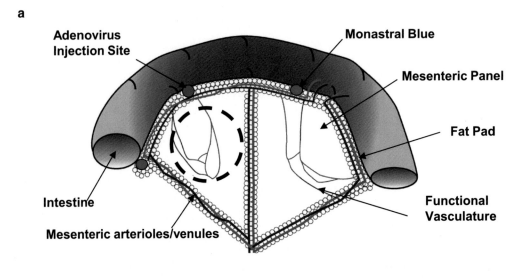

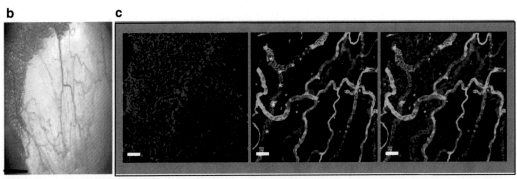

Fig. 1 (**a**) Schematic representation of a section of rat mesentery shows the intestine, fat pad, membranous mesenteric panels, and blood vessel network. The adenovirus injection site (*green circle*) is labeled with Monastral blue tattoos (*blue circles*) in adjacent panels. (**b**) Intravital imaging of perfused vessels in the mesentery and (**c**) immunofluorescent staining for isolectin and DAPI shows the vasculature present the mesenteric window (indicated as *broken circle* in **a**). *Asterisk* indicates lymphatic vessels (indicated by differential junctional lectin staining). Scale bar = (**b**) 1 mm; (**c**) 40 μm

6. Throughout the procedure (which should last approximately 2–5 min) superfuse the exposed tissue with mammalian Ringer, kept at a constant 37 ± 1 °C. Temperature is maintained by connecting the superfusate to a heat exchange coil fed by a thermostatically controlled water bath.

7. Inject approximately 5×10^8 Plaque forming units (PFU, in a volume of approximately 25 μL) of adenovirus into the surrounding fat pad using a 30-gauge needle and Hamilton syringe. Tattoo surrounding mesenteric panels with 0.6 % w/v Monastral blue in mammalian Ringer to enable the same panel to be located on day 7 or 14. The following viruses have been used for the angiogenesis assay: adenovirus-cytomegalovirus (Ad-CMV)–VEGF$_{165}$, titer 8×10^8 PFU/mL; Ad-CMV-Ang-1, titer 8×10^8 PFU/mL; Ad-CMV-eGFP (enhanced green fluorescent protein), titer 5×10^8 PFU/mL; and Ad-CMV-eNOS, titer 6×10^8 and Ad-VEGF-C (5×10^8) In experiments requiring multiple viruses to be used, 25 μL of each virus is used but injected as a mixture.

8. Gently return the gut/mesentery into the abdominal cavity and suture the muscle, and skin.

9. Analgesia is provided by an intramuscular injection of buprenorphine.

10. Allow animals to recover on 100 % O_2 before being transferred to a warmed, clean animal cage.

3.2 Immuno-fluorescence on Whole-Mount Mesentery

1. Repeat surgery (day 7 or 14) as before to find same panel and identify the panel is imaged as before and fixed in vivo with 4 % paraformaldehyde (in 1× phosphate-buffered saline [PBS], pH 7.40) for 5 min (*see* **Note 3**).

2. Sacrifice rat by cervical dislocation, and cut out the mesenteric panel of interest.

3. Remove fat pads and snap freeze for later protein analysis and stored at −20 °C. Wash the panel in mammalian Ringer and refix for a maximum of 1 h at room temperature in the same fixative (*see* **Note 4**).

4. Wash panels with 0.5 % PBX, for 1 h. Changing solutions every 10 min at room temperature.

5. Block panels in 1 % bovine serum albumen (BSA)—0.5 % PBX, or 1.5 % normal goat serum for 1 h.

6. Incubate mesentery overnight at 4 °C with 10 μg/mL biotinylated *Griffonia simplicifolia* isolectin IB4 and mouse monoclonal antibodies to either Ki-67 for dividing cells, NG2 for pericytes, or α-smooth muscle actin for smooth muscle, *see* Fig. 2. Rabbit anti-mouse lymphatic vessel endothelial hyaluronic acid receptor (LYVE-1, Fig. 3) can be used to identify lymphatic vessels (*see* **Note 5**).

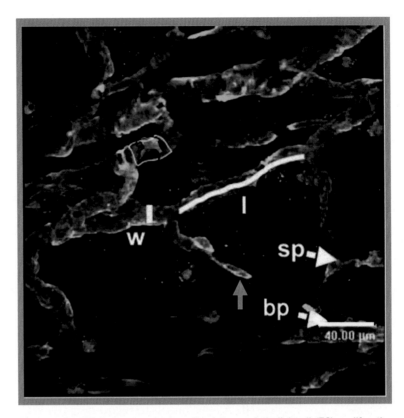

Fig. 2 (**a**) Growth factor expression stimulates endothelial cell (EC) proliferation. Fluorescent staining of mesenteries after injection of adenovirus-enhanced green fluorescent protein (Ad-EGFP) (control) or Ad-growth factors. Isolectin IB4-TRITC (*red*) stains endothelial cells ECs; Hoechst 33324 (*blue*) stains all mesenteric nuclei and antibodies to Ki67-AF488 (*green*) to detect proliferating cells. Overlaying of the stack images is used to calculate the number of proliferating endothelial cells (PECs). Images are triple stained with TRITC-streptavidin and biotinylated GSL isolectin IB4 (EC, *red*). Alexa Fluor 488-labeled goat anti-mouse immunoglobulin G and mouse monoclonal anti-Ki67 antibody (proliferating cells, *green*) and overlay. (**b**) Immunohistochemical analysis of mural cell (pericyte/vSMC) demonstrates presence of NG2+ and αSMA positive mural cells. Scale bar = 40 μm

7. Wash panels as before and add secondary antibodies in block solution. Incubate for 2 h on a rocker at 4 °C.

8. Wash panels a further six times, and if appropriate DAPI or Hoechst 33324 (1 μM) can be added to stain mesenteric nuclei.

9. Carefully manipulate the panels using fine forceps and flatten onto a glass slide Mount in Vectashield (Vector Lab) and place a coverslip on the tissue.

10. Image five random sections of each panel are imaged, at 40× magnification.

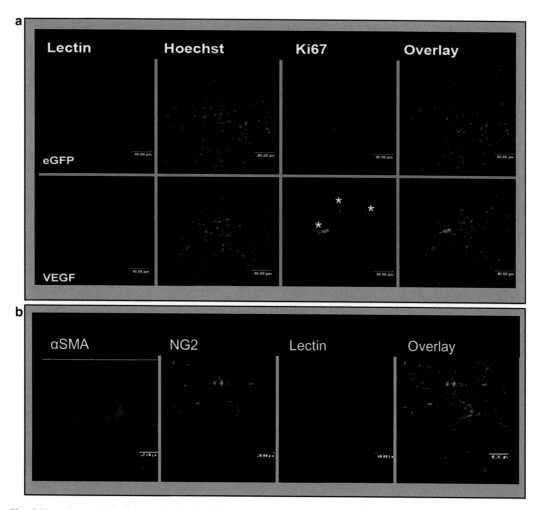

Fig. 3 Vascular endothelial growth factor-C treated mesenteries stained with IB4 (**a**), Ki67 (**b**) and LYVE 1 (**c**). The overlay (**d**) shows that vascular endothelial growth factor-C induced proliferation of lymphatic vessels as well as blood vessels. Vascular endothelial growth factor-C induced filopodial extensions both at the blind ends of the lymphatic system (**e**), and along the length of the vessel (**f**, *inset*). Proliferation was seen both in the cells at the end of the sprouts and in the cells behind (**g**). Scale bar 40 μm, inset 20 μm

3.3 Microvessel Analysis

1. Use images obtained using intravital microscopy to generate an indirect assessment of information regarding neovessel perfusion. The red blood cells flowing through the microvessel bed contrast with the clear mesothelial layer surrounding the vessels. Consequently, fractional vessel area (FVA) is measured as the area of flowing blood vessels per area of mesenteric tissue.

2. All microvessel analysis is carried out using Openlab 3.1 (Improvison) or ImageJ (NIH). Measure the vessel area using the Wand tool, with a pixel threshold of 32. The FVA is a

measure of the percentage of vessel area per mesenteric panel. The angiogenesis index (AI) is expressed as the following equation:

$$AI\ (\%) = \left[(FVA\ Day\ 7 - FVA\ Day\ 1)/FVA\ Day\ 1\right] * 100$$

3. Using an image analysis program and the acquired z-stack images , make measurements of vessel diameter, vessel length (distance of vessel not broken by a branch or sprout point), vessel number, sprout point number, branch point number, and PEC number (*see* Fig. 2).

4. Take the mean value from five panels is and the density (/mm²) can then be calculated.

5. The fractional pericyte area (FPA) is calculated as the percentage of the vessel covered by pericyte (NG2 positive). Staining for α-smooth muscle actin is used to confirm the presence of vascular smooth muscle cells (vSMCs, Fig. 2b); therefore, the density of vSMC-positive vessels can also calculated as well as the fractional smooth muscle area. *See* Fig. 4 for an illustration of the analyses.

6. Lymphatic vessels can be identified as LYVE-1+ vessels and analyzed as before.

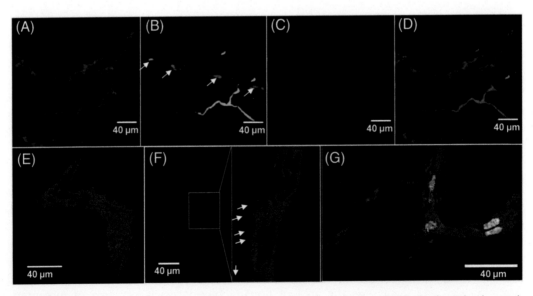

Fig. 4 Measurement of microvascular parameters on stained mesenteries. *Red* isolectin-stained vessels, *green* NG2-stained pericytes. Parameters of width (measured line by w), length (measured line by l), sprouts (sp), branches (bp), and pericyte coverage (area of pericyte/area of endothelial cell). Note that the pericytes can be seen right up to the tips of the sprouts (*grey arrowhead*). Images obtained at 40× magnification using Z-stacked confocal images

4   Conclusion

The mesenteric angiogenesis assay is a highly sensitive, flexible, and subtle assay for investigation of blood vessel growth in vivo in a physiological vascular bed. The experiments are not onerous or technically difficult and do not require special equipment. They do require substantial analysis time, but that results in a very useful analysis of growth characteristics and can be linked to physiological data.

5   Notes

1. Ringer salts can be measured and stored. Add immediately prior to use. Do not autoclave after the addition of glucose containing salts.

2. Heat solution but do not boil. The PFA wont dissolve unless the solution is made basic, through the addition of 1 M NaOH. pH must be corrected after this.

3. Ice-cold methanol can also be used, but is not ideal for in vivo fixation.

4. 24-well plates are ideal for this purpose.

5. Centrifuge antibodies before use. Use 250–500 µL per mesenteric panel.

References

1. Wang WY et al (2004) An adenovirus-mediated gene-transfer model of angiogenesis in rat mesentery. Microcirculation 11(4):361–375

2. Rissanen TT et al (2003) VEGF-D is the strongest angiogenic and lymphangiogenic effector among VEGFs delivered into skeletal muscle via adenoviruses. Circ Res 92(10):1098–1106

3. Vajanto I et al (2002) Evaluation of angiogenesis and side effects in ischemic rabbit hindlimbs after intramuscular injection of adenoviral vectors encoding VEGF and LacZ. J Gene Med 4(4):371–380

4. Visconti RP, Richardson CD, Sato TN (2002) Orchestration of angiogenesis and arteriovenous contribution by angiopoietins and vascular endothelial growth factor (VEGF). Proc Natl Acad Sci U S A 99(12):8219–8224

5. Baluk P et al (2004) Regulated angiogenesis and vascular regression in mice overexpressing vascular endothelial growth factor in airways. Am J Pathol 165(4):1071–1085

6. Detmar M et al (1998) Increased microvascular density and enhanced leukocyte rolling and adhesion in the skin of VEGF transgenic mice. J Invest Dermatol 111(1):1–6

7. Sundberg C et al (2001) Glomeruloid microvascular proliferation follows adenoviral vascular permeability factor/vascular endothelial growth factor-164 gene delivery. Am J Pathol 158(3):1145–1160

8. Thurston G et al (2000) Angiopoietin-1 protects the adult vasculature against plasma leakage. Nat Med 6(4):460–463

9. Cursiefen C et al (2004) VEGF-A stimulates lymphangiogenesis and hemangiogenesis in inflammatory neovascularization via macrophage recruitment. J Clin Invest 113(7): 1040–1050

10. Ziche M et al (1997) Nitric oxide synthase lies downstream from vascular endothelial growth factor-induced but not basic fibroblast growth factor-induced angiogenesis. J Clin Invest 99(11):2625–2634

11. Glass CA, Harper SJ, Bates DO (2006) The anti-angiogenic VEGF isoform VEGF165b transiently increases hydraulic conductivity,

probably through VEGF receptor 1 in vivo. J Physiol 572(Pt 1):243–257

12. Bates DO (1998) The chronic effect of vascular endothelial growth factor on individually perfused frog mesenteric microvessels. J Physiol 513(Pt 1):225–233

13. Dietrich HH (1989) Effect of locally applied epinephrine and norepinephrine on blood flow and diameter in capillaries of rat mesentery. Microvasc Res 38(2):125–135

14. Takano H et al (2004) Spreading dilatation in rat mesenteric arteries associated with calcium-independent endothelial cell hyperpolarization. J Physiol 556(Pt 3):887–903

15. Benest AV et al (2006) VEGF and angiopoietin-1 stimulate different angiogenic phenotypes that combine to enhance functional neovascularization in adult tissue. Microcirculation 13(6):423–437

16. Cristofaro B et al (2010) Neurotrophin-3 is a novel angiogenic factor capable of therapeutic neovascularization in a mouse model of limb ischemia. Arterioscler Thromb Vasc Biol 30(6):1143–1150

17. Benest AV et al (2008) VEGF-C induced angiogenesis preferentially occurs at a distance from lymphangiogenesis. Cardiovasc Res 78(2):315–323

18. Stone OA et al (2009) Critical role of tissue kallikrein in vessel formation and maturation: implications for therapeutic revascularization. Arterioscler Thromb Vasc Biol 29(5):657–664

19. Benest AV et al (2008) Arteriolar genesis and angiogenesis induced by endothelial nitric oxide synthase overexpression results in a mature vasculature. Arterioscler Thromb Vasc Biol 28(8):1462–1468

20. Schreihofer AM, Hair CD, Stepp DW (2005) Reduced plasma volume and mesenteric vascular reactivity in obese Zucker rats. Am J Physiol Regul Integr Comp Physiol 288(1): R253–R261

21. Sweat RS, Stapor PC, Murfee WL (2012) Relationships between lymphangiogenesis and angiogenesis during inflammation in rat mesentery microvascular networks. Lymphat Res Biol 10(4):198–207

Chapter 25

In Vivo Models of Muscle Angiogenesis

Stuart Egginton

Abstract

Angiogenesis is an important determinant of tissue function, from delivery of oxygen and other substrates to removal of waste products, in health and disease (e.g., adaptive or pathological remodelling). The phenotype and functional responses of endothelial cells are conditioned by systemic humoral signals and local environmental factors, including the haemodynamic forces that act upon them. Here we describe some interventions that have been helpful in unraveling the integrative nature of the complex in vivo response, and quantitative assessment of angiogenesis in muscle.

Key words Capillary growth, Skeletal muscle, Rats, Mice, Shear stress, Muscle strain, Electrical stimulation, VEGF

1 Introduction

In principle, endothelial cell (EC) cultures can be used to measure responses relevant to angiogenesis (e.g., migration and proliferation), to determine gene expression in response to putative signals, and to quantify intracellular, surface or secreted proteins (e.g., adhesion receptors, cytokines, proteases). In the circulation ECs are continually exposed to shear stress (luminal viscous drag applied by flowing blood), to compression by blood pressure (pulsatile distortion), and to tension from strain in the extracellular matrix (e.g., cyclic deformation during muscle activity). Intracellular signalling pathways and gene expression respond to such changes and modify EC functions including proliferation, apoptosis, adhesion, motility, and matrix deposition or degradation [1]. However, modifications of angiogenic pathways have not been systematically analysed, although responses to cytokines or growth factors will be modified if applied to EC in the presence of mechanical stress rather than to undisturbed cells [3]. In contrast to the majority of studies on chemical factors in pathological angiogenesis and in vitro models, the physical environment is known to play an important role in controlling capillary growth in normal cardiac and skeletal muscle

Stewart G. Martin and Peter W. Hewett (eds.), *Angiogenesis Protocols*, Methods in Molecular Biology, vol. 1430, DOI 10.1007/978-1-4939-3628-1_25, © Springer Science+Business Media New York 2016

in vivo [2, 4]. The control of in vivo angiogenesis is complex and a reductionist approach to dissect potential mechanisms is tempting, but although in vitro models are often described as 'angiogenesis assays' they can at best only partially represent the phenomenon. For example, use of Matrigel as a substrate may be better than flat monocultures but the subsequent tube formation is not a unique property of endothelium, and the mechanism of lumen formation is different to that found in tissues [2]. While there has been considerable progress in this approach, e.g., flow-based studies are a more physiologically relevant modelling of angiogenesis, there is a long way to go before in vitro studies can reproduce the situation in vivo [3]. In addition, each step of the angiogenic cascade described in reviews based primarily on tissue culture experiments is known to have exceptions in observed capillary growth [4].

In vivo, ECs continually experience haemodynamic forces: shear stress due to flowing blood, compressive pressure, and circumferential tension from the cardiac cycle, a pressure wave travelling along the wall. In addition to circulating humoral factors and local metabolites, in striated muscle capillaries are also subjected to stretch and compression during a duty cycle. ECs are sensitive to the shear stress applied [5] with rapid responses, e.g., production of reactive oxygen species and/or nitric oxide, and prolonged adaptations involving changes in gene expression. The physical environment is therefore recognized as an important modifier of a range of endothelial functions including proliferation, apoptosis, motility, adhesion, and matrix deposition or degradation; changes in which may directly induce angiogenesis. In animal models, skeletal muscle capillaries exposed to hyperaemia and hence increased wall shear stress, or to a sustained strain, showed similar angiogenic responses [6, 7]. Metabolic consequences of systemic hypoxaemia or local hypoxia can stimulate capillary growth, but is dependent on regional phenotype and/or requires an additional stimulus such as muscle activity [8, 9]. The responses are reliant on key chemical mediators, both pro- and anti-angiogenic growth factors, but with different time courses [10, 11].

Angiogenesis is a complex, multifactorial process that is regulated in different ways among vascular beds and according to the stimulus, often involving proliferation and migration of EC to form new capillaries from pre-existing vessels. Angiogenesis is essential for development of the vasculature and in wound repair, and is highly regulated allowing capillary growth when required or capillary rarefaction when not. Many diseases are associated with excessive angiogenesis (e.g., rheumatoid arthritis, diabetic retinopathy, gastric ulcers), or insufficient angiogenesis (e.g., cardiac hypertrophy, peripheral vascular disease), making it a major target for therapeutic intervention [12, 13]. However, there are relatively few data on angiogenesis in normal adult organs where, apart from the female reproductive tract, capillary growth is very rare [14].

Irrespective of the cause of angiogenesis, it is assumed there is a common sequence of events initiated by proteolytic breakage of the basement membrane surrounding a capillary, followed by migration of normally quiescent EC into the interstitium, accompanied by modification of the extracellular matrix and mitosis at the base of the column of migrating EC. These abluminal sprouts elongate and develop lumen, eventually fusing with existing vessels or other sprouts to create functional anastomoses. However, despite repetition in most reviews, this paradigm is not universal [15, 16].

Endothelial cell proliferation and/or migration seen in vitro are not necessarily synonymous with angiogenesis in vivo, and care in interpretation is needed if the ability to isolate putative mechanisms is not to be achieved at the expense of contextual complexity. For example, different patterns of angiogenesis may occur in vivo when the physical environment, both inside and outside vessels, changes. During development, angiogenesis by intussusceptive growth has been described, where extracellular material penetrates and divides vessels [17], perhaps driven as much by growth and remodelling of surrounding tissue as by endothelial cell-specific stimuli. Muscle activity represents a complex mix of cellular, chemical, and physical influences on angiogenesis. However, some reductionism is possible: ECs are sensitive to mechanical strain imposed by the surrounding tissue, and stretch of skeletal muscle during the early phases of overload provides a uniquely sprouting phenotype, whereas intravascular mechanical stimuli such as high shear stress as a result of hyperaemia produce a unique phenotype of longitudinal splitting [4]; the latter two forms of capillary growth exhibit both common and differential signalling responses [10, 18], with the distinct growth patterns involving perivascular cells in different ways [16]. In this chapter three in vivo models of muscle angiogenesis are described, along with techniques to assess various parameters.

Animal models of exercise usually involve treadmill or wheel running, with apparent differences in outcomes between approaches and species. For example, control of duration/intensity is difficult to normalise using running wheels, whereas the treadmill is imposed rather than voluntary so may include a stress component. Angiogenesis appears to be easier to induce in rats than mice on treadmill training, but direct comparisons are difficult due to likely different workloads imposed. Here an alternative method of muscle activity, that of indirect electrical stimulation by implanting electrodes in the vicinity of a motor nerve, is described [19, 20]. This allows more targeted recruitment of muscle and better-defined recruitment parameters.

Endothelial sprouting can also be induced by unilateral muscle overload [7]. This procedure involves the removal (surgical ablation, or extirpation) of the majority of the *m. tibialis anterior* (TA) from the hindlimb of the animal, which leads to a compensatory overload of the synergist *m. extensor digitorum longus* (EDL) and *m. extensor hallucis proprius* (EHP). The consequence is stretch of the EDL leading to an increase in sarcomere length of ~20 % [21]. Longitudinal splitting angiogenesis can be induced by oral ad libitum administration of prazosin [6]. Prazosin is an α_1 adrenoreceptor antagonist and acts as a vasodilator by blocking tonic sympathetic tonus on the resistance vessels, predominantly within skeletal muscle, mainly affecting the arteriolar side of the vasculature. It has little or no effect on capillary diameter and therefore increases blood flow and shear stress in capillaries [14].

2 Materials

2.1 Animals

All work must be performed in accordance with local and national policies and guidelines governing animal work.

1. Animals should be housed at ~21 °C with a 12 h:12 h light:dark cycle, kept with littermates in an enriched environment, and given standard laboratory feed pellets and tap water ad libitum. *Mice*: Male C57Bl/6 or C57Bl/10 mice should be 6–8 weeks old, and weigh ~25 g for optimal results (*see* **Note 1**).

2. *Rats*: Male Sprague-Dawley or Wistar rats should be used when they are 200–250 g (*see* **Notes 1** and **2**).

2.2 Surgery

1. Isoflurane: For induction of anesthesia use 5 % Isofluorane (Novartis) delivered at 5 L/min O_2 flow rate, and maintain with 2.5 % at 2 L/min O_2 (*see* **Note 3**).

2. Buprenorphine analgesic: 2.5 mL/kg buprenorphine (Temgesic; National Veterinary Services) delivered subcutaneously.

3. Duplocillin LA antibiotic: Dilute Duplocillin LA antibiotic (NVS) 1:10 with sterile water.

4. Scalpels: No. 10 blades for surgical procedures.

5. Monofilament sutures: Use 6-0 Vicryl (Ethicon) for mice; 4-0 Vicryl for rats (*see* **Note 4**).

6. Teflon-coated multi-stranded stainless steel wires (e.g., Clark or AM Systems).

7. Neurotech Multichannel Stimulator (Bio-Medical Research).

8. Food-safe silicone sealant (e.g., Dow Corning 744).

2.3 Vasodilator Treatment

1. Prazosin solution: Dissolve 50 mg/L Prazosin (*see* **Note 5**) in tap water, and gently warm while stirring overnight (solubility is very low). Add a small amount (suggested 1–2 g/L) of sugar to the solution to mask the bitter taste.

2.4 Tissue Sampling

1. Liquid nitrogen.

2. 2-Methylbutane (isopentane).

3. Optimal Cutting Temperature embedding matrix (OCT).

4. Cork discs, 20 mm diameter.

5. Poly-L-lysine-coated slides (e.g., polysine™; VWR International).

6. Cryostat microtome.

2.5 Biochemical Assays

1. Lysis buffer: 4-(2-hydroxyethyl)-1-piperazineethanesulfonic acid (HEPES) buffer (975 μL), containing 10 μL Protease Inhibitor Cocktail (Sigma), 10 μL phenylmethylsulfonyl fluoride (PMSF), 5 μL sodium vanadate solution.

2. ELISA kits and reagents as required, e.g., mouse and human VEGF protein and PGF1α (Cambridge Biosciences, UK).

3. Bradford or similar protein assay reagents.

2.6 Alkaline Phosphatase

1. Stock buffer: Add 1.7 g magnesium sulfate ($MgSO_4 \cdot 7H_2O$) and 3.8 g sodium borate ($NaBO_2 \cdot 4H_2O$) to distilled water and make up to a total volume of 1 L. Store at 4 °C (*see* **Note 6**).

2. Working buffer: Add 7–10 mg nitroblue tetrazolium salt (NBT), 2 mg 5-bromo-4-chloro-3-indoxyl phosphate, toluidine salt (BCIP) to the stock buffer and make up to 10 mL. Sonicate to dissolve and adjust to pH 9.3 with boric acid (3.09 g/L).

2.7 Lectin Staining

1. Paraffin wax pen.

2. Electron microscopy grade 4 % (w/v) formaldehyde (e.g., TAAB, Berkshire, UK).

3. Lectin solution: *Griffonia simplicifolia* Lectin I labelled with 5 μL/mL fluorescein- or rhodamine-labelled *Griffonia simplicifolia* Lectin I.

4. Phosphate buffered saline (PBS): 137 mM NaCl, 2.7 mM KCl, 10 mM, Na_2HPO_4, 2 mM KH_2PO_4, and adjust pH to 7.4 with HCl.

5. Fluorochrome-safe mountant (Vectashield; Vector Labs, UK) with and without 4′,6-diamidino-2-phenylindole (DAPI).

2.8 α-Smooth Muscle Actin Staining

1. Formalin fixative: 10 % buffered neutral formalin solution. Add 100 mL of formaldehyde solution (37–40 %), to 900 mL distilled water containing 4.0 g NaH_2PO_4, 6.5 g Na_2HPO_4 (anhydrous).

2. Tris-Buffered Saline (TBS): Prepare 10× stock TBS solution containing 0.5 M Tris Base and 9 % (w/v) NaCl at pH 7.6. Store at room temperature. Dilute the stock 1:10 with distilled water before use and adjust pH to 7.6 if necessary.

3. Blocking solution: Add 10 % (v/v) normal goat serum to TBS containing 1 % (w/v) bovine serum albumin (BSA). Three milliliters is sufficient for ten slides.

4. Anti-alpha smooth muscle actin antibody (e.g., rabbit monoclonal, Abcam): Dilute to 1:500 in TBS with 1 % BSA—6 μL primary antibody solution in 3 mL TBS with 1 % BSA is enough for ten slides.

5. Horse radish peroxidase (HRP) labelled secondary antibody (e.g., goat anti-rabbit; Abcam); diluted to 1:1000 on the slides in TBS with 1 % BSA.

6. 3,3-Diaminobenzidine (DAB): Add two DAB tablets (10 mg) to 5 mL of water, vortex and filter.

2.9 BrdU Pulse Labelling

1. In Situ Cell Proliferation Kit, FLUOS (Roche, Mannheim, Germany).

2.10 Proliferating Cell Nuclear Antigen (PCNA) Staining

1. Formalin fixative (*see* Subheading 2.8, **item 1**).

2. Washing buffer: Add 500 μL of 6 % (v/v) Triton X-100, 666 μL 7.5 % (w/v) BSA in water to 3.83 mL PBS.

3. Rabbit anti-PCNA antibody (Santa Cruz); 1:100 (5 μL antibody in 500 μL washing buffer).

4. Donkey anti-rabbit antibody: CY2 conjugated Donkey anti-rabbit antibody (Jackson ImmunoResearch Laboratories) dilute 1:50 with rhodamine conjugated GSL-1 lectin (1:100; Vector) (10 μL antibody, 5 μL lectin in 500 μL washing buffer).

5. Aquamount/glycerol jelly.

2.11 TUNEL Staining

1. Formalin fixative (*see* Subheading 2.8, **item 1**).

2. Proteinase K solution: Add 1 μL TUNEL staining kit solution (Trevigen 2 TdT-DAB in situ apoptosis detection kit—Trevigen Inc., Gaithersburg, MD, USA) in 50 μL of water per slide.

3. Quenching solution: Add 10 mL of 30 % hydrogen peroxide solution to 90 mL of methanol.

4. Labelling reaction mix: 1 μL TdT-dNTP kit solution, 1 μL Mg^{2+} kit solution, 1 μL TdT enzyme kit solution, 50 μL TdT labeling buffer (kit solution diluted 1:9 in dH_2O).

5. DAB solution: Add 50 mL PBS, 250 μL DAB stock, 50 μL 30 % hydrogen peroxide.

6. Nuclear counter stain: Add 0.5 g of methyl green (ethyl violet free, Sigma) to 100 mL 0.1 M sodium acetate buffer, pH 4.2.

7. Histomount mounting solution.

2.12 Histological Quantification	1. Standard light microscope system equipped with a digital camera (*see* **Note 7**).
	2. Image analysis software (*see* **Note 8**).

2.13 Fibre Type Composition

1. Acid preincubation buffer: 0.2 M sodium acetate adjusted to pH 4.5 containing 50 % (v/v) acetic acid.

2. Alkali preincubation buffer: Dissolve 3.445 g sodium metaborate and 2.36 g calcium nitrate $(Ca(NO_3)_2)$ in 500 mL of distilled water and adjust the pH to 10.3 using 0.1 M sodium hydroxide.

3. Glycine/calcium chloride buffer: Add 3 g glycine and 2.94 g calcium chloride in distilled water, adjust the pH to 9.4 using 0.1 M potassium hydroxide and make up to 1 L.

4. Incubation buffer: Add 160 mg of adenosine 5-triphosphate (ATP) to 100 mL of glycine/calcium chloride buffer and adjust to pH 9.4 with 0.1 M KOH.

5. 1 % cobalt chloride (or nitrate) and 1 % ammonium sulphide in distilled water.

3 Methods

3.1 Muscle Stimulation

1. Surgery should be conducted under aseptic conditions with the use of inhalation anesthetic (*see* **Note 3**). Five to seven animals are required in each treatment groups. *Control groups*: Animals without any intervention are used as controls. The untreated contralateral limbs of animals receiving surgery may also serve as controls (*see* **Note 9**). Sham control animals should have the electrodes implanted, but not receive stimulation (*see* **Note 10**).

2. In the test group of animals implant seven-stranded stainless steel wires close to the lateral popliteal nerve to indirectly stimulate ankle flexors. Shave the site prior to making the incision (*see* **Note 11**).

3. Strip the insulation from the end of the electrodes and wrap them around a 25 G needle to produce a coil, which aids intermuscle location (located deep between the vastus lateralis and the tibialis anterior, which requires minimal dissection of covering muscle), and fix in place with sutures.

4. Feed the electrodes under the skin onto the back of the animals and exteriorise through a piece of Velcro® attached to the skin at the nape of the neck. Attach the electrodes to an external stimulator. When not in use the electrodes should be coiled up inside a Velcro® pocket to prevent them being damaged by the animals.

5. The animals should be recovered within 20–30 min of anesthetic removal and given systemic analgesic and topical antibiotic twice a day for the first 2 days post-surgery.

6. Stimulate the extensor digitorum longus and tibialis anterior muscles for 2–28 days, starting ~24 h after surgery at 10 Hz, pulse width 0.3 ms, and sufficient voltage (3–6 V) to produce palpable contractions using an external stimulator [19].

7. Alternatively, in-house constructed miniaturised implantable stimulators may be used when located subcutaneously on the back of the animal. Home Office guidelines suggest these should be <15% of the animals body mass (easy to achieve <10% with rats, less so with mice due to battery size). These are either encased in heat-shrink plastic and sealed with silicone sealant, or coated with hypo-allergenic beeswax. The stimulator is operated through a superficial magnetic reed-switch allowing it to be activated 24 h following surgery. A similar approach has been used by Greene and colleagues [20]. The limitation for their use is battery weight versus life; lithium coin batteries give ~1 week life and allow stimulators to be implanted at ~10 % body weight in mice.

3.2 Muscle Overload

1. Anaesthetise animals using isoflurane delivered in oxygen to maintain blood oxygen saturation during cardiorespiratory depression, and maintain anaesthesia using a lower dose throughout the duration of the procedure (*see* Subheading 2.2, **item 1**).

2. Systemic buprenorphine analgesic 2.5 mL/kg is administered subcutaneously, usually by scruffing the skin at the neck (*see* Subheading 2.2, **item 2**).

3. Shave the hindlimb (*see* **Note 11**) and disinfect the site with 70 % ethanol spray. After a few seconds wipe away alcohol with a sterile cotton bud or tissue; surface evaporation causes cutaneous vasoconstriction that minimises bleeding on incision.

4. Locate the TA tendon, usually visible under the skin at the instep. It can usually be found by feeling for the femur with the back of the scalpel blade, and running down the anterolateral surface. The outline of the TA can be seen as a slight depression on leg extension.

5. Using moderate pressure, with the scalpel make an incision from the tendon to about midway up the TA. Make a single incision of 1–2 cm through the skin. Free the skin and carefully clear away any fascia overlaying the muscle with blunt dissection using strabismus scissors.

6. Free up the tendon from surrounding connective tissue (ensure the surface is glistening white), being careful to avoid

damage to any blood vessels in close proximity (the ankle is well vascularised).

7. Slide a pair of curved forceps under the tendon and cut with scissors or scalpel, leaving a generous portion still attached to the muscle (*see* **Note 12**).

8. Holding the cut tendon, lift the muscle free from the underlying tissue, blunt dissect the inner surface from the underlying EDL, and free the lateral edges. With one clean cut slice through the muscle with a scalpel 2/3–3/4 along its length.

9. Replace the cut end of the muscle and apply pressure to the site for a few moments in order to stop any bleeding.

10. Apply one drop of topical antibiotic, leaving for about 10 s before blotting away excess.

11. Using monofilament sutures close the wound with 4–5 interrupted stitches (*see* **Note 13**). Clean the wound carefully, as any dried blood invites the animal to nibble at sutures; an anaesthetic cream may dampen irritation but is not usually necessary.

12. Allow the animal to recover in a warmed cage, with jelly feed to aid rehydration. Check on them regularly—the only issue noted is occasional biting of sutures that require repair within the first 12 h. Thereafter they feed and drink normally, with minor gait impairment lasting 36–48 h. Normal climbing activity returns earlier.

3.3 Chronic Vasodilatation

1. Replace normal drinking water of animals with the prazosin solution (*see* **Note 14**) and allow access ad libitum (*see* **Note 15**).

2. Replace with fresh Prazosin solution every 2–3 days, as the drug will precipitate (evident as a cloudy appearance of the water).

3.4 Tissue Sampling

1. For immunohistochemistry and protein and mRNA quantification kill the animals by stunning and cervical dislocation.

2. Immediately dissect the muscles and divide into two. Process as described below for biochemical and histological analysis in **items 3** and **4** respectively. For the EDL aim to sample from the central half of the muscle.

3. Snap-freeze one half of the muscle in liquid nitrogen for biochemical analysis.

4. Coat the other half of muscle sample with OCT for structural support and cryoprotection. Orientate the muscle perpendicularly on a cork disc by leaning the muscle against a pin or soft tissue (e.g., piece of liver). Carefully invert and drop the disc into liquid nitrogen-cooled isopentane bath (*see* **Note 16**).

5. Keep discs in liquid nitrogen for short-term storage before sectioning; for long-term storage wrap in aluminum foil packets to aid sorting and place at –80 °C.

6. Cut ~10 μm sections (8 μm is better, 12 μm is acceptable) on a cryostat microtome at –20 °C, depending on the rigidity of the sections obtained this may be varied by ±2 °C only.

7. Pick up sections on poly-L-lysine-coated slides for better adhesion during subsequent immunohistochemistry and allow to air-dry for 30–60 min before staining or re-freezing and storage at –20 °C (*see* **Note 17**).

3.5 Protein Quantification

1. Powder frozen tissue using a chilled pestle and mortar under liquid nitrogen and suspended in ice-cold lysis buffer.

2. Centrifuge at $5000 \times g$ for 20 min at 4 °C.

3. Assay samples for the protein of interest in duplicate by ELISA, according to manufacturer's instructions (5–6 animals/group).

4. Determine total tissue protein concentration by Bradford assay performed in duplicate at the same time as the ELISA.

3.6 Alkaline Phosphatase Staining

Alkaline phosphatase is an enzyme found at high levels in capillary endothelium, but may also stain terminal arterioles and motor nerves; it is not suitable for all tissues or species (e.g., not found in human muscle, lower staining in mouse muscle). This staining protocol uses the phosphatase activity to cleave BCIP, which becomes a strong reducing agent, then acts on NBT to form an insoluble dye which deposits on the capillary wall.

1. Air-dry cryostat sections (30 min if fresh or taken from freezer).

2. Fix sections in precooled acetone (4 °C or less) for 5 s.

3. Air dry (5–10 min).

4. Stain the slides with working buffer for 45 min to 1 h at room temperature.

5. Wash in distilled water.

6. Mount with aquamount and a cover slip. Seal with glue or nail varnish if sections are to be kept for long periods.

3.7 Lectin Staining

Lectins are carbohydrate-binding proteins from plant extracts; those isolated from *Griffonia* (*Bandeiraea*) *simplicifolia* seeds (GSL-1/BSL-1) identify ECs in mouse and rat tissues. They bind to galactose residues, which are found at high concentration on the proteoglycans in the glycocalyx lining blood vessels and laminin glycoproteins, which are major components of the basal lamina that surround the vasculature. The choice of which lectin to use is species-dependent [22], and other variables, e.g., use *Ulex europaeus* lectin, UEA-I, for humans but staining may be influenced by blood group.

1. Air-dry cryostat sections fully (30 min if fresh or taken from freezer).

2. Draw around sections with a paraffin wax pen, and allow to dry completely (~15 min). This produces a shallow well for incubating tissue, and helps to reduce problems with evaporation while minimizing incubation volume.

3. Fix in ice-cold acetone (4 °C) for 15 s. Alternatively, fix with formaldehyde, followed by 3×5 min PBS washes.

4. Incubate for 1 h at room temperature using fluorescein- or rhodamine-labelled lectin.

5. Wash 3×5 min in PBS.

6. Wash in distilled water.

7. Mount using 1:10 Vectashield plus DAPI:Vectashield without DAPI (to normalise fluorescent intensity of different structures).

3.8 α-Smooth Muscle Actin

During vascular remodelling, haemodynamic control needs to be preserved and so estimating the size of microvascular units (number of capillaries fed by individual arterioles) is an important readout. The process of arteriolisation has been shown to accompany, or even precede angiogenesis. Alpha actins are used to label vascular smooth muscle cells, and identify arterioles and venules on the basis of size and layer thickness.

1. *Day 1*: Fix in formalin fixative for 10 min.

2. Wash the slides 4×5 min in TBS + 0.025 % Triton X-100 with gentle agitation.

3. Incubate in blocking solution for 2 h.

4. Drain slides for a few seconds.

5. Incubate with antibody to α-smooth muscle actin overnight at 4 °C.

6. *Day 2*: Wash 2×5 min in TBS + 0.025 % Triton X-100 with gentle agitation.

7. Wash 2×5 min TBS + 0.025 % Triton X-100 with gentle agitation.

8. Incubate the slides in 0.3 % hydrogen peroxide in TBS for 15 min.

9. Wash 2×5 min TBS + 0.025 % Triton X-100 with gentle agitation.

10. Apply HRP secondary antibody and incubate for 1 h at RT.

11. Wash 3×5 min TBS.

12. Develop with DAB for 10 min at RT.

13. Wash in running tap water for 5 min.

14. Mount in glycerol.

3.9 Bromode-oxyuridine (BrdU) Pulse Labelling

Understanding the relationship between cellular proliferation and angiogenesis is important; e.g., the longitudinal splitting form is efficient in that it requires little mitotic activity, while proliferative activity in perivascular or mural cells gives insight into local regulatory mechanisms. Incorporation of an identifiable nucleoside during mitosis has long been used to label the site of proliferation, with BrdU (a synthetic thymidine analog) often preferred. Co-localisation of BrdU with capillaries can provide a measure of cell proliferation associated with the capillaries (though not necessarily ECs), and that associated with interstitial cells or myonuclei.

1. Animals are injected i.p. with BrdU labeling reagent 16 h prior to tissue sampling. Other time points can be used to assess the kinetics of cell turnover.

2. Cryosections are obtained as above (*see* Subheading 3.4).

3. Immunostaining is also performed using primary anti-BrdU antibodies as described above (*see* Subheading 3.8) with the addition of 1:100 rhodamine *Griffonia simplicifolia* lectin-1 to allow co-localisation of BrdU-labelled cells with vascular structures.

4. Slides are mounted using 1:10 Vectashield + DAPI:Vectashield to permit confirmation of nuclear localisation of BrdU labelling.

3.10 PCNA Staining

For archived material, an alternative approach to BrdU labeling is to retrospectively stain for factors upregulated during the cell cycle. PCNA (a.k.a. cyclin) is a DNA polymerase co-factor expressed by cells undergoing mitosis, and so can act as a measure of cell proliferation.

Air-dry cryostat sections (30 min).

1. Fix in formalin fixative for 5 min.

2. Wash in PBS for 3 × 5 min.

3. Block with washing buffer at RT for 30 min.

4. Incubate with primary antibody for 1 h (usually 1 % (*see* Subheading 2.10, **item** 3).

5. Wash in PBS 3 × 5 min.

6. Incubate with secondary antibody (usually 1/200) and lectin for 1 h. Keep sections covered and out of direct light.

7. Wash in PBS 3 × 5 min.

8. Wash in distilled water.

9. Mount with Aquamount/glycerol jelly if required.

3.11 TUNEL Staining

1. An important balance to angiogenesis during remodelling is vascular pruning. During apoptosis, endonucleases digest DNA to form short double-stranded DNA fragments that can

be detected by attaching biotinylated nucleotides to the 3′-OH ends using a terminal deoxynucleotidyl transferase (TdT) enzyme, visualised using a streptavidin-horseradish peroxidase conjugate followed by the substrate DAB which forms a brown precipitate when broken down by peroxidase. Apoptotic cells must be distinguished from necrotic cells morphologically. Air-dry cryostat sections fully (30 min if fresh or taken from freezer).

2. Draw round sections with a paraffin wax pen, and allow to dry (~15 min).

3. Fix for 10 min in formalin fixative.

4. Wash slides for 2×5 min in PBS.

5. Blot slides dry, then immediately add 50 μL of proteinase K solution and cover.

6. Incubate for 15 min and then wash 2×2 min in dH$_2$O.

7. Place slides into 50 mL of quenching solution for 5 min.

8. Wash 2×1 min in PBS.

9. Incubate slides in 50 mL TdT labeling buffer for 5 min.

10. Blot slides dry, then immediately add 50 μL labelling reaction, mix and cover.

11. Incubate slides in a 37 °C incubator in a humidified chamber for 2 h.

12. Transfer slides into 50 mL TdT stop buffer (kit solution diluted 1:9 in dH$_2$O) and incubate for 5 min.

13. Wash slides 2×2 min in PBS.

14. Blot slides dry and then add 50 μL streptavidin-HRP detection solution onto each sample (1 μL kit solution in 50 μL dH$_2$O per slide) and cover. Incubate at RT for 10 min.

15. Wash slides 2×2 min in PBS.

16. Incubate slides in DAB solution for 5 min.

17. Wash slides in dH$_2$O and transfer to fresh water.

18. Counterstain nuclei by transferring slides to 50 mL methyl green solution for 5 min.

19. Wash slides by sequentially dipping in 1-butanol until the sample turns from blue to mainly green, dip once more in fresh 1-butanol, then dip ten more times in two changes of xylene.

20. Dry xylene from the back of the slide. Leave xylene on the surface of the slide to aid the mounting process.

21. Place two drops of Histomount on the samples and gently lower the coverslip.

22. Leave to harden overnight. Store out of direct light.

3.12 Quantification of Capillary Supply

1. Once capillary visualisation is complete, three to four images per section are photographed and analysed (*see* **Note 18**).

2. Photomicrographs can be taken using ×10 or ×20 objectives, depending on the field of view required (e.g., 0.14–0.20 mm^2); ×40 is better for co-localisation studies.

3. Some pre- or post-processing may be required to compensate for the dynamic range of capillary staining intensities, or to adequately visualise fibre boundaries (try equalisation or logarithmic filters).

4. For each image place one to four sampling squares (regions of interest, ROI) on the field of view (Fig. 1), and count the number of capillaries and fibers using an unbiased counting rule (Fig. 2).

5. The top and left sides of the sampling square are designated 'inclusion lines' with the bottom and right sides designated 'exclusion lines.' Any capillary or fibre exclusively touching a line of inclusion is counted in the analysis; any touching a line of exclusion is excluded from the analysis.

6. It is best if the counting frame is systematically oriented on the same circumscribed regions, relative to the major axis and the section boundaries, to provide a systematic-random sampling. This is statistically preferable to simple random sampling, which will increase data variance and tend to obscure any localised response of specific muscle regions. Location is especially important in the outer cortex of the TA muscle, where apparent capillarisation/fibre composition is very sensitive to the sample position.

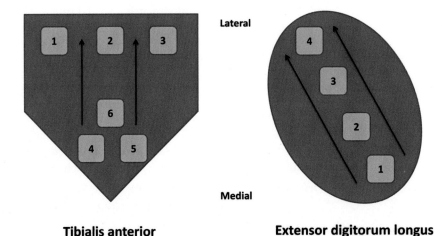

Tibialis anterior   **Extensor digitorum longus**

Fig. 1 Sampling ROI for individual muscle cross sections. *Arrows* show the gradation in oxidative capacity, decreasing in the medial-lateral axis (*see* Ref. 8)

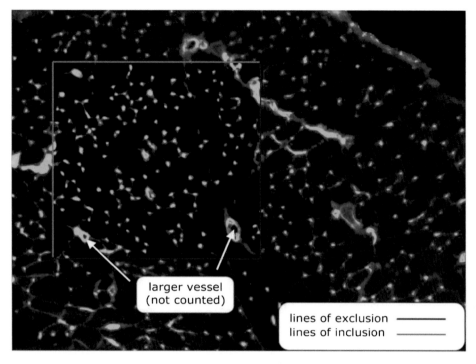

larger vessel
(not counted)

lines of exclusion ———————
lines of inclusion ———————

Fig. 2 Image showing a cross section of muscle stained with FITC-labelled *Griffonia simplicifolia* (GSL-1) illustrating the protocol for unbiased sampling for estimating capillary supply in skeletal muscle

7. A fibre is defined as an isolated nonstained, convex area surrounded by collagen IV or laminin.

8. A capillary is defined as an ALP/lectin/CD31-positive structure <8 µm in diameter. They are seen as either a circle in cross section or elongated structures when sectioned obliquely, in which case individual branches are counted as single capillaries.

9. The size of the ROI is important in limiting 'edge effect' errors, the most common being too few fibres (ideally >50) or capillaries (>100) included.

10. There are a large number of indices used to describe the adaptive nature of the microcirculation, most of which have severe limitations in their utility (for detailed explanation, *see* Ref. 23).

11. An unbiased counting rule [24] should be used to estimate the capillary density (CD or $N_A(c,f)$ in stereological notation; expressed as number of capillaries per mm^2 of section (ideally fibre area), the capillary to fiber ratio (C:F, $N_N(c,f)$) is the number of capillaries/total number of fibers), and the average fiber cross-sectional area (FCSA; total area of fibres/number of fibres) (*see* **Note 19**).

3.13 Fibre Type Composition

Muscle fibre types may be detected using enzymatic methods, exploiting differential pH stability of different myosin ATPases [25]. For example, in sections preincubated at pH 4.55 dark staining indicates Type I (slow oxidative), light staining indicates Type IIa (fast oxidative glycolytic), and intermediate staining intensity indicates Type IIb (fast glycolytic) fibre types. Intrinsic mATPase activity may be revealed by omitting the preincubation step. It does not easily allow discrimination between Type IIb and Type IId/x/c fibres whose oxidative capacity is between Type IIa and IIb, which may be assessed by staining for succinic dehydrogenase activity (note differences in notation between rodent and human fibre types). This approach is increasingly replaced by monoclonal anti-MHC antibodies allowing for targeted multiple fibre staining (*see* Ref. 26).

1. Preincubate sections: 2–4 min in either acid or alkali preincubation solution.

2. Wash in distilled water (dH_2O).

3. Wash in glycine/$CaCl_2$ buffer.

4. Incubate sections: 15 min in incubation solution at room temperature.

5. Wash three times in dH_2O.

6. Incubate in 2 % $CaCl_2$ three times for 30 s each.

7. Wash three times in dH_2O.

8. Incubate in 1 % $CoCl_2$ for 3 min.

9. Wash five times in dH_2O.

10. Incubate in 1 % ammonium sulphide for 3 min in a fume cupboard.

11. Wash five times in dH_2O.

12. Mount with Aquamount/glycerol jelly and coverslips.

13. The muscle cross-sectional area occupied by the three main fibre types may be quantified using standard stereological point-counting methodology [24]. Type I or Type IIa fibres from fast or mixed muscles are about half the size of Type IIb fibres in adult rats, and consequently their total oxidative capacity is overestimated if only numerical frequency is quantified. In addition, muscle tension development is proportional to FCSA; thus a quantitative estimate of their relative area may be a better index of functional capacity.

3.14 Power Calculations

Based on previously published data, significant differences can be seen in capillary supply indices and most ultrastructural data between two experimental groups using $n = 4$ animals; however if submaximal responses are induced this will require a larger sample size. Our previous data has a median average standard deviation (SD) of 0.12 for C:F which requires $n = 6$ to give 80 % power to means separated by 0.2, or roughly half the response to our mechanical stimuli. The most subtle ultrastructural change we would expect to see is changes in pericyte coverage. Our previously published data has a median average SD of 2.1, which therefore requires $n = 7$ to have 80 % power in a mean difference of 3, again half the response in our previous model. Power calculation for microarray analysis is a rather contentious topic, but based on a power of 85 % and $P < 0.05$, previous experience gives a pooled variance term $S = 0.15$, then using $D/S \times N(2/n)$ gives a requirement for 6–8 animals/experimental group. Data are usually presented as mean $\pm$ SEM, with statistical analysis by one-way ANOVA using PLSD or Bonferroni post-hoc analysis.

4 Notes

1. Males are almost always used in order to avoid interference in vascular remodeling that may occur during the estrous cycle in females.

2. Sprague-Dawley rats are fast growing with relatively aerobic muscles, whereas Wistar rats tend to be smaller at similar age and have slightly less aerobic muscles.

3. We originally used halothane (Fluothane®, ICI), but now use Isoflurane (IsoFlo®) routinely.

4. It is preferable to use absorbable sutures for internally and nonabsorbable for cutaneous sutures.

5. Prazosin was originally a gift from Pfizer. It is now commercially available from Fluka, Tocris, and Sigma-Aldrich.

6. The stock buffer can be stored for several months at 4 °C.

7. Any standard microscope system is adequate for acquisition of photomicrographs. We used a Zeiss Axioskop 2 plus fluorescent microscope, with image capture using an Axiocam MRc digital camera and Axiovision software (Karl Zeiss).

8. Analysis of digital images can be performed using commercial image processing packages, but we routinely use ImageJ (NIH).

9. In animals receiving surgery there is some gait adjustment following surgery that means the untreated contralateral limb muscles are not exactly the same as those from naïve control animals, but serve to normalise inter-animal variation.

10. Sham controls serve to normalize for surgical trauma, but are probably only important for the earliest time points as for example edema resolves within 2–3 days of surgery.

11. Removing the fur avoids allergic reaction to fine hair, and allows clean wound closure with minimal ingress of bacteria etc.

12. Leaving a small stump of the TA may be preferable to complete removal as it can cause less bleeding, surgical trauma, and/or inflammation (the latter consequences may affect angiogenesis), but there is no noticeable difference in extent of final capillary supply.

13. Continuous sutures may unravel if the animal chews them.

14. Prazosin has been the vasodilator drug of choice for our group, although other compounds may have similar effects.

15. Each mouse receives approximately 175 μg per day, based on the average water consumption monitored throughout the experiment. The rat dosage should be scaled according to body mass and drinking rates.

16. Note this avoids prolonged freezing due to volatilisation of liquid nitrogen alone, but the wetting agent must therefore be treated carefully. To avoid freezing artifacts the isopentane should be close to its freezing point, which can be ascertained through an increase in viscosity and clouding whilst stirring, and preferably has solid bits floating in it; for larger tissue samples this is essential.

17. Avoid condensation on removal from the freezer as freezing artifacts can also occur on temperature reversal and re-storage.

18. More sections may be analyzed provided they are adequately separated to avoid the same microvascular units being sampled twice (e.g., 100–200 μm apart), but this rarely provides additional accuracy in the absence of gross structural heterogeneity.

19. I would recommend these three indices as a minimum: CD offers insight into the functional capacity of the microcirculation, C:F is a robust index of angiogenic activity under most conditions, while FCSA is a direct determinant of CD so cannot be ignored. Numerical or areal fiber type composition can offer additional insights (*see* Subheading 3.13).

Acknowledgements

Development of the ideas in this chapter has been supported by the BBSRC, BHF, MRC, and Wellcome Trust; discussions with many colleagues and students have been extremely helpful.

References

1. Lekes PI (1999) Endothelium and mechanical forces. Harwood Academic Publishers, London

2. Egginton S, Gerritsen M (2003) Lumen formation: in vivo versus in vitro observations. Microcirculation 10:45–61

3. Nash GB, Egginton S (2005) Modelling the effects of the haemodynamic environment on endothelial cell responses relevant to angiogenesis, in angiogenesis assays: a critical appraisal of current techniques. In: Staton CA, Lewis C, Bicknell R (eds) Angiogenesis assays—a critical appraisal of current techniques. John Wiley & Sons, London

4. Egginton S, Zhou A-L, Brown MD et al (2001) Unorthodox angiogenesis in skeletal muscle. Cardiovasc Res 49:634–646

5. Barakat A, Lieu D (2003) Differential responsiveness of vascular endothelial cells to different types of fluid mechanical shear stress. Cell Biochem Biophys 38:323–343

6. Zhou A, Egginton S, Hudlicka O et al (1998) Internal division of capillaries in rat skeletal muscle in response to chronic vasodilator treatment with alpha1-antagonist prazosin. Cell Tissue Res 293:293–303

7. Zhou AL, Egginton S, Brown MD et al (1998) Capillary growth in overloaded, hypertrophic adult rat skeletal muscle: an ultrastructural study. Anat Rec 252:49–63

8. Deveci D, Marshall JM, Egginton S (2001) Relationship between capillary angiogenesis, fiber type, and fiber size in chronic systemic hypoxia. Am J Physiol Heart Circ Physiol 281:H241–H252

9. Milkiewicz M, Hudlicka O, Verhaeg J et al (2003) Differential expression of Flk-1 and Flt-1 in rat skeletal muscle in response to chronic ischaemia: favourable effect of muscle activity. Clin Sci (Lond) 105:473–482

10. Williams JL, Weichert A, Zakrzewicz A et al (2006) Differential gene and protein expression in abluminal sprouting and intraluminal splitting forms of angiogenesis. Clin Sci (Lond) 110:587–595

11. Olfert IM, Birot O (2011) Importance of anti-angiogenic factors in the regulation of skeletal muscle angiogenesis. Microcirculation 18:316–330

12. Folkman J (1995) Angiogenesis in cancer, vascular, rheumatoid and other disease. Nat Med 1:27–31

13. Carmeliet P, Jain RK (2011) Molecular mechanisms and clinical applications of angiogenesis. Nature 473:298–307

14. Hudlická O, Brown M, Egginton S (1992) Angiogenesis in skeletal and cardiac muscle. Physiol Rev 72:369–417

15. Egginton S (2009) Invited review: activity-induced angiogenesis. Pflügers Arch 457:963–977

16. Egginton S (2011) Physiological factors influencing capillary growth. Acta Physiol (Oxf) 202:225–239

17. Djonov V, Baum O, Burri PH (2003) Vascular remodeling by intussusceptive angiogenesis. Cell Tissue Res 314:107–117

18. Williams JL, Cartland D, Hussain A et al (2006) A differential role for nitric oxide in two forms of physiological angiogenesis in mouse. J Physiol 570:445–454

19. Hudlicka O, Egginton S, Brown MD et al (1994) Effect of long-term electrical stimulation on vascular supply and fatigue in chronically ischemic muscles. J Appl Physiol 77:1317–1324

20. Linderman JR, Kloehn MR, Greene AS (2000) Development of an implantable muscle stimulator: measurement of stimulated angiogenesis and poststimulus vessel regression. Microcirculation 7:119–128

21. Egginton S, Hulicka O, Brown MD et al (1985) Capillary growth in relation to blood flow and performance in overloaded rat skeletal muscle. J Appl Physiol 85:2025–2032

22. Kirkeby S, Mandel U, Vedtofte P (1993) Identification of capillaries in sections from skeletal muscle by use of lectins and monoclonal antibodies reacting with histo-blood group ABH antigens. Glycoconj J 10:181–188

23. Egginton S (1990) Morphometric analysis of tissue capillary supply. Adv Comp Environ Physiol 6:73–141

24. Egginton S (1990) Numerical and areal density estimates of fibre type composition in a skeletal muscle (rat extensor digitorum longus). J Anat 168:73–80

25. Brooke MH, Kaiser KK (1970) Muscle fiber types: how many and what kind? Arch Neurol 23:369–379

26. Bloemberg D, Quadrilatero J (2012) Rapid determination of myosin heavy chain expression in rat, mouse, and human skeletal muscle using multicolor immunofluorescence analysis. PLoS One 7:e35273

Chapter 26

Use of the Hollow Fiber Assay to Evaluate Agents That Target the Tumor Neovasculature

Patricia A. Cooper and Steven D. Shnyder

Abstract

In vivo preclinical assays are required to screen potential agents that target the tumor vasculature. Here a hollow fiber-based assay for the quantification of neovasculature in the presence or absence of an agent that potentially targets tumor neovasculature is described. The neovasculature is developed as a consequence of the presence of tumor cells encapsulated in hollow fibers, which are transplanted sub-cutaneously in the dorsal flanks of mice.

Key words Hollow fiber assay, Angiogenesis, Vascular-targeting agents, Cancer therapy, Tumor neovasculature, Preclinical screening, CD-31

1 Introduction

The progression of novel agents that target tumor vasculature into the clinic is reliant on demonstration of activity in suitable preclinical assays. With these assays, the analytical endpoint is either the detection of new vessel growth in an area where there is minimal vascularization (e.g., matrigel or sponge assays [1, 2]) or the differentiation of neovasculature from existing vasculature (e.g., dorsal skin chamber [3]). In vivo assays curently used in this field have distinct features and limitations (as reviewed in [4]), and thus it is not ideal when screening to rely on a single assay. Hence, there is a requirement for further development of novel in vivo screening assays.

As the basis for developing such an assay, we have adapted the hollow fiber assay that is currently used as part of the drug-screening programme at the National Cancer Institute (NCI) in the USA [5] for studies on vasculature. In the NCI assay, human tumor cell lines are loaded into biocompatible PVDF hollow fibers (HF) with an internal diameter of 1 mm and a 500 kDa molecular weight exclusion, which allows for free passage of macromolecules and drugs, while restricting passage of cells. The HF are cut into

Stewart G. Martin and Peter W. Hewett (eds.), *Angiogenesis Protocols*, Methods in Molecular Biology, vol. 1430,
DOI 10.1007/978-1-4939-3628-1_26, © Springer Science+Business Media New York 2016

2 cm lengths which are heat-sealed at both ends, and using aseptic surgical procedures, these are then transplanted subcutaneously and intraperitoneally into anesthetized mice. Animals are treated with the test agent for 4 consecutive days from day 3 and the studies terminated on day 7 or 10, with cells then analyzed for viability using the MTT cell viability assay, with the amount of viable cells in HF from the treated animals compared with untreated controls.

The rationale for adapting this assay is from initial studies carried out in this laboratory where assay times were extended beyond the standard 6-day period, and extensive vascular networks were observed surrounding subcutaneously implanted fibers [6].

We have subsequently modified the NCI assay to make it more suitable for investigating anti-angiogenic and other vascular targeting strategies by only transplanting subcutaneously and reducing the size and number of fibers transplanted in order to minimise inflammation surrounding the fibers due to damage inflicted by the fiber seals on the surrounding tissue [7]. We have also incorporated immunohistochemical analysis of the neovascularization surrounding the fibers as an end-point for quantifying agent effect [7, 8].

2   Materials

2.1   Hollow Fiber Preparation and Loading

1. Polyvinylidene fluoride (PVDF) hollow fibers with an internal diameter of 1.0 mm and a molecular weight cut-off point of 500 kDa (Spectrum Medical Inc.) are used. These are supplied dehydrated in 180 mm lengths and in different colors to enable different cell lines to be placed in the same animal.

2. 70 % ethanol solution in distilled water.

3. Sterile 5 ml and 10 ml syringes with 21-gauge blunt needles.

4. 600 mm glass chromatography column or similar.

5. Stainless steel tray (approximate dimensions: 220 mm × 150 mm).

6. Autoclavable stainless steel box and lid (approximate dimensions: 250 mm × 50 mm × 40 mm).

7. Sterile distilled water.

8. Autoclave tape.

9. Autoclavable bags for sterilising box containing fibers and instruments.

10. Tumor cell lines obtained from the European Collection of Cell Cultures or ATCC Cell Biology Collection (see Note 1).

11. Tissue culture medium, e.g., RPMI 1640 supplemented with 10 % or 20 % fetal calf serum (FCS), 2 mM sodium pyruvate, and 2 mM L-glutamine, stored at 4 °C.

12. Cell culture consumables including: 75 cm² culture flasks, 10 ml pipettes, 6-well plates, 30 ml polypropylene tubes.

13. Hanks' Balanced Salt Solution (HBSS), stored at room temperature.

14. 0.25 % Trypsin-EDTA solution, stored at –20 °C.

15. Hemocytometer.

16. Polystyrene box containing ice.

17. "Fireboy plus" rechargeable mobile Bunsen burner with CV360 butane cyclinder.

18. Sterilized demarcated stainless steel tray (approximate dimensions: 450 mm × 140 mm). Demarcations every 1.5 cm.

19. Sterilized instrument tray (approximate dimensions: 220 mm × 150 mm) and instruments including: smooth-jawed needle holders, scissors, and forceps.

2.2 In Vivo Transplantation of Hollow Fibers and Treatment

1. Mice (*see* **Note 2**).

2. Polystyrene box to transport fibers to operating theatre.

3. Anesthesia equipment.

4. Sterile instruments—scissors, forceps × 2, 3 mm diameter trocar.

5. Histoacryl tissue adhesive. Store at 4 °C.

6. Autoclips.

7. Sterile 1 ml syringe plus either a 25-gauge needle for i.p., or a 26-gauge needle for i.v. administration of compound.

2.3 Removal and Processing of Hollow Fibers

1. CO_2 euthanasia system.

2. Dissecting board with pins.

3. Instruments—scissors and forceps.

4. Compact digital camera and stand.

5. Stiff card cut to 22 mm × 22 mm.

6. Processing cassettes, 28 mm × 32 mm.

7. Zinc fixative. Store at room temperature.

8. 50 ml polystyrene sample pots.

9. Graded ethanols—70 %, 90 %, and 100 %.

10. Histoclear.

11. Paraffin wax.

2.4 Assessment of Effects on the Vasculature

1. Xylene.

2. Distilled water.

3. 50 ml Coplin jars.

4. 1 % hydrogen peroxide solution. Made immediately before use by adding 3.33 ml of 30 % hydrogen peroxide solution (stored at 4 °C) to 96.67 ml of distilled water.

5. Phosphate Buffered Saline (PBS), 0.01 M, pH 7.4.

6. Wax pen.

7. Normal rabbit serum (Vector Laboratories Inc.). Made by adding 150 µl of stock solution to 9.95 ml of PBS. Make up fresh before every run-through.

8. Purified rat anti-mouse CD-31 (PECAM-1) monoclonal antibody (BD Pharmingen). Store stock at 4 °C. Dilute 1:100 in normal rabbit serum prior to use.

9. Secondary antibody: Biotinylated rabbit anti-rat IgG (H + L) (Vector Laboratories Inc.). Store stock at 4 °C. Dilute 1:400 in PBS prior to use.

10. ABC-peroxidase reagent kit (Vector Laboratories Inc.). Components provided in dropper bottles, store at 4 °C. Make up 30 min before use by adding 1 drop of "bottle A" plus 1 drop of "bottle B" to 5 ml of PBS, mixing thoroughly between each addition.

11. Peroxidase substrate kit (Vector Laboratories Inc.). Components provided in dropper bottles, store at 4 °C. Make up immediately before use by adding two drops of "buffer, pH 7.5" plus 4 drops of "DAB", plus two drops of "hydrogen peroxide" to 5 ml of distilled water, mixing thoroughly between each addition.

12. Sheet of white paper.

13. 10 % bleach in tap water solution for disposal of DAB solutions.

14. Harris's hematoxylin solution. Store at room temperature.

15. Scott's tap water substitute. Made by dissolving 3.5 g of sodium bicarbonate and 20 g of magnesium sulphate in 1 L of distilled water. Store at room temperature.

16. DPX mountant for microscopy.

17. 22 × 50 mm glass coverslips.

3   Methods

The evaluation of the effects of agents on the neovasculature surrounding hollow fibers loaded with tumor cells comprises of four subsequent procedures: the rehydration and sterilization of the hollow fibers followed by loading of the tumor cells; the transplantation of the fibers into recipient mice and treatment with agents; the removal of the fibers and processing for analysis, and the analysis itself.

3.1 Hollow Fiber Preparation and Loading

1. Ensure that any closed, cut ends of the fiber are opened by applying gentle pressure, and flush each fiber through with 70 % ethanol using the syringe attached to a blunt needle.

2. Place the fibers in a tray containing more 70 % ethanol, and when all fibers have been flushed through, transfer the fibers to the glass chromatography column. Fill with more 70 % ethanol so that the fibers are well-covered and sealed.

3. Store vertically for at least 72 h (*see* **Note 3**).

4. In order to sterilize the fibers, first transfer them to a tray.

5. Place 100 ml of distilled water into the autoclavable box.

6. Flush each fiber through with fresh 70 % ethanol using the syringe and needle.

7. Flush each fiber through with distilled water, and with the syringe and needle still attached, place the fiber in the box filled with distilled water and flush to eliminate trapped air from the fiber. Remove the syringe and needle (*see* **Note 4**).

8. Once the required number of fibers have been rehydrated, put the lid on the box and secure with autoclave tape.

9. Carefully place the box in an autoclave bag, seal and label, and put into an autoclave.

10. Autoclave at 131 °C for 5 min (*see* **Note 5**).

11. Once fibers have been sterilized, they can be stored at 4 °C in their container until required.

12. Prepare conditioned medium by decanting the medium from one of the flasks of cells to be harvested into a 30 ml polypropylene tube and centrifuge at $700 \times g$ for 10 min. Decant the supernatant into a second tube and label as conditioned medium.

13. Use monolayers of tumor cells grown in 75 cm² flasks approaching confluence. Wash twice in HBSS, followed by trypsinization in the trypsin/EDTA solution at 37 °C for between 2 and 10 min (dependent upon the cell line), and resuspend the cells in conditioned medium.

14. Count the cells using a hemocytometer and adjust to the required cell density with conditioned medium (*see* **Note 6**).

15. Store the cell suspension on ice while the class II cabinet is prepared for loading the fibers. Spray the cabinet with 70 % ethanol and wipe. Lay out the equipment for loading the cells into the hollow fibers as suggested in Fig. 1a.

16. Position the Bunsen burner to the rear, right-hand side of the cabinet (*see* **Note 7**). Place the demarcated tray at the front of the cabinet with the instrument tray to the right of it, and the small steel tray to the left of it.

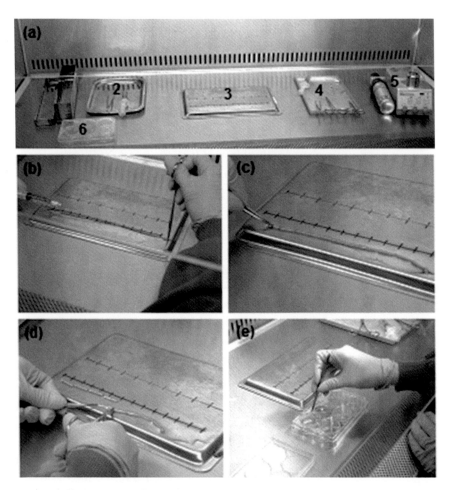

Fig. 1 Loading of cells into hollow fibers. (**a**) Suggested layout of equipment in the class II cabinet: *1*, box containing sterilized fibers; *2*, small tray for instruments in use; *3*, demarcated tray; *4*, instrument tray; *5*, "Fireboy plus" rechargable Bunsen burner; *6*, 6-well plate containing medium. (**b**) Heat-sealing one end of the fiber after loading with cells. (**c**) Fiber following heat-sealing at 1.5 cm intervals along its length. (**d**) Dividing the fiber into segments by cutting across the heat seals. (**e**) Placing the fiber segments into a 6-well plate

17. Add 2 ml of growth medium containing 20 % FCS to each well of a 6-well plate, and also fill a 30 ml polypropylene tube with the medium, and store both at 37 °C.

18. Open the packaging on the syringe and needles without touching the inside of the package.

19. Bring the tube containg the medium into the cabinet and loosen the lid. Using the sterile 10 ml syringe, take up approximately 7–8 ml of medium, attach a sterile blunt needle, and place on the small steel tray.

20. Mix the tube containing the cell suspension by inversion, bring into the cabinet and loosen the lid. Using the sterile 5 ml syringe, take up approximately 3–4 ml of medium, attach a sterile blunt needle, and place on the instrument tray.

21. Place the box containing the sterilized fibers on the furthest left of the cabinet and open the lid.

22. Manipulate the fiber using one set of forceps and flush through with fresh medium onto the demarcated tray.

23. While holding the fiber with a second set of forceps, flush the fiber through with cell suspension onto the demarcated tray (*see* **Note 8**). Keep the syringe and needle attached to the fiber.

24. Heat the needle holders in the Bunsen flame for 3 s and clamp the loose end of the fiber to heat-seal (*see* Fig. 1b). Dispense further cell suspension into the fiber until it is evidently full and heat-seal the needle end. Remove the needle and syringe and place on the instrument tray.

25. Lay the cell-filled fiber along the demarcated tray, lining up the first sealed end with one of the marks on the tray, and cover the fiber with medium. Heat-seal at 1.5 cm intervals (*see* Fig. 1c), and then using scissors, cut across the seals and separate the segments (*see* Fig. 1d). For ease of handling, each segment should have a tail of 2 mm (*see* **Note 9**).

26. Wash off any cells from the outside of the fiberes by immersing the fibers in a series of wells of a 6-well plate filled with medium.

27. Place the washed segments into fresh 6-well plates and incubate at 37 °C in a 5 % CO_2 atmosphere overnight before proceeding to implantation.

3.2 In Vivo Transplantation of Hollow Fibers and Treatment

All animal procedures adhere to guidelines issued by the UK National Cancer Research Institute Guidelines for the Welfare of Animals [9].

1. In order to minimise fluctuations in temperature during transport to the operating theatre, the 6-well plates which hold the fibers are transported in an insulating polystyrene box.

2. Following induction general inhalation anesthesia with 2 % isofluorane in a 2 % O_2 atmosphere, anesthesia is maintained at the same rate using a nose cone, with the recipient mouse placed on its abdomen.

3. A small incision is made in the skin near to the left leg on the left dorsal flank and a fiber is placed sub-cutaneously using a trocar. The incision is sealed with tissue glue and an autoclip applied (*see* **Note 10**). The procedure is repeated to load another fiber in the right flank, with the fibers lying as seen in Fig. 2a.

4. Mice are then transferred to a box containing a heating pad maintained at 37 °C and are monitored until they have recovered from the anesthesia. They are then returned to their cages and frequently monitored for any deleterious effects until the time of sacrifice.

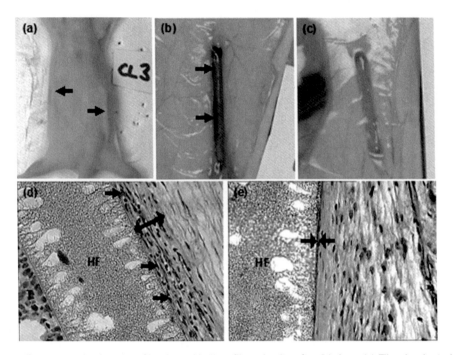

Fig. 2 (**a–c**) Representative images of implanted hollow fibers in situ after 21 days. (**a**) Fiber implanted subcutaneously in the right and left dorsal areas with no visible signs of inflammation (*arrows*). (**b**) image of a fiber loaded with cells exposed in situ following dissection. The area directly adjacent to the fiber is heavily vascularized (*arrows*). (**c**) Similar image of an unloaded fiber showing lack of vascularization surrounding the fiber in comparison. (**d, e**) Photomicrographs at 21 days of sections of tissue surrounding implanted fibers immunostained with α-CD31 antibody. (**d**) Fiber (HF) loaded with MCF-7 mammary tumor cells (T) with large numbers of vascular profiles positive for CD31 (*arrows*) in a thick layer of granulation tissue (*double-headed arrow*). (**e**) Area next to an unloaded fiber (HF) with minimal granulation tissue (*arrows*) and negligible CD31 immunopositivity is seen

3.3 Removal and Processing of Hollow Fibers

1. At various times following implantation (studies have been carried out up to 32 days), mice are sacrificed by CO_2 inhalation and are pinned out onto a dissecting board with their dorsal flank exposed.

2. An incision is made in the skin near the base of the left fiber which is then extended parallel with the fiber towards the center of the body, and then up along the top of the fiber, such that there is at least a 5 mm margin of skin surrounding the fiber on three sides. The skin flap is then carefully pulled back and pinned to the dissecting board. The same procedure is repeated for the right fiber.

3. The dissection board is then placed on the photographic stand and an image of the fibers in situ captured using a digital camera (*see* Fig. 2b and c) to show the extent of the neovascularization.

4. The skin flap surrounding the fiber is then totally dissected and laid flat on a piece of card and placed into a histology process-

ing cassette. This in turn is placed into an histology pot containing zinc fixative.

5. The samples are then fixed overnight, and then processed through a graded alcohol series and Histoclear before paraffin wax embedding using the automatic tissue processor.

3.4 Assessment of Effects on the Vasculature

All incubations in this section are carried out at room temperature unless otherwise stated, with 100 μl of solution applied to each section.

1. 5 μm thick sections of the fiber in longitudinal profile are cut using a microtome and collected onto APES-coated slides (*see* **Note 11**). Sections are then de-waxed in xylene and hydrated using a graded alcohol series and then finally distilled water.

2. Slides are then placed in Coplin jars containing freshly made 1 % H_2O_2 solution in distilled water for 30 min in order to block any endogenous peroxidase activity.

3. Following a wash in PBS for 5 min, slides are carefully blotted dry with a tissue, and a circle drawn around the section using a wax pen, which ensures that solutions remain over the area of the section when applied, and thus that a smaller volume of reagent can be used.

4. NRS is then applied as a blocking serum for 20 min, after which the excess liquid is blotted off by carefully tilting the slide and allowing the edge of a folded tissue to come into contact with the solution to remove the majority of it.

5. The α-CD31 primary antibody is then applied to the sections and these are incubated for 90 min (*see* **Note 12**), followed by three washes of 5 min each in PBS.

6. The biotinylated rabbit α-rat IgG (H + L) secondary antibody is then applied for 30 min. At this stage, the ABC peroxidase reagent is made up in a 30 ml polypropylene tube.

7. After three washes of 5 min each in PBS, the ABC peroxidase reagent is applied for 30 min.

8. Slides are then washed a further three times in PBS. During the third wash, the DAB substrate is prepared.

9. The DAB substrate is then applied to the slides, which are placed on a piece of white paper and are closely observed until brown staining is evident (approximately 3 min). Once color is observed, the reaction is stopped by gently running distilled water over the whole of the section using a wash bottle, with the waste collected into a 500 ml beaker containing 10 % bleach solution (*see* **Note 13**).

10. The sections are then placed in a Coplin jar filled with distilled water for 5 min.

11. Sections are then counterstained with Harris's hematoxylin for 5 min followed by washing in running tap water for 2 min, immersion in Scott's tap water substitute for 1 min, a further rinse in tap water for 30 s, and then dehydration through the graded series of alcohols and xylene before mounting the sections with DPX mounting medium and addition of a 22 mm × 50 mm coverslip.

12. Sections are then analyzed by capturing digital images from a light microscope with a 40× magnification objective lens, which is connected to a CCD video camera, using image capture software (*see* Fig. 2d and e).

13. Using a 21 cm × 17 cm graticule overlayed on the image on the PC monitor, measurements are taken in the area towards the center of the fiber, with the area of granulation tissue surrounding the fiber, and the amount of CD-31 immunolabelling of the neovasculature recorded. Twenty fields are measured for each section, and three sections are evaluated for each fiber.

14. A figure for the mean percentage density of neovasculature in the granulation tissue surrounding the fiber can then be calculated by dividing the mean number of CD31-positive intersects on a section by the mean area of the granulation tissue, and multiplying by 100.

15. Statistical analysis of differences in neovascularization surrounding treated and control fibers can then be carried out using a one-way ANOVA.

---

4   Notes

1. Cell lines are chosen for their ability to stimulate angiogenesis, e.g., high VEGF expression. This can be confirmed by ELISA analysis of VEGF levels in hollow fiber experiments in vitro of the same duration as the in vivo experiments.

2. We have demonstrated that for the longer implantation times required for studying vascular effects, human tumor cell lines must be placed in immunocompromised nude mice, whereas if murine tumor cell lines are used, then they can be placed in either immunocompetent or immunocompromised mice [10].

3. Fibers can be stored in this way indefinitely as long as they do not dry out. If this occurs, then the procedure must be repeated.

4. When the fibers are being rehydrated with water, the trapping of air bubbles may occur. To prevent this happening, ensure that the fiber is fully immersed in the autoclavable box before removing the needle.

5. Fibers melt at 143 °C and should not be autoclaved above 131 °C.

6. Cell density depends upon the growth characteristics of the cell line to be used, but the typical range is for densities between 2×10^5 and 1×10^7 cells per ml.

7. Positions are for a right-handed person. Reverse if left-handed.

8. Keep the fiber as near to the horizontal as possible while loading and sealing.

9. Seals must be properly formed or else they may break open resulting in leakage of cells. It is important to practice not only the sealing procedure, but also the segmentation, as both procedures require steady hands and good eyesight. It may be helpful to employ a magnifying lens during this stage.

10. Once the incision wound has healed, the clips can be removed. This normally occurs around 7 days postimplantation.

11. Discard initial sections until the full lumen of the hollow fiber is exposed. Then take serial sections at 200 μm intervals.

12. For the immunostaining procedure, control sections are included in the protocol where: (a) NRS is applied rather than the primary antibody; (b) in addition to omitting the primary antibody, PBS is then applied instead of the secondary antibody.

13. DAB is a known carcinogen, therefore all unused DAB solution should be poured into 10 % bleach which can then be disposed of by pouring down the drain in the presence of a large excess of running tap water.

Acknowledgments

The author would like to thank Shofiq Al-Islam for his additional technical assistance. This work was supported by Cancer Research UK Programme Grant C7589/A5953, and Yorkshire Cancer Research Programme Grants R00248 and R02222.

References

1. Passaniti A, Taylor RM, Pili R et al (1992) A simple, quantitative method for assessing angiogenesis and antiangiogenic agents using reconstituted basement membrane, heparin, and fibroblast growth factor. Lab Invest 67:519–528

2. Mahadevan V, Hart IR, Lewis GP (1989) Factors influencing blood supply in wound granuloma quantitated by a new in vivo technique. Cancer Res 49:415–419

3. Algire GH (1945) An adaptation of the transparent chamber technique to the mouse. J Natl Cancer Inst USA 4:1–11

4. Mriouah J, Boura C, Thomassin M et al (2012) Tumor vascular responses to antivascular and antiangiogenic strategies: looking for suitable models. Trends Biotechnol 30:649–658

5. Hollingshead MG, Alley MC, Camalier RF et al (1995) In vivo cultivation of tumor cells in hollow fibres. Life Sci 57:131–141

6. Phillips RM, Pearce J, Loadman PM et al (1998) Angiogenesis in the hollow fibre tumor model influences drug delivery to tumor cells: implications for anticancer drug screening programs. Cancer Res 58:5263–5266

7. Shnyder SD, Hasan J, Cooper PA et al (2005) Development of a modified hollow fibre assay for studying agents targeting the tumor neovasculature. Anticancer Res 25: 1889–1894

8. Hasan J, Shnyder SD, Clamp AR et al (2005) Heparin octasaccharides inhibit angiogenesis in vivo. Clin Cancer Res 11:8172–8179

9. Workman P, Aboagye EO, Balkwill F et al (2010) Guidelines for the welfare and use of animals in cancer research. Br J Cancer 102:1555–1577

10. Shnyder SD, Cooper PA, Scally AJ et al (2006) Reducing the cost of screening novel agents using the hollow fibre assay. Anticancer Res 26:2049–2052

Chapter 27

Studying Vascular Angiogenesis and Senescence in Zebrafish Embryos

Emma Ristori, Sandra Donnini, and Marina Ziche

Abstract

The zebrafish is an excellent animal model to study the formation of the vertebrate vascular network. The small size, the optical translucency, and the ability to model endothelial-specific fluorescent transgenic lines in the zebrafish embryo had facilitate, in the past 10 years, the direct visualization of vessels formation and remodeling. Furthermore, zebrafish is an excellent disease model such as for cancer and neurodegenerative diseases. Cerebral amyloid angiopathy (CAA) is a human neurovascular degenerative disease, caused by Amyloid β (Aβ) peptides deposition around brain microvessels, and characterized by vascular brain degenerative changes. By using the zebrafish model, we investigated the effect of Aβ peptides treatment in vessel formation during embryogenesis. We showed that the defects in the vascular remodeling and senescence can be detected, respectively, via staining for alkaline phosphatase activity and β-galactosidase or cyclin-dependent kinase inhibitor p21 expression. We demonstrated that treating zebrafish embryos with these oxidative peptides reduces angiogenesis and promotes premature vascular senescence. In this chapter, we will describe the methods to reveal both angiogenesis and senescence defects upon Aβ peptides treatment of the zebrafish embryos.

Key words Vascular imaging, Cerebral amyloid angiopathy, Amiloid β peptides, Vascular senescence, β-Galactosidase, In situ hybridization

1 Introduction

Zebrafish (*Danio rerio*) is a small tropical freshwater fish that in the past decade has became a favorite vertebrate organism to study angiogenesis in vivo. Zebrafish embryos develop externally and are optically clear, providing noninvasive and high-resolution methods to observe blood vessels morphogenesis at every stage of the embryonic development using transmitted light or fluorescent imaging techniques. The husbandry of the adults is inexpensive and easy to maintain long-term [1], while the embryonic development and the organogenesis are very rapid, between 24 and 72 of hours postfertilization (hpf) [2]. Moreover, the zebrafish shares many conserved gene pathways with other vertebrates, making this organism a useful model system to study vascular development and

Stewart G. Martin and Peter W. Hewett (eds.), *Angiogenesis Protocols*, Methods in Molecular Biology, vol. 1430,
DOI 10.1007/978-1-4939-3628-1_27, © Springer Science+Business Media New York 2016

function [3]. Finally, several forward and reverse genetic screens, using vascular transgenic lines (Table 1), were able to identify fundamental genes driving vascular remodeling.

Angiogenesis is essential in normal developmental processes and is critical for the progression of several disease states including cancer and age-associated cerebrovascular diseases, including cerebral amyloid angiopathy (CAA) [14, 15]. The CAA is an age-associated cerebrovascular disease, characterized by deposit of amyloid peptides (Aβ) around brain vessels, and, as a consequence, by a massive brain blood vessels degeneration [15, 16]. Loss of vascular function associated with an accelerated endothelial cell senescence is one of the early consequences of the vascular Aβ peptides deposition during CAA [15, 17]. In zebrafish, the vascular senescence can be analyzed by β-galactosidase activity, a histochemically detectable biomarker in organismic aging [18], or measuring the expression of the messenger RNA for p21, a cyclin-dependent inhibitor that induces cell cycle arrest [19–21].

In this chapter, we will show that the treatment with Aβ peptides of wild type strain (AB) or transgenic line (*fli1*:EGFP)[y1] embryos results in angiogenesis defect of the subintestinal vessels (SIVs) as well as the premature vascular β-galactosidase activity and p21 expression. By measuring these senescence markers, we demonstrated that these oxidative peptides promote senescence at an early stage of vascular development, a harbinger of vascular clinical symptoms in adult [15].

Table 1
Zebrafish transgenic lines used for vascular imaging

Transgenic strain	Reference
Tg(fli1a:EGFP)[y1]	[4]
Tg(fliia:nEGFP)[y7]	[5]
Tg(fliia:EGFP–cdc42wt)[y48]	[6]
Tg(mTie2:GFP)	[7]
Tg(kdrl:G-RCFP)	[8]
Tg(kdrl:memCherry)[s896]	[9]
Tg(kdrl:EGFP)[s843]	[10]
Tg(fli1a: DsRed)	[11]
Tg(fli1a:EGFP; kdrl:ras-cherry)	[12]
Tg(flt1:YFP, kdrl:mCherryRed)	[12]
Tg(stabilin:YFP)[hu4453]	[12]
Tg(gata1:DsRed)[sd2]	[13]

2   Materials

2.1 General Materials	Zebrafish AB strain and *tg* (*fli1*:EGFP)[y1] strain (*see* **Note 1**). Fish water (Instant Ocean, 0.01 % Methylene Blue). Petri dishes. Pronase solution (100 mg/ml; Protease from Streptomyces griseus, Sigma). 1-Phenyl-2-thiourea (PTU, Sigma). Sterile distilled water. Dulbecco's phosphate-buffered saline (PBS). Paraformaldehyde (PFA). Absolute Ethanol. Glycerol (>99 %). Inverted microscope equipped with a digital camera for image capture and processing, i.e., Leica microscope DM6000B equipped a Leica DCF480 digital camera and the software LAS (Leica 85 Application Suite) (Leica, Germany). Planachromatic objectives.

2.2   Whole-Mount Alkaline Phosphatase Staining

1. Rinse Buffer: 10 ml 10× PBS, 5 ml 10 % Triton-X100, 1 ml normal horse serum, 84 ml distilled water, makes 100 ml, scale up or down as needed.
2. Staining Buffer: 1 ml 5 M NaCl, 2.5 ml 1 M $MgCl_2$, 5 ml 1 M Tris–HCL pH 9.0–9.5, 500 μl 10 % Tween-20, 41 ml distilled water, makes 50 ml, scale up or down as needed.
3. Staining Solution: 10 ml staining buffer, 45 μl NBT, 35 μl BCIP, scale up or down as needed.
4. 4-Nitro Blue Tetrazolium, NBT, 100 mg/ml in 70 % dimethylformamide.
5. BCIP X-Phosphate or 5-Bromo-4-Chloro-3-indolyl-phosphate, 50 mg/ml in dimethylformamide.

2.3   SA-b-Gal Staining

4 % Paraformaldehyde/phosphate-buffered saline. SA-β-galactosidase detection kit (Millipore-Chemicon). 30–50–70 % EtOH in water. 0.1 % NaN_3. 70 % Glycerol in water. LR White resin (London Resin Company). Reichert Ultra-cut "E" (*see* **Note 2**).

2.4   p-21 In Situ Hybridization

p21 plasmid or genomic p21. PCR master mix (Promega). Trizma base. Ethylenediaminetetraacetic acid (EDTA) (*see* **Note 3**). Boric acid. Agarose. Transcription buffer (Promega). dl-Dithiothreitol (DTT). DIG RNA labeling mix (UTP). RNasein. T3 RNA-Polymerase. T7 RNA-Polymerase. RNase free DNAse I. NaOH. RNAlater. Methanol (MeOH). NaCl. KCl. $MgSO_4$. $Ca(NO_3)_2$. HEPES. H_2O_2 (*see* **Note 4**). KOH. Tween 20 (*see* **Note 5**). Proteinase K (10 mg/ml). Formamide. Citric acid trisodium salt. Citric acid monohydrate. Heparin sodium salt. tRNA from wheat germ type V, lyophilized powder. Phenol solution saturated with 0.1 M citrate buffer, pH 4.3. Chloroform. Sodium acetate. Sheep serum. Albumin from bovine serum (BSA). Sheep anti-dioxygenin-AP Fab fragments (Roche Diagnostics). HCl. $MgCl_2$. Nitro blue tetrazolium (NBT) (*see* **Note 6**). 5-Bromo 4-chloro

3-indolyl phosphate (BCIP) (*see* **Note** 7). 99.8 % *N,N*-dimethylformamide anhydrous. Na$_2$HPO$_4$·12H$_2$O. NaH$_2$PO$_4$. Stop solution (mix 1× PBS, pH 5.5, with 1 mM EDTA, and 0.1 % Tween 20 (vol/vol) and store at room temperature). 10× PBS, pH 5.5, stock solution (*see* **Note 8**). PBT (mix 1× PBS, and 0.1 % Tween 20 (vol/vol) and store at room temperature). 20× SSC stock solution (dissolve 175.3 g of NaCl and 88.2 g of citric acid trisodium salt in 1 L of water and store at room temperature). Blocking buffer (mix 1× PBT, 2 % sheep serum (vol/vol), BSA 2 mg/ml, and store at 4 °C). Alkaline Tris buffer (100 mM Tris–HCl, pH 9.5, 50 mM MgCl, 100 mM NaCl and 0.1 % Tween 20 (vol/vol)). Microcon YM-50 columns (Millipore). Sigmaspin post-reaction purification columns (Sigma-Aldrich).

3   Methods

3.1   Aβ Peptides Treatment and Visualization of the Vascular Defects

Embryos from two different strains, AB and Tg(*fli1*:EGFP)^{y1}, were treated with Aβ peptides (the natural Aβ1–40 variant (WT), and the mutant Dutch Aβ, E22Q [17]) from the 12-somite stage (somitogenesis) until 7 days postfertilization (dpf; larval stage).

1. Collect and raise AB and Tg(*fli1*:EGFP)^{y1} zebrafish embryos at 28 °C in Petri dishes.

2. Dechorionate embryos with pronase solution (100 μg/ml) at 6 hpf.

3. Treat with 0.003 % phenylthiourea (PTU) at 24 hpf to prevent pigmentation (*see* **Note 9**).

4. Administer from 18 hpf (12 somites stage) until 7 dpf (larval stage), Aβ peptides dissolved to 5 mM in anhydrous dimethyl-sulfoxide (DMSO) and further diluted (1–25 mM) in fish water (*see* **Note 10**).

5. Fresh aliquots are administrated every 24 h.

6. Control embryos are maintained in 0.1 % DMSO in fish water.

7. Images are taken with a Leica MZFLIII epifluorescence stereomicroscope equipped with a DFC 480 R2 digital camera and LAS Leica imaging software (Leica) and processed using Adobe Photoshop (Adobe Systems).

We observed that the treatment with a concentration of 2.5 μM of Aβ peptides affects vascular development in zebrafish; indeed the intersegmental vessels (ISVs) in tg(*fli1*:EGFP)^{y1} embryos result slightly disorganized (Fig. 1).

3.2   Whole-Mount Alkaline Phosphatase Staining

To analyze the angiogenic defects upon Aβ peptides treatment, we modeled the growth of the subintestinal venous vessels (SIVs) using endogenous alkaline phosphatase (AP) staining [22]. AP

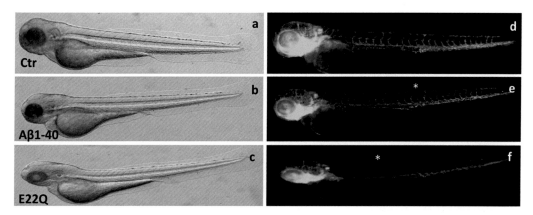

Fig. 1 Transgenic zebrafish tg (*fli1*:EGFP)^{y1} at 72 hpf following repeated Aβ peptides (Aβ1–40 and E22Q, both 2.5 μM) treatment. Stereomicroscopic images (3.2×) of whole fish bodies, both brightfield (*left panels* **a, b, c**) and fluorescence (*right panels* **d, e, f**). *Asterisks* indicate disorganized intersegmental vessels

staining is an accepted method for staining endothelial cells, as these cells express relatively high levels of endogenous AP activity [23]. In zebrafish, endogenous AP activity is not detectable in 24 hpf embryos, but is weakly detectable by 48 hpf and strongly at 72 hpf. Staining vessels by endogenous AP activity is useful for easy and rapid visualization of the vasculature in many specimens, but provides less resolution than many of the other methods. We found a decreased extension and reduction of vessel diameter in the SIVs plexus throughout the dorsal–ventral axis in both the Aβ1–40 and E22Q peptide-treated embryos (Fig. 2).

1. Treat AB zebrafish embryos with Aβ peptides as previously described.

2. Fix embryos at 72 hpf in 4 % paraformaldehyde for 2 h at room temperature.

3. Wash fixed embryos 5 × 10 min at room temperature in rinse buffer or leave washing in rinse buffer at 4 °C for up to several days. If doing the latter, wash again at RT for 10 min before going on to the next step.

4. Wash 2 × 5 min in staining buffer.

5. Stain in 1 ml of staining solution. Color development takes about 5–30 min.

6. To stop reaction, wash three times in rinse buffer without horse serum, then fix in 4 % paraformaldehyde for 30 min, and store at 4 °C.

7. Mount in agarose-coated Petri dishes and take images under a Leica MZFLIII epifluorescence stereomicroscope equipped with a DFC 480 R2 digital camera and LAS Leica imaging software.

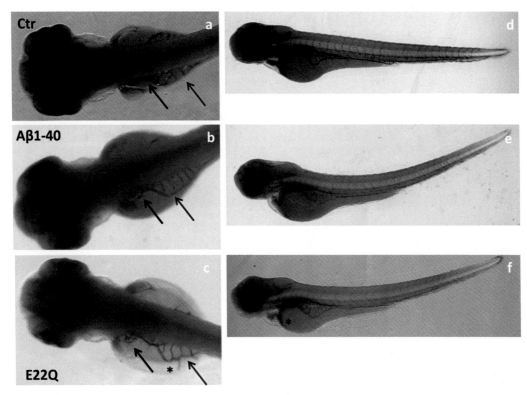

Fig. 2 Alkaline phosphatase activity in SIV basket at 72 hpf, following Aβ peptide exposure (2.5 μM). SIV basket in dorsal view (10×, **a**, **b**, **c**). *Red arrows* indicate SIV basket; *black arrows* indicate renal plexus. SIV basket (lateral view, 10×, **d**, **e**, **f**) at 72 hpf. *Asterisk* indicates SIV spikes protruding toward the yolk, together with pericardial edema overshadowing the SIV basket (E22Q)

It is better to avoid putting the embryos in methanol (this destroys endogenous AP activity). If embryos have been placed in methanol, some AP activity can be reconstituted by washing embryos in PBT overnight or even over a weekend before starting the staining procedure, although staining will be weaker than in non-methanol-exposed embryos.

3.3 Senescence Associated to the Quantification of the β-Galactosidase (β-Gal) Activity in Zebrafish

β-Gal activity is a marker of senescence validated in a previous work on zebrafish mutants [18]. We used a β-Gal assay at 7 dpf to directly analyze whether Aβ peptides were inducing senescence in treated AB larvae. This assay reveals that staining, detectable in control group, increased with the Aβ treatment (Fig. 3).

1. Treat AB embryos with Aβ peptides as previously described.

2. Fix zebrafish larvae (7 dpf) at room temperature for 2 h in 4 % paraformaldehyde/phosphate-buffered saline. Alternatively, embryos can be left in fixing solution overnight at 4 °C.

3. Wash in 1× PBS 3 × 15 min.

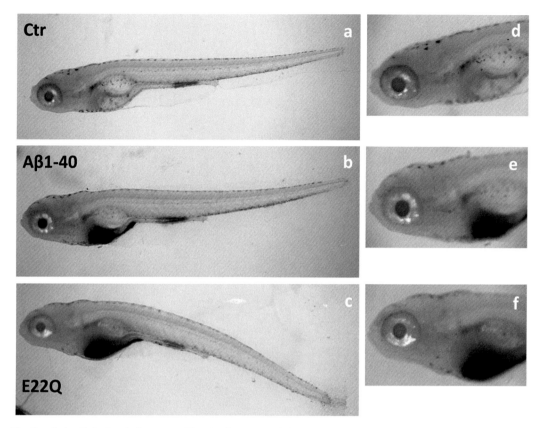

Fig. 3 β-Gal activity in whole-mount AB zebrafish at 7 dpf, following Aβ-repeated treatment at 2.5 μM (images at 3.2×, **a, b, c**). On the right, images of β-Gal activity on zebrafish trunk (lateral view, 10×, **d, e, f**)

4. Make up the senescence staining mixture as recommended by the vendor (Chemicon). We have used the SA-Galactosidase Detection Kit from Millipore/Chemicon.

5. Add 1 ml to embryos and incubate for 4 h at 37 °C.

6. Wash in 1× PBS 3×15 min.

7. Store embryos at 4 °C in 1× PBS and 0.1 % NaN$_3$ or in 70 % glycerol.

8. Take images with a Leica microscope DM6000B equipped with a DFC 480 R2 digital camera and LAS Leica imaging software.

9. Quantify SA-β-Gal activity using a selection tool in Adobe Photoshop for blue color range, according to Kishi et al. [18].

3.4 In Situ Hybridization (ISH) of p21

Whole-mount ISH was carried out as described by Thisse et al. [24]. We found a striking increase of p21 signal in the rostral head region of Aβ-peptide-treated fish (2.5 μM), in contrast with control embryos (Fig. 4). The up-regulation of p21 expression suggests that p53-p21 is the prevailing pathway involved in the Aβ-driven senescence.

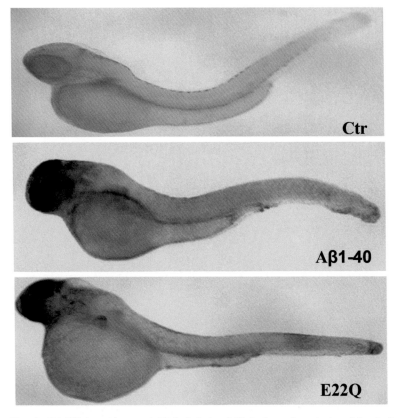

Fig. 4 p21 ISH in embryos at 72 hpf. Lateral (4×) view of control and treated embryos (Aβ1–40 and E22Q at 2.5 μM), showing p21 mRNA expression in whole body, with enrichment in the head

3.4.1 Preparation of the p21 Probe

When working with RNA, use appropriate tips with filter RNAse/DNase free and wear gloves.

1. For the p21 probe [also known as **cdkn1a**], set up 100 μl PCR in a 0.5 ml sterile tube as follow:

Template DNA	0.5 μl	(10–100 ng)
Forward primer (500 ng/ml)	0.5 μl	(250 ng)
Reverse primer (500 ng/ml)	0.5 μl	(250 ng)
PCR master mix (2×)	50 μl	(1×)
Sterile water up to	100 μl	

p21 forward: *5'-ATGCAGCTCCAGACAGATGA-3'*,
    p21 reverse: *5'-CGCAAACAGACCAACATCAC-3'* (*see* **Note 11**).

2. Run the PCR using the following conditions: 4 min at 95 °C, 40 cycles as follow: 30 s at 95 °C, 30 s at 55 °C, and 1 min at 72 °C; 7 min at 72 °C.

3. Add the 100 µl PCR to a Microcon YM-50 column and add 400 µl of sterile water. Centrifuge for 15–20 min at $1000 \times g$ at room temperature.

4. Place the Microcon column into a new microfuge tube, add 20 µl of sterile water, vortex briefly, and then turn the Microcon column upside down. Spin for 1 min at $1000 \times g$ at room temperature to recover the DNA.

5. Check the quality, quantity, and size of the PCR amplification product by loading 1/20 of the preparation on a 1 % (wt/vol) agarose gel in 1× TBE buffer.

6. Add to a microfuge tube:

Template DNA from PCR	2.5 µl	200 ng
5× Transcription buffer	1 µl	1×
DTT (0.1 M)	0.5 µl	10 mM
DIG-RNA labeling mix (10×)	0.5 µl	1×
RNAsein (40 U/ml)	0.25 µl	10 U
T3 or T7 RNA polymerase (20 U/ml)	0.25 µl	5 U

Mix and incubate for 2 h at 37 °C (*see* **Note 12**).

7. Add 2 µl of RNase-free DNase I and 18 µl of sterile water. Mix and incubate for 30 min at 37 °C.

8. Stop the reaction by adding 1 µl of sterile 0.5 M EDTA and 9 µl of sterile water.

9. Place a Sigmaspin post-reaction purification column on top of a microfuge tube. Centrifuge for 15 s at $750 \times g$.

10. Break the base of the column and discard the lid. Spin for 2 min at $750 \times g$.

11. Place the column on a new microfuge tube. Add the RNA template on top of the resin. Centrifuge for 4 min at $750 \times g$. Discard the column.

12. Add 1 µl of sterile EDTA 0.5 M and 9 µl of RNAlater to the sample.

13. Visualize 1/20 of the synthesized RNA on 1 % (wt/vol) agarose gel in 1× TBE buffer after 30 min of electrophoresis at 230 V.

3.4.2 p21 In Situ Hybridization

1. Treat AB embryos with Aβ peptides as previously described.

2. Fix embryos (72 hpf, after dechorionation) for 2 h at room temperature in 4 % paraformaldehyde/phosphate-buffered

saline, then wash twice in PBS-Tween, 5 min each, at room temperature.

3. Dehydrate the embryos in 100 % methanol (MeOH) for 15 min at room temperature and store at –20 °C until processed (*see* **Note 13**).

4. Rehydrate embryos into successive dilutions of methanol in 1× PBS: 5 min in 75 % (vol/vol) methanol; 5 min in 50 % (vol/vol) methanol; and 5 min in 25 % (vol/vol) methanol.

5. Wash four times, 5 min per wash, in 100 % PBT.

6. Permeabilize the embryos by digestion with proteinase K (10 µg/ml in 1× PBS) at room temperature (40 min) (*see* **Note 14**).

7. Stop the proteinase K digestion by incubating the embryos for 20 min in 4 % (wt/vol) paraformaldehyde in 1× PBS.

8. Transfer the embryos from to 1.5 ml sterile Eppendorf tubes (up to 50 embryos per tube).

9. Prehybridize the embryos for at least 3 h at 62 °C in hybridization buffer [50 % formamide, 5× SSC, 50 µg/ml heparin, 500 µg/ml tRNA, 0.1 % Tween 20, 1 M citric acid pH 6.0].

10. Discard the hybridization buffer and replace with 400 µl of hybridization buffer containing 200 ng of antisense DIG-labeled RNA probe. Hybridize overnight at 62 °C (*see* **Note 15**).

11. Incubate for 10 min the embryos with 200 µl of hybridization buffer (without RNase-freetRNA and heparin) warmed at 62 °C.

12. Gradually change the hybridization buffer to 2× SSC: wash once in each of the following: 75 % hybridization buffer, 50 % hybridization buffer, 25 % hybridization buffer, and 100 % 2× SSC. Perform washes in a 62 °C water bath with gentle shaking, 10 min each wash.

13. Wash twice, for 30 min/wash, in 0.2× SSC at 62 °C.

14. Gradually replace 0.2× SSC with PBT: wash once at room temperature, in agitation, in 200 µl of the following solutions: 75 % 0.2× SSC, 50 % 0.2× SSC, 25 % 0.2× SSC and 1× PBT, 10 min each wash.

15. Incubate the embryos for 3–4 h at room temperature in blocking buffer under agitation.

16. Incubate in 200 µl of anti-DIG antibody solution diluted at 1/10,000 with blocking buffer overnight at 4 °C with gentle agitation.

17. Discard the antibody solution and wash the embryos briefly in PBT.

18. Wash six times, 15 min/wash, in PBT at room temperature with gentle shaking.

19. Briefly dry the embryos on a sheet of absorbent paper.

20. Incubate the embryos at room temperature three times, 5 min/wash, in alkaline Tris buffer with agitation.

21. Transfer the embryos to 1.5 ml microfuge tubes.

22. Remove the alkaline Tris buffer and replace with 0.7 ml staining solution prepared fresh and kept in the dark.

23. Transfer the embryos to 12 well ceramic plates. Monitor the color reaction periodically under a dissecting microscope, lit from above. Keep the embryos in the dark between checks.

24. Stop the reaction by transferring the embryos to 1.5 ml Eppendorf tubes containing 1 ml stop solution.

25. Transfer embryos to a six-well plate containing 100 % glycerol. Shake gently overnight at room temperature in the dark.

26. Mount the embryos in 100 % glycerol and observe microscopically.

27. Take images (Fig. 4).

4 Notes

1. All animal experiments should be performed in accordance with the relevant authorities' guidelines and regulations.

2. Methacrylate resins as LR White are hydrophilic and tend to wet the block surface because they attract water. Dry the block face with filter paper, and if necessary, use an antistatic device (gun) to eliminate electrostatic charging.

3. To prepare 0.5 M EDTA, dissolve 186.1 g of EDTA in 800 ml of water. Add 15 g of NaOH pellets and make up to 1 l by adding water when all pellets are dissolved. EDTA may behave as a mild irritant for the eyes, skin, and respiratory system.

4. Hydrogen peroxide solution should be prepared fresh immediately before use.

5. Stock solution is prepared by diluting 200 ml of Tween 20 with 800 ml of sterile water. After complete homogenization of the solution, store at room temperature. Protect the solution from light.

6. Dissolve 50 mg of NBT in 0.7 ml of N,N'-dimethylformamide and 0.3 ml of sterile water and store at −20 °C for max 6 months.

7. Dissolve 50 mg of BCIP in 1 ml of N,N'-dimethylformamide anhydrous and store at −20 °C for max 6 months.

8. Dissolve 10.8 g of Na_2HPO_4, 65 g of NaH_2PO_4, 80 g of NaCl, and 2 g of KCl in 1 l of water.

9. PTU solution prevents the formation of melanin pigments and greatly facilitates visualization of the final signal. PTU is an inhibitor of tyrosinase, an enzyme required for melanin synthesis. PTU affects early development: do not treat embryos before the end of gastrulation. Further, a reduction of cell viability in catecholaminergic neuronal cells has been reported after PTU treatment. These effects can be avoided by using the hydrogen peroxide for dechorionation.

10. Aβ (1–40) peptides, WT or bearing the Duch mutation—were synthesized at Espikem (University of Florence, Italy). In all cases, peptides were purified by reverse phase high pressure liquid chromatography, eluting as a single peak, and their respective molecular masses corroborated by mass spectrometry. Peptides were first dissolved to 1 mM in cold (4 °C) hexafluoro-isopropanol (HFIP,) in a chemical fume hood to break down β-sheet structures and disrupt hydrophobic forces in aggregated Aβ. HFIP was allowed to evaporate, and the resulting clear peptide films were vacuum-dried. Before use, HFIP-treated peptides were dissolved to 5 mM in anhydrous dimethyl sulfoxide (DMSO) and further diluted (1–25 μM) in fish water.

11. In addition to the use of antisense RNA probes, sense (control) RNA probes of the corresponding gene should be performed to provide indications about the background signal that may appear. Primers should generate linear DNA containing one T3 or T7 RNA polymerase promoter, which should be located at 3′ (for antisense probes) or 5′ (for sense probes) of the cDNA or exon of interest. The sequences to include the T3 or T7 RNA polymerase promoter in the appropriate primer are T3: 5′-*CATTAACCCTCACTAAAGGGAA*-3′ or T7: 5′-*TAATACGACTCACTATAGGG*-3′ (reverse primer for antisense probes, forward primer for sense probes). Sequences should be located at the 5′ extremity of the primer. Be careful not to contaminate the PCRs. Use sterile tubes and filter tips and wear gloves. Alternatively to PCR amplification, cDNAs in plasmids can be linearized using restriction enzymes that have a unique site located 5′ (for antisense probes) or 3′ (for sense probes) to the insert. Purification of linear DNA can be achieved by phenol/chloroform extraction followed by ethanol precipitation.

12. In this chapter, we describe the approach using the synthesized p21 RNA probe tagged with dioxygenin uridine-5′-triphosphate. Following hybridization, the transcript is visualized by immunohistochemistry using an anti-digoxygenin antibody conjugated to alkaline phosphatase.

13. This step is necessary for permeabilization of embryos even if you don't want to store them. These embryos can be stored at −20 °C for several months. Use MeOH and not ethanol because ethanol causes a higher background.

14. This step permeabilizes the embryos and allows the RNA probe to penetrate. It is important to use the times indicated for proteinase K treatment. There are different times of treatment for each developmental stage. Under-digestion would not allow the probe to get in, whereas over-digestion will alter the morphology of the embryo.

15. Do not use excessive amounts of RNA probe. This increases background labeling. This high hybridization temperature and the percentage of formamide in the hybridization buffer ensure high stringency of hybridization and decrease the occurrence of cross-hybridization.

Acknowledgments

The work was supported by Istituto Toscano Tumori (ITT). We are grateful to Dr. Stefania Nicoli for her scientific support.

References

1. Brand M, Granato M, Nüsslein-Volhard C (2002) Keeping and raising zebrafish. In: Nüsslein-Volhard C, Dahm R (eds) Zebrafish: a practical approach. Oxford University Press, Oxford, pp 7–37

2. Dahm R (2002) Atlas of embryonic stages of development in the zebrafish. In: Nüsslein-Volhard C, Dahm R (eds) Zebrafish: a practical approach. Oxford University Press, Oxford, pp 219–236

3. Rubinstein AL (2003) Zebrafish: from disease modeling to drug discovery. Curr Opin Drug Discov Devel 6:218–223

4. Lawson ND, Weinstein BM (2002) In vivo imaging of embryonic vascular development using transgenic zebrafish. Dev Biol 248:307–318

5. Yaniv K, Isogai S, Castranova D, Dye L, Hitomi J, Weinstein BM (2006) Live imaging of lymphatic development in the zebrafish. Nat Med 12:711–716

6. Kamei M, Saunders WB, Bayless KJ, Dye L, Davis GE, Weinstein BM (2006) Endothelial tubes assemble from intracellular vacuoles in vivo. Nature 442:453–456

7. Motoike T, Loughna S, Perens E, Roman BL, Liao W, Chau TC, Richardson CD, Kawate T, Kuno J, Weinstein BM et al (2000) Universal GFP reporter for the study of vascular development. Genesis 28:75–81

8. Cross LM, Cook MA, Lin S, Chen JN, Rubinstein AL (2003) Rapid analysis of angiogenesis drugs in a live fluorescent zebrafish assay. Arterioscler Thromb Vasc Biol 23:911–912

9. Chi NC, Shaw RM, De Val S, Kang G, Jan LY, Black BL, Stainier DY (2008) Foxn4 directly regulates tbx2b expression and atrioventricular canal formation. Genes Dev 22:734–739

10. Jin SW, Beis D, Mitchell T, Chen JN, Stainier DY (2005) Cellular and molecular analyses of vascular tube and lumen formation in zebrafish. Development 132:5199–5209

11. Geudens I, Herpers R, Hermans K, Segura I, Ruiz de Almodovar C, Bussmann J, De Smet F, Vandevelde W, Hogan BM, Siekmann A et al (2010) Role of dll4/notch in the formation and wiring of the lymphatic network in zebrafish. Arterioscler Thromb Vasc Biol 30:1695–1702

12. Hogan BM, Bos FL, Bussmann J, Witte M, Chi NC, Duckers HJ, Schulte-Merker S (2009) Ccbe1 is required for embryonic lymphangiogenesis and venous sprouting. Nat Genet 41:396–398

13. Traver D, Paw BH, Poss KD, Penberthy WT, Lin S, Zon LI (2003) Transplantation and in vivo imaging of multilineage engraftment in zebrafish bloodless mutants. Nat Immunol 4:1238–1246

14. Carmeliet P, Jain RK (2000) Angiogenesis in cancer and other diseases. Nature 407:249–257

15. Solito R, Corti F, Fossati S, Mezhericher E, Donnini S, Ghiso J, Giachetti A, Rostagno A, Ziche M (2009) Dutch and Arctic mutant peptides of beta amyloid (1–40) differentially affect the FGF-2 pathway in brain endothelium. Exp Cell Res 315:385–395

16. Revesz T, Holton JL, Lashley T, Plant G, Frangione B, Rostagno A, Ghiso J (2009) Genetics and molecular pathogenesis of sporadic and hereditary cerebral amyloid angiopathies. Acta Neuropathol 118:115–113

17. Donnini S, Solito R, Cetti E, Corti F, Giachetti A, Carra S, Beltrame M, Cotelli F, Ziche M (2010) Aβ peptides accelerate the senescence of endothelial cells in vitro and in vivo, impairing angiogenesis. FASEB J 24:2385–2395

18. Kishi S, Bayliss PE, Uchiyama J, Koshimizu E, Qi J, Nanjappa P, Imamura S, Islam A, Neuberg D, Amsterdam A, Roberts TM (2008) The identification of zebrafish mutants showing alterations in senescence-associated biomarkers. PLoS Genet 4:e1000152

19. Kurz DJ, Decary S, Hong Y, Trivier E, Akhmedov A, Erusalimsky JD (2004) Chronic oxidative stress compromises telomere integrity and accelerates the onset of senescence in human endothelial cells. J Cell Sci 117:2417–2426

20. Erusalimsky JD, Skene C (2009) Mechanisms of endothelial senescence. Exp Physiol 94:299–304

21. Kurz DJ, Decary S, Hong Y, Erusalimsky JD (2000) Senescence-associated (beta)-galactosidase reflects an increase in lysosomal mass during replicative ageing of human endothelial cells. J Cell Sci 113:3613–3622

22. Isogai S, Horiguchi M, Weinstein BM (2001) The vascular anatomy of the developing zebrafish: an atlas of embryonic and early larval development. Dev Biol 230:278–301

23. Childs S, Chen JN, Garrity DM, Fishman MC (2002) Patterning of angiogenesis in the zebrafish embryo. Development 129:973–982

24. Thisse C, Thisse B, Schilling TF, Postlethwait JH (1993) Structure of the zebrafish snail1 gene and its expression in wild-type, spadetail and no tail mutant embryos. Development 119:1203–1215

INDEX

Stewart G. Martin and Peter W. Hewett (eds.), *Angiogenesis Protocols*, Methods in Molecular Biology, vol. 1430,
DOI 10.1007/978-1-4939-3628-1, © Springer Science+Business Media New York 2016

Printed in the United States
By Bookmasters